GURDJIEFF'S HYDROGENS

VOLUME 1

THE RAY OF CREATION

by Robin Bloor

KARNAK PRESS

KARNAK PRESS
Austin, Texas

GURDJIEFF'S HYDROGENS
VOLUME 1: THE RAY OF CREATION

Author: Robin Bloor
Editor: Paula Schmidt

Copyright © 2021 by Robin Bloor, All Rights Reserved.

Without limiting the rights under copyright reserved above, no part of this publication may be reproduced, stored in or introduced into a retrieval system, or transmitted, in any form or by any means (electronic, mechanical, photocopying, recording or otherwise), without the prior written permission of both the copyright owner and the above publisher of the book. Please do not participate in or encourage piracy of copyrighted materials in violation of the author's rights. Purchase only authorized editions.

The author has made every effort to provide accurate Internet addresses in this work at the time of publication, neither the publisher nor author assumes any responsibility for errors or changes that occur after publication. Further, the publisher and author have no control over and do not assume responsibility for third-party websites or their content.

Second Edition May, 2022

ISBN 978-0-9966299-5-9

Printed in the United States of America

In right knowledge the study of man must proceed on parallel lines with the study of the world, and the study of the world must run parallel with the study of man.

~ *George Gurdjieff*

Gurdjieff's Hydrogens

Volume 1

The Ray of Creation

Contents

An Author's Introduction — 1
   The Gurdjieff Legacy — 2
   Objective Science — 2
   How to Best Use this Book — 3
   The Third Obligolnian Striving — 5

The Apes of Objective Science — 7
   The Fundamentals of Contemporary Science — 7
   Points to Ponder Concerning Contemporary Science — 10
   The Hypnotic Power of Mathematics — 14
   The Genesis of Standard Models — 18
   The Cosmology of Contemporary Science — 19
   Darkness — 31
   The Cosmic Microwave Background Radiation — 33
   The Birth of The Elements — 36
   The Evolution of Galaxies — 38
   The Big Bang Objections — 38
   The Quantum Mechanical Morass — 41
   The Wave/Particle Problem — 43
   Philosophical Discord — 45
   The Abuse of Probability and Statistics — 46
   Fundamental Forces — 46
   The Large and The Small — 48
   Mach and Weber — 50
   What is a Field? — 52

The Threshold of Objective Science — 55
   Fundamental Concepts of Objective Science — 58
   The Ray of Creation — 62
   The Law of Three — 64
   The Law of Seven — 66
   The Modern Musical Scale — 72
   Materiality, Atoms, Forces, and Laws — 74
   The Lateral Octave From The Sun — 80
   Three Octaves Within The Ray of Creation — 81
   Knowledge and Information — 82

The Ladder of Hydrogens	85
The Four Elements: Earth, Water, Air, and Fire	86
Points of Stability	89
Hydrogen, Carbon, Oxygen and Nitrogen	91
The Table of Hydrogens	94
The Trogoautoegocratic Perspective	100
The Reduced Hydrogens Scale	103
Hydrogens as Substances	104
The Essence of Plasma	115
What is Plasma?	115
Waves: Propagation & Radiation	117
The Three States of Plasma	123
An Excursion into Fundamental Physics	125
The Varieties of Energy	129
Energy and Changes in Materiality	134
Planck's Constant	138
Plasma and Static Electricity	139
Electricity and Magnetism	142
Emanations and Radiations	146
Form in Plasma	149
The Diagram of Everything Living	153
Cosmic Units, Life-Forms	153
The Step Diagram	157
The Enneagram of a Cosmos	160
The Step Diagram and The Law of Three	162
The Lateral Octave	164
Earth, Water, Air and Fire	166
The Domain of Earth	169
The Cycle of Rock	170
Kernel Life	175
Metals Life	178
Minerals Life	180
Earth in Overview	183
The Domain of Water	185
The Earth's Water Cycle	186
The Domain of Water	187
Minerals Life	188

Soil	191
Plants Life	194
Invertebrates	199
Water In Summary	200

The Domain of Air — 203

The Cycle of Air	204
The Domain of Air	207
Invertebrates	207
Vertebrates	213
Nature—the Great Survivor	219
The Emotional Center	225
Man	229
Air in Summary	230

The Domain of Fire — 233

The Plasma Cycle	234
Domain of Fire	237
Man, Organic Life and Plasma	245
Angels and Archangels	250
The Second Lateral Octave	258
Feeding in a Different Way	263
The Sun Absolute and the Eternal Unchanging	270
Fire in Overview	272

Plasma Cosmology — 273

Birkeland Currents	274
The Electric Sun	281
The Birth of Suns and Planets	285
The SAFIRE Project: An Electric Sun	287

Planets, Earth and Moon — 291

The Planets	292
Objective Science and the Moon	298
Feeding The Moon	303
The Growing Earth	309

A New Model of The Universe — 319

The Narrative	320
The Living Universe and the Trogoautoegocrat	323
Plasma and Plasma Structures	324

The Unimportance of Humanity	328
The Genesis of Nature	330
A Tale of Four Elements	331
Mother Nature	339
The Possibility of Failure	342
Man	343
The Trogoautoegocrat and Plasma Beings	345
Further Questions	347
Bibliography	**349**
About the Author	**351**

Illustrations

1. Redshift of Quasar 3C 273 — 27
2. Quantum Mechanics' Standard Model — 42
3. The Double Slit Experiment — 44
4. The Thought Experiments — 51
5. The Ray of Creation — 62
6. The Law of Three — 65
7. Inner Octaves — 67
8. Octaves, Notes and Stopinders — 68
9. Involution — 69
10. Evolution — 71
11. The Ray, Atoms and Laws — 75
12. The Ray of Creation From Below — 76
13. The Ray of Creation From Above — 78
14. The Lateral Octave — 80
15. The Three Octaves — 82
16. Worlds and Laws — 86
17. Earth, Water, Air and Fire — 88
18. The Ray of Creation and Law of Three — 93
19. The Three Octaves — 95
20. The First Class of Hydrogens — 96
21. Three Octaves of Hydrogens: From the Absolute to the Moon — 97
22. The Reduced Scale — 103
23. The Periodic Table and the Halogens — 106
24. Water Molecule — 108
25. The Periodic Table and Life — 112
26. Ions as Particles, Atoms or Molecules — 116
27. Types of Wave — 118
28. Newton's Cradle — 118
29. The Movement of Waves on Water — 119
30. The Spectrum of Electromagnetic Radiation — 121
31. An Electromagnetic Wave — 122
32. The Three States of Plasma — 123
33. A Pendulum — 132
34. Atoms and Materiality — 135
35. Six Levels of Materiality — 136
36. Coulomb's Law — 140
37. Capacitors with a Vacuum or Dielectric Material — 141
38. Metalloid Staircase — 142
39. Magnetic Fields and Electric Fields — 143
40. A Wire and Fields — 144
41. Current Through Two Adjacent Wires — 145
42. Electromagnetic Field — 147
43. Simple Dynamo — 148
44. Plasma Crystallized in Zero Gravity — 149
45. The Fourth Phase of Water — 150
46. Plasma Double Layer — 151
47. The Step Diagram — 158
48. Food Octaves on The Enneagram — 161
49. The Step Diagram: Alternate Squares — 163
50. The Lateral Octave From The Sun — 165
51. Organic Life on Earth and the mi-fa Interval — 166
52. Earth, Water, Air, Fire — 167
53. The Earth Triplet — 169
54. The Earth's Core — 170
55. The Earth's Cycle of Rock — 171
56. The Periodic Table — 178
57. The Cycle of Water — 186

58. The Water Triplet	187
59. Plants	194
60. The Cycle of Air	205
61. The Triplet of Air	207
62. The Taxonomy of Life	218
63. The Four Bodies of Man	228
64. The Plasma Cycle	235
65. The Fire Triple	237
66. The Ionic Structure of ATP	246
67. The Ray and the Two Lateral Octaves	259
68. The Higher Realm	270
69. Birkeland Currents	275
70. The Compressed Central Cylinder and its Ions	276
71. Z-pinch Plasmoid	277
72. Birkeland Currents Connecting Galaxies	278
73. The Solar System	280
74. The Electric Sun	281
75. Hertzsprung-Russell (HR) Graph	283
76. The Birth of Stars	285
77. The SAFIRE Reactor Chamber	288
78. Moon and Earth	300
79. The Moon, Earth and its Magnetotail	302
80. Earth's Growth based on the Size of Land Life	312
81. Earth's Growth based on Continental Drift	313
82. The Ray The Ray, Laws, Atoms and Elements	321
83. The Enneagram	322
84. The Three States of Plasma	324
85. Plasma Structures	325
86. The Z-pinch at a Galactic Center	326
87. The Lateral Octave	329
88. The Step Diagram	332
89. Nature as a Living Being	339
90. The Flow of Plasma to the Moon	341
91. The Ray and the Two Lateral Octaves	346

Tables

1. The Just and Equally Tempered Scales Compared — 73
2. The Hydrogens with Examples — 105
3. The Inner Ansapalnian Octave — 110
4. A Comparison of Atoms — 137
5. Key to *Figure 60*: The Cycle of Air — 204
6. Planetary Atmospheres — 206
7. Parallels Between Man and The Solar System — 239
8. The Planets and The Endocrine Glands — 241
9. The Speed of Evolution — 244
10. The "Average" Hydrogens — 250
11. Man's Food and Elimination — 254
12. Hydrogens of the Four Bodies — 266
13. The Psychic Hydrogens of the Center of Gravity Squares — 267
14. The Psychic Hydrogens of the Transition Squares — 268
15. The Solar System's Large Moons — 308
16. The Step Diagram and Relationships of Life-Forms — 333
17. Planetary Atmospheres — 339

An Author's Introduction

"I've always felt that the ideas of the Work were to be used. Not just to be 'bees in amber.'"

~Rina Hands

Please read this chapter. It will be easier to profit from this book if you do. The book intends to help the reader gain a deeper knowledge of objective science than can be acquired simply by reading *In Search of the Miraculous* by P D Ouspensky and deducing what you can from Gurdjieff's obtuse descriptions of objective science in various parts of *An Objectively Impartial Criticism of the Life of Man or Beelzebub's Tales to His Grandson* (*The Tales*).

The descriptions of objective science that haunt the pages of *In Search of the Miraculous* are over 100 years old. And while the explanations of objective science in *The Tales* are only 90 or so years old, they are far less accessible to the average reader. There needs to be a modern context within which these ideas can be pondered and comprehended. This book seeks to provide it.

Our Dubious Inheritance

Each of us possesses a readily available collection of low-quality information. Some of it is the residual imprint of our education. It also includes information about contemporary science fed to us by modern media, and books we have read. We formed our personal world view from this material and, most likely, embellished it with personal beliefs that have been neither pondered nor tested.

The author hopes that if such a dubious inheritance sullies the reader, he or she will gradually turn away from and discard that influence as they make their way through the pages of this book. Ideally, they will restate or redefine their "first principles"—the basic principles they employ to understand the world. Objective science provides that opportunity, but the reader has to choose it.

The Gurdjieff Legacy

The legacy Gurdjieff handed down to us has multiple pillars. One pillar is the Gurdjieff Movements—remarkable dances whose virtue becomes obvious if and only if you participate in them. These may be objective dances. Another pillar is the music Gurdjieff composed with Thomas De Hartmann. Part, but not all of it, was created for the movements. You can assess its value by listening to live performances. It is less easy to assess whether this music is "objective" in its own right, but when combined with the movements, experience suggests that it is.

The third pillar of Gurdjieff's legacy is group activities—the days Gurdjieff groups spend working together. The form of these activities was inherited, at least in part, from Gurdjieff's organized work activities at The Prieuré. Participants in such events can decide for themselves whether their impact is objective.

And then there are Gurdjieff's written works. Readers who have made an effort to read them in the ways Gurdjieff advised tend to conclude that they are indeed objective literature.

The fifth and final pillar of Gurdjieff's legacy is his articulation of objective science. Objective science is the foundation of the psychological techniques and methods that Gurdjieff taught. It may be that it is the foundation of everything he taught, but he never said that.

The only direct source for objective science are Gurdjieff's books and lectures including P D Ouspensky's account of Gurdjieff's teaching in *In Search of the Miraculous*. Aside from that, useful details can be gleaned from the writings of some of his pupils. We provide a bibliography of what we regard to be useful sources in the Appendix to this book. Gurdjieff never claimed to have invented any of it.

Objective Science

The Merriam Webster dictionary's definition of "objective," as it applies to human activity, is as follows:

> "expressing or dealing with facts or conditions as perceived without distortion by personal feelings, prejudices, or interpretations."

Gurdjieff's use of the word "objective" and the meaning he gives it does not agree with that definition. In *In Search of the Miraculous*, when discussing subjective and objective art, he insists that objective art is mathematical. The artist plans and calculates his work to have the impression he wants to convey on everyone who observes it—according to their level of understanding. So, in respect of people on the same level, they will receive the same impression, with mathematical certainty. It is as exact as a car manual might be. All readers who can understand will understand the same thing.*

Objective science is objective in the same way. People understand according to level, to their ability to understand. Some of the scientific concepts of objective science are difficult to understand. Transforming those words into knowledge and understanding requires personal effort—a distinctly different personal effort than is required in any area of subjective science. None of it can be understood by rote learning, and the ability to describe it does not prove any understanding.

To make matters a little more complicated, Gurdjieff only provides fragments rather than a complete corpus, and deliberately so. It has the character of a jigsaw puzzle with some of the pieces missing. You are required to create those pieces yourself.

How to Best Use this Book

This book is intended primarily for people interested in or involved in the Gurdjieff Work. They are likely to have a familiarity with some of the ideas and information the book offers. It focuses particularly on three things: the Ray of Creation, the *Hydrogens*, and the Diagram of Everything Living (commonly referred to as the Step Diagram).

However, whether you fall into this category or not, you may find it helpful to refer to *In Search of the Miraculous* where appropriate. For that reason, we provide relevant references to pages of that book, so readers can consult it if they wish.

This book is organized and proceeds according to the following order:

The first chapter, entitled *The Apes of Objective Science*, discusses modern science and provides an impartial critique of its methods and beliefs. The subsequent two chapters, *The Threshold of Objective Science* and *The Ladder of Hydrogens*, provide a detailed definition of what objective science is, followed by descriptions and discussions of the Ray of Creation, the Law of Three, the Law of Seven and the ladder of Hydrogens, focusing on substances as defined by objective science.

If you think of this book as an octave (and it inevitably is), then chapter four, *The Essence of Plasma*, intends to fill an interval. The problem it addresses is

---

* *In Search of The Miraculous* by P D Ouspensky, p296

that most readers will know very little about electricity or the state of matter that modern science calls plasma. They will know more after reading this chapter. It is difficult, perhaps impossible, to understand objective science without any knowledge in this area.

In chapter five, *The Diagram of Everything Living*, the question we examine is: what is life on Earth? This chapter provides objective science's perspective. The four chapters that follow, *The Domain of Earth*, *The Domain of Water*, *The Domain of Air*, and *The Domain of Fire*, expand on this perspective.

The subsequent two chapters, *Plasma Cosmology*, and *Earth and Moon*, scale up from considering life in the solar system to the life of the macrocosm: moons, planets, suns and galaxies. You could say that this is where the book ends. However, to complete the octave, the author provides a summary chapter entitled *A New Model of The Universe*—a new narrative to describe the real world.

A Note About Typography

The terminology used in objective science can be a potential source of confusion because some of the words used add an additional meaning to an already familiar word. For that reason we have applied a specific typographical style to diminish possible confusion, as follows:

- When discussing the Ray of Creation we capitalize the first letter of the name of the note as follows: Absolute, Sun Absolute or All Worlds, All Suns or Milky Way, Sun, Planets, Earth, Moon.
- When referring to the four elements, Earth, Water, Air and Fire, we use small caps in the following way: EARTH, WATER, AIR, FIRE.
- When describing substances as types of Hydrogen, Carbon, Oxygen and Nitrogen, we italicize and capitalize as follows: *Hydrogen, Carbon, Oxygen, Nitrogen*.
- When writing about the various squares in the Step Diagram, we also italicize and capitalize as follows: *Kernel, Metals, Minerals, Plants, Invertebrates, Vertebrates, Man, Angels, Archangels, Eternal Unchanging, Sun Absolute*.
- When referring to the notes of a specific octave, do, re, me, fa, sol, la and si, we print them bolder and italicized as follows: ***do, re, mi, fa, sol, la, si***. We treat octave intervals in the same way: ***mi-fa, si-do***.

Standard typography is used for all other words.

Author's Introduction

The Third Obligolnian Striving

Ponder this assertion:

In respect of science, all evidence is circumstantial.

It is as true of objective science as it is of contemporary science. This book is aware of that. Objective science is a set of theories of the outer world that can be falsified by contradictory evidence but can only be affirmed by circumstantial evidence. This book provides both a narrative and circumstantial evidence for those theories.

As regards the inner world of man, the responsibility is with the reader. The author hopes the reader has taken to heart the ancient words of Hermes Trismegistus:

"As above, so below."

As stated on page 386 of *The Tales*, the third obligolnian striving is as follows:

… the conscious striving to know ever more and more concerning the laws of World-creation and World-maintenance.

People in the Gurdjieff Work often regard the five obligolnian strivings as direct injunctions.

Sadly, in this media-driven age, the pursuit of the third striving is complicated by the dubious assertions and ceaseless stream of "news" emanating from the scientific establishment and its servants in the media. People are suggestible and we too are prey to that weakness. If we are to pursue the third striving honestly, on one hand we must ignore the siren songs of subjective science, yet on the other we need to analyze its activity and peruse the data. It is not without value.

Objective science requires a modern context within which it can be pondered. The aim of this book is to provide such a context.

GURDJIEFF'S HYDROGENS: THE RAY OF CREATION

Chapter 1

THE APES OF OBJECTIVE SCIENCE

—※—

"I sought great human beings, but found only the apes of their ideals."
~ *Friedrich Nietzsche*

In the author's experience of the Work, he has repeatedly noticed and experienced a definite tendency in Work groups to accord credibility to the theories of contemporary science. He even once heard a definite opinion that "In Gurdjieff's time, science was less advanced and so Gurdjieff's scientific understanding was clearly 'off the mark.'"

This chapter disputes that opinion without an atom of compromise. For the sake of brevity, it confines itself just to modern physics and it proposes that modern physics, even in terms of its own approach to knowledge, is misguided and wrong-headed in respect of the theories it espouses. It is "off the mark," by a country mile.

The Fundamentals of Contemporary Science

The paragraph below, taken from Wikipedia, briefly describes the scientific method:

> *The scientific method is a body of techniques for investigating phenomena, acquiring new knowledge, or correcting and integrating previous knowledge. To be termed scientific, a method of inquiry is based on empirical or measurable evidence subject to specific principles of reasoning.*

We can formulate it as a series of easy to understand actions, as follows:

- **Observation:** Some phenomenon that is deemed worthy of investigation is observed in order to arrive at an explanation that can be expressed as a set of principles.

- **Problem statement**: A statement of the phenomenon is made as accurately as possible, perhaps in the form of a question such as: How does A react with B to produce C?
- **Prior evidence:** Prior validated evidence relating to the phenomenon (if any exists) is examined and used if necessary as reference material.
- **Hypothesis:** A hypothesis is derived both from existing evidence and the formulation of the problem statement. The general rule here is that the hypothesis must be falsifiable.
- **Prediction:** A set of unambiguous and well-defined predictions representing the logical consequences of the hypothesis are formulated.
- **Experiment(s):** The predictions of the hypothesis are empirically tested with measured results being obtained.
- **Analysis:** An analysis of the outcome of the experiments is conducted in an effort to prove the hypothesis wrong. If the hypothesis is not negated by the experiments, the outcome of the experiments can be regarded as support for the hypothesis.
- **Reformulation:** If the hypothesis is disproved then it may be reformulated and another iteration of prediction, experiment and analysis may take place.

So, scientists observe the natural world, formulate hypotheses, test them experimentally and then adjust the hypotheses if necessary in response to experimental outcomes. If the hypothesis is general enough, and has sufficient experimental support, it becomes a theory that is taken to apply to many contexts. If there is enough scientific consensus for a long enough time, the theory may even be accepted as a "law," implying that it holds unfailingly in a set of well-delineated contexts.

So, for example, we have Boyle's Law, which states:

The absolute pressure exerted by a given mass of an ideal gas is inversely proportional to the volume it occupies if the temperature and amount of gas remain unchanged within a closed system.

Note that in the statement of a scientific law, the context is precisely defined (in this example by the terms: absolute pressure, ideal gas, closed system). This theory acquired the status of "law" by being repeatedly proved in all appropriate contexts. Science rarely proclaims something to be a law. For example, Einstein's general theory of relativity is widely regarded as correct at large scale and has been validated to some degree (in the sense of predicting experimental outcomes), but is still only accorded the status of theory.

Science is essentially collaborative. One individual could formulate a hypothesis in a given area and carry out many experiments that (in his opinion) unquestionably proved his hypothesis. On its own this counts for nothing: the scientist may be incompetent, he may be competent but have made an error in the design of his experiments, he may have failed to account for some factor that could impact the results and so on.

Consequently, within the scientific community, these hypotheses and experiments are subject to peer review by other scientists working in the same field. Hypotheses and the results of experiments are shared via the publication of papers, articles in scientific magazines and by presentation at scientific conferences. Comments and criticisms ensue and, over time, a general consensus emerges as to what is regarded as true, or likely to be true, in any scientific field.

In areas of science that attract the interest of the general public, information is disseminated by way of articles in magazines, newspapers and documentary television programs. Information is also disseminated through the education system, as various theories and "accepted truths" are included in school and university curricula.

The body of scientific theory, knowledge and information gradually expands over time, with some theories being adjusted and others being abandoned in favor of new ones. Occasionally some scientific hypotheses and experiments prove to be revolutionary, provoking a whole area of science to be rethought and reconstructed.

In some areas, science has become an expensive activity because of the cost of equipment needed to carry out well-designed experiments. This is the case, for example, in many areas of physics, chemistry and materials science. Here, funding is provided by governments and commercial interests, some of whom hope to profit from their donations. While this can at times exert an influence on science, it is rarely a malign influence.

Direct political influence in various eras has interfered with scientific activity, most obviously as occurred when the Roman Catholic Church tried and failed to enforce a biblical world view in contradiction to the ideas of Galileo and Copernicus. Science in Nazi Germany and Soviet Russia was deflected for a while by political interference, but this eventually faded when the political weather changed. In recent times oil interests have interfered politically with climate science, but this also is now fading. In such situations, the inherent idealism of science—the search for the truth—is difficult to suppress indefinitely.

This is not to suggest that science is truly impartial. Some theories are established, become popular within the scientific community and eventually represent vested interests that the community defends against any contrary view. Sometimes a scientific idea becomes so offensive to the scientific establishment that, as Gurdjieff describes in *The Tales*, the one who proposed it is "pecked to death," within the scientific community, and at times in the court of public opinion. This was the fate of Mesmer, and more recently the fate of both Immanuel Velikovsky and Wilhelm Reich. This is not to imply that the theories of these individuals were correct; only that they suffered the process of being "pecked to death."

The Gurdjieff Work itself has received the occasional "peck" from the representatives of the scientific establishment, usually being dismissed as mystical claptrap and Gurdjieff himself being described as a charlatan or worse. This is to be expected. Contemporary science tends to denigrate the mystical.

Points to Ponder Concerning Contemporary Science

The question to ask is: How useful is contemporary science as a source of knowledge? There are good reasons to be cautious about its various theories and proclamations. Consider the following.

Suggestibility

As normal human beings we are suggestible. Throughout our life we have received many suggestions that originated with contemporary science. Mostly, we believed these suggestions without question.

Most of the things we think we know we simply accepted "in good faith." For example, ask anyone who is not in the Work about "how life came to be" and you will normally be treated to a mishmash of ideas that revolve around Darwinian evolution. Most likely the person describing these ideas will not have studied the field at all and will simply have accepted "in good faith" what they were taught at school, or encountered in the media, or have read about in books and magazines. They are unlikely to provide a critical view. Even if they do provide a critical view, it is most likely that their critique will reflect something they read or heard, rather than their own thinking.

To consider the opinions of another to have any validity, one needs to know their source. If they themselves are not the source, one needs to determine who was the original source and consider how they arrived at their opinion. An agreement with the opinion requires either a review of how the original opinion was arrived at, or one's own analysis that arrives at the same opinion via a different route.

Contemporary science has no unity

We might believe that the body of scientific thought constitutes a unity, allowing perhaps for the reality that this body of thought is gradually evolving. However, it has no unity; it is a consensus reflecting the opinion of those highest in the scientific hierarchy. You can see this. You might think otherwise, if you peruse Wikipedia, which is a large conscientiously maintained scientific information source. However, if you choose any particular theory at random and google the topic, you will usually discover opposing theories and dissenting opinion. Wikipedia has evolved into the "official" mouthpiece of science.

There is no individual in any scientific field who is the acknowledged "master." Even in fields where one individual's work and opinions are dominant for a while, it is unlikely that he or she is conversant with all hypotheses and experimental results in their own field. And their expertise beyond their own field is likely to be thin or non-existent.

Even if we assume that such atypical individuals have achieved genuine knowledge in their field, it is not our knowledge. We, who have never delved deeply into their scientific domain, can only accept their theories and proclamations "on faith." And unless there is evidence to the contrary, it will be prudent to assume that they are normal human beings endowed with the usual failings. It would not be prudent to assume that they know anything "for certain."

"Scientific truth" is and will forever be the aggregation of many 'I's.

Are scientific experiments truly repeatable?

The repeatability of experiments is a supremely important criterion for accepting any scientific hypothesis. Some experiments certainly are repeatable. If you mix a given amount of silver nitrate with a given amount of sodium chloride at a specific temperature, you will produce a precipitate of a given amount of silver chloride. You always get the same result. So it is with some scientific experiments. However there are also many notable failures to repeat "discovered" phenomena.

One famous area of disputed claims is the ESP research of J B Rhine, which suggested experimental "proof of telepathy." It was never independently verified. Once you enter the area of scientific psychology, you encounter the problem of the experimenter unwittingly influencing the experiment, and the additional problem that one group of subjects is not necessarily equivalent to another. Rhine's experiments may have suffered from both of these failings. Perhaps J B Rhine and his methods were at fault, and perhaps not.

As Heraclitus noted "No man ever steps in the same river twice, for it's not the same river and he's not the same man."

The point is that repeatability is not easy to establish, because all the factors that influence the outcome of an experiment may not be known.

Where you do not have repeatability, the scientific method rules itself out—in theory. In practice that important criteria is not always enforced. Some experiments, notably those carried out in the Large Hadron Collider (LHC) that is buried beneath the France-Switzerland border, escape the "rule of repeatability" because there is only one LHC and the demands for its use far outstrips availability.

Even if you successfully "exactly repeat" an experiment with this equipment, until someone builds another equivalent LHC, you cannot know for sure that there wasn't some subtle fault in the experimental equipment.

The problem of "the closed system"

It is usually the case that a scientific hypothesis is expressed in terms of cause and effect, in the sense that a particular action in a particular situation causes a particular result. The problem in proving such a hypothesis is that the scientist needs to design an experiment that eliminates all extraneous influences. A closed system needs to be created which includes only the relevant components. However since the scientist cannot know everything that must be eliminated—since he is dealing to some extent with the unknown—it is difficult to be certain that an experimental design creates a genuinely closed system.

More to the point, the truth is that there are no fully closed systems. The only truly closed system in the universe is the universe itself and even that it a conceptual assumption. It is also worth noting that almost all the experiments that have been carried out since the dawn of science have been carried out on the planet Earth.

All, including those carried out in orbit around Earth or in its vicinity, are proven only in this locality. If there is some influencing factor in this locality that does not generally apply throughout the universe, then the generality of all of science is in question.

The statistical problem

Where scientists cannot create a closed system, they will attempt to verify a hypothesis statistically. If an experiment does not always provide the same result, but when repeated produces the result a statistically significant number of times, the hypothesis is deemed to be supported if not proven. Common examples of this are found in the field of medicine.

When searching for the cause of a particular disease, epidemiologists will conduct experiments to try to identify the responsible pathogen. If you review such experiments, you will find that there is normally a control group of people in the locality under investigation, who show no symptoms of the disease. Their health is compared to a group suffering from the disease. If the pathogen is isolated, it will normally be found, by test, to exist in the bodies of most of the infected group—but not all of them. In the control group it will be found to be absent in the bodies of most, but not all. This is a strong sign that the identified pathogen is the cause.

You might protest the fact that the pathogen cannot be isolated in every one of the infected group, and that it can be found in one or two of the "uninfected" group. But the human body is a very complex system and there can be great variability from one such system to another. The few in the uninfected group, who show signs of the pathogen, may have very robust immune systems and antibodies that can cope with the pathogen. On the other side of the line, those who showed no evidence of the pathogen, but had symptoms of the disease, may have been affected by undetectable levels of the pathogen.

In any event, with epidemiology, that is merely the beginning of the story. The next steps are to proceed from these results to identify how infection by the pathogen occurs (by contagion, by insect bite, etc.) and to find ways to prevent transmission. Where such campaigns are successful it is clear that the pathogen has been nailed.

The point is that the statistics only demonstrated a correlated association. Such associations do not prove causation at all, they only indicate the possibility of causation. Nevertheless, such statistical data is often imputed to demonstrate causation, even among scientists. The fault is not in statistics itself, but in its abuse.

The book *Spurious Correlations** presents many excellent and amusing examples of correlations that clearly have no direct relation to causation. They include:

---
* *Spurious Correlations* by Tyler Vigen

- Figures from 1999 to 2009 demonstrate a 99.79% correlation between US spending on science, space and technology and US suicides by hanging, strangulation and suffocation.
- Figures from 1996 to 2008 demonstrate a 95.23% correlation between Math doctorates awarded in the US and the amount of uranium stored at US nuclear power plants.
- Figures from 1999 to 2009 demonstrate a 95.45% correlation between US crude oil imports from Norway and US drivers killed in a collision with a railway train.

At above 95%, all of these are very high correlations, demonstrating how slippery correlation can be in any scientific context. And yet, contemporary science cannot proceed without using statistical correlation. If a scientist can present high correlation along with a convincing explanation of why A causes B, the hypothesis is likely to be given credence. Contemporary science is obliged to walk this line.

Scarcity of data

In some areas of scientific study there is insufficient data to offer strong support to any theory. If we take cosmology as an example, the field of study is handicapped because we can only make observations of the universe from the Earth or from satellites within the solar system. The accurate data that has been gathered is also confined to a relatively short period of time—a few hundred years at best—less than the blink of an eye in the life of the universe.

Similarly, paleontology, the study of ancient life, is restricted to what can be discovered via the fossil record. Data is also confined to the specific times when fossils were created. And the fossil record from any given era is only a minuscule snapshot of that time. This leaves huge gaps in the evidence for any theory in this field.

Other areas of science are not so constricted. For example, with modern instruments, zoologists can examine both living and recently dead specimens of a species in fine detail and gather very large amounts of data to support or oppose any specific theory.

The Hypnotic Power of Mathematics

Mathematics is not a science per se. It is a very useful related discipline that provides scientists and engineers with extremely useful tools—statistics being just one of them. Nevertheless, the point needs to be understood at the outset that mathematics does not and cannot prove anything in relation to reality.

The Apes of Objective Science

Albert Einstein said, famously;

> "As far as the laws of mathematics refer to reality, they are not certain; and as far as they are certain, they do not refer to reality."

He got it exactly right. There is only one area where mathematics aligns almost completely with reality. That is in the act of counting. You might argue philosophically that, say, "three apples" in the real world only embody the concept of "threeness," but it is splitting hairs to distinguish that from their embodying the concept of "appleness." When there are three apples, there are three apples. The counting numbers—the natural numbers, as mathematicians call them, can reasonably be taken to denote a reality of the real world.

Beyond that, when we talk in terms of negative numbers, real numbers, irrational numbers, or complex numbers, we are manipulating abstract concepts that cannot be demonstrated to exist in reality. They are inventions of the mind of man that can, nevertheless, be put to good use to model reality. And that's just for starters. We can introduce algebra, geometry, calculus and all the various fields of mathematics, most of which can also be put to excellent use in modeling reality.

Contemporary science and engineering employ mathematics to model reality and, time and again, the models turn out to be so close to reality that it predicts the real world accurately. In some instances the very laws that science proposes can be expressed mathematically—as for example with Newton's laws of motion.

Indeed, Newton serves as an excellent example of the productive use of mathematics, since his gravitational theory and its associated equations are pretty much all you need to spray space shots around our solar system. He formulated it more than 200 years before the first space shot.

And none of that proves Newton's gravitational theory correct. In fact nowadays his gravitational theory is regarded as incorrect and has been superseded by Einsteinian gravitation. The mathematics did not and never could prove the theory correct, but it created a very close-to-accurate model of reality.

The Map is not the Territory

There are several things to be concerned about with mathematics. The first is to note that in terms of the models it can create, mathematics can be divided roughly into two parts: discrete mathematics and continuous mathematics.

Discrete mathematics is the study of mathematical structures made up of separate components, individual items, like quanta. So the objects studied in discrete mathematics, such as integers and statements in logic, have distinct, separated values.

In contrast, continuous mathematics deals with objects that vary smoothly, without gaps. An example of this continuity is the simple equation $y = 2x$. The two linked variables, y and x, in the equation are continuous.

We can ask the question: Is reality continuous or discrete?

The evidence from quantum mechanics is that reality is discrete. So, for example, a spectrometer viewing the light emissions from a particular substance shows us spectral lines rather than a continuity of wavelengths. This, incidentally, accords with the objective science view of reality. However the way our minds model the world is necessarily that way. We see, for example, an iron bar. It is a thing and hence discrete. However we see it and conceive of it as continuous from end to end. We can imagine that it is composed of atoms—discrete things—but we do not know for sure. Even if we possessed a microscope that was powerful enough, and could clearly see the atoms, we could not know for sure whether the space between them was really empty rather than containing some kind of material or energy.

While it is the case that discrete and continuous mathematics can be used together in some contexts, a mathematical model of a real situation always depicts the world as one or the other, discrete or continuous. If the model works well, it will be valued. For example, the mathematical models that were used to calculate artillery tables were valued by the military because they were very accurate within practical parameters, but they were not perfect.

We adopt the same attitude to the problem of infinity—a concept that emerges in both continuous and discrete mathematics. We require the concept, for example, to establish calculus. We cannot demonstrate infinity in the real world, we can only presume it. Nevertheless, if we're careful in using it, we can employ it productively in mathematics, and employ the mathematical models we create productively. Mathematics can be right within its own context of modeling and mapping. However the map is not the territory.

Extrapolation

Mathematically, extrapolation is where you extend a method (say a formula or a technique) outside the range of proven real world applicability, and logically deduce that it applies to all areas outside the range. Mathematics even has a specific kind of proof (the inductive proof) that works by extrapolation.

This is fine in the domain of mathematics, as it does not need or even care for real world confirmation. It is correct axiomatically and thus an inductive proof applies all the way to infinity. All mathematical extrapolation is valid for the mathematical map, but the map is not the territory.

As soon as you apply extrapolation in science you are on shaky ground. Consider, for example, an activity as fundamental to geology as estimating the age of rocks. Such ages are calculated on the basis of radioactive decay. For example, the element Uranium 238 decays to become Thorium, which in turn decays until it becomes Lead. There are many steps to this process. The geological dating of a specimen can be achieved by estimating the original content of Uranium 238 and all the elements and associated isotopes in the rock sample when the rock was formed. The rock's age is deduced according to the quantities of those elements and isotopes. The known proven-in-the-laboratory pattern of decay of Uranium 238 is applied. This is an extrapolation.

The accuracy of the calculation suffers from three problems:

1. The estimate of the original content of U238 and the elements and radioisotopes it decays into could be incorrect.
2. The rock could have been contaminated during its long life in a way that altered the ratios.
3. The normal (predictable) process of radioactive decay could have been accelerated or decelerated by unusual conditions some time during the lifetime of the rock.

One example of an anomaly is sufficient to demonstrate this problem. Radioactive dating on recent (roughly 50-year-old) lava flows at Mt. Ngauruhoe, New Zealand, have yielded a rubidium-strontium "age" of 133 million years, a samarium-neodymium "age" of 197 million years, and a uranium-lead "age" of 3.9 billion years. In each case, the dating method gives a wildly incorrect result and, as you can see, they are not even close to agreeing among themselves.

But what is the geologist to do? There are no better methods for estimating the age of rocks. It may even be that some of these estimates are correct in some cases. However, there is good reason to doubt.

Cognitive Bias and Mathematical Manipulation

In formulating hypotheses and proposing scientific models of real-world events, scientists almost always encounter the fact that their carefully designed experiments do not produce the hoped-for result, but may provide something that is close to the hoped-for result. In this area we encounter the problem of "cognitive bias."

The term "cognitive bias" describes errors in thinking processes caused by holding on to individual preferences in the face of contrary evidence. This could be described as "unintentional dishonesty," in that the individual affected by it is completely unaware of their bias. Where it crops up in scientific experimentation (outside of psychology, where it is an area of study), it is described as "confirmation bias." It is the tendency to interpret experimental results in a way that confirms one's cherished hypotheses or even pre-existing beliefs. In science pre-existing beliefs are often just fashionable theories.

The scientific method is supposed to eliminate such bias by the process of peer review. Other experts in the field review the published results produced by a specific scientist or scientific team and offer critiques. However, peer review is only effective if the reviewers are not also suffering from the same confirmation bias.

As we review some of the theories of modern physics in the coming pages, we will encounter the existence of "adjustable parameters." We can explain by example:

Consider the trajectory of a ball thrown at an upward angle through the air. It will follow a parabolic curve almost exactly, rising in the air at first and then falling. Its position in the air at any point will depend on the initial upward angle of its trajectory and the time elapsed since it was thrown. If there were no other forces affecting the ball it would move in a perfect parabola. However, the resistance of the air to the movement of the ball inevitably distorts the parabola.

If we adjust the mathematical equation by adding an "adjustable parameter," we can compensate for the air resistance. Adding a fixed parameter might do the trick, but air resistance can vary. It will be different at sea level than on a high mountain, and hence the parameter will need to be adjusted, for context.

This does not mean that the theory of the parabolic movement is incorrect, just that we need to adjust the model. The scientific problem is not that adjustable parameters are necessarily wrong—they may not be. But if you cannot assign a cause to the adjustable parameters in a model, then the model is clearly suspect. You can usually make inconvenient results look respectable by resorting to adjustable parameters.

The Genesis of Standard Models

The problem of geological dating highlights a common occurrence with contemporary science. If any new phenomenon is observed in a given area of science, it is soon accompanied by a theory to explain it, no matter how little empirical support the theory has. The motivation to invent a theory is considerable. It's likely that the interesting observation occurred as the result of some hypothesis the scientist was investigating, so it is logical for the scientist to explain the hypothesis when he documents the observations. His reputation within the scientific community will be greatly enhanced if he is recognized as the scientist who invented the accepted theory. If he simply reports the phenomenon, he will be asked for an explanation anyway.

And there are questions that are natural for anyone to ask, and science feels obliged to provide an answer, even when there can be little certainty. How old is the Earth? How did life develop on Earth? Why does the Earth have a magnetic field? When did mankind first appear on Earth?

Providing some theory or other—the best theory available or the accepted theory—always seems preferable to "I don't know." Contemporary science abhors "I don't know"—it abhors a theoretical vacuum.

In time, as various theories are threaded together to generate a common narrative, a "Standard Model" emerges, which becomes the scientific worldview in a given area. This is only to be expected. Psychologically, men develop formatory attitudes about many things. Hence it is to be expected that societies of mechanical men will develop a consensus world-view. There are Standard Models in astrophysics and in quantum mechanics. The theory of evolution provides the Standard Model in botany and zoology.

These Standard Models are absorbed into the education system in most countries and nowadays educational documentaries, regularly broadcast on television, spread the narratives to the general public. They are happily accepted without question and usually without discussion.

This is no different to what happened in previous eras. In the Middle Ages, when the Roman Catholic Church was the arbiter of knowledge in Europe, its Standard Models were spread in a similar manner. Man's suggestibility ensures the success of this.

The Cosmology of Contemporary Science

Currently physics abides by a Standard Model of Cosmology for astrophysics (the macro scale), and an entirely different Standard Model for quantum mechanics (the micro scale). Although attempts have been made to reconcile these two Standard Models, it has so far proved fruitless. Quantum mechanics does not scale up, and astrophysics does not scale down.

Gurdjieff's Hydrogens: The Ray of Creation

It is useful to examine these models from the perspective of objective science, primarily to demonstrate that objective science and contemporary science do not agree at all.

The Standard Model of Cosmology is referred to popularly as The Big Bang Theory, but among physicists it is usually named the Lambda Cold Dark Matter model. It can be summarized as follows:

- The universe originated in a rapid expansion of energy from nothing. This original "nothing" is commonly referred to as a "singularity"—a term borrowed from mathematics. In mathematics, a singularity is a point at which a given mathematical object is not defined, usually due to such problems as "dividing by infinity." In terms of astrophysics the original "nothing" was a gravitational or spacetime singularity—a point where the gravitational field was infinite, or at least exceedingly large.

- The expansion from this "nothing" is described in terms of Einstein's general theory of relativity. Consequently, the "nothing" from which expansion began was not a "nothing" located somewhere in empty space. There was no empty space. In theory the space came into existence because of the expansion; it was created by the expansion. This expansion from a "nothing" occurred around 13.7 to 13.8 billion years ago, and the universe has been expanding ever since. The act of expansion caused cooling of the energy within the expanding space.

- As a consequence, at some point after the beginning, the energy condensed into atoms of hydrogen and helium (very light elements). The expansion gave rise to cooling and the cooling continued. There were thus gas clouds. Under the influence of gravity these gas clouds condensed, forming stars and eventually, planets.

- The process of planet formation involved a large number of supernovas. Contemporary science has just one explanation for the formation of heavy elements. They are believed to be formed by nuclear fusion within stars. Once a star explodes in a supernova, heavy elements of every variety are scattered into the surrounding space and many of these heavy atoms eventually participate in planet formation.

- Some stars do not explode in a supernova event, but condense so completely that they become "Black Holes." They become regions of spacetime with such a strong gravitational field that nothing (no particles or radiations of any kind) can escape from inside.

- The stars grouped themselves into galaxies, the galaxies we observe today. Thus the universe consists of galaxies composed of billions of stars and the stars (or at least many of them) are accompanied by planets, which in turn may have moons.
- Currently this Standard Model includes the idea of "dark matter." Dark matter is non-luminous matter in the sense that it does not emit light and hence is undetectable directly. Dark matter is estimated to constitute roughly 27 percent of the whole universe. Additionally the Standard Model includes the idea of dark energy. This is non-luminous energy and it is estimated to constitute about 70 percent of the whole universe. Thus only around 5 percent of the universe is believed to be luminous (in the sense of being visible to telescopes and other scientific instruments).
- At the farthest reaches of this expanding universe, the extreme edge of its expansion consists of thermal radiation left over from the time soon after the expansion began when hydrogen atoms first formed. This is called the Cosmic Microwave Background Radiation (CMBR).
- The universe is expanding, but it is not expanding at a constant speed, the expansion is accelerating.

This is currently the dominant model. It is not the only scientific model and it is not without its critical opponents within the scientific community. It is very likely that it will be updated with new details or superseded as time passes. It may even have been updated since we wrote these words.

"Evidence" for The Big Bang Theory

In the absence of evidence, the Standard Model of Cosmology may seem bizarre if you have never encountered it. However, those who adhere to it seem comfortable with it and cite evidence for it in four main areas.

1. **The Recession of Galaxies.** One interpretation of measured galactic observations is that the galaxies are moving away from each other. This is taken to demonstrate, if we go back in time, that they originated from a common point.
2. **CMBR.** The existence of Cosmic Microwave Background Radiation with a black body spectrum and a temperature of 2.725 K is observed in every area of the sky (between stars). This is interpreted to be evidence of expansion.
3. **Galactic Evolution.** Galaxies appear to evolve in the sense that those

furthest away appear to be in earlier stages of evolution (astrophysics models galactic evolution).

4. **Light Elements.** The abundance of the "light elements," hydrogen and helium, observed throughout the universe, is deemed to support the Standard Model, although it would be more accurate to say it is simply consistent with it.

Objective science agrees that there was a moment of creation when the current universe came into existence from a previous state. It disagrees with the proposition that it began with a singularity. Its proposition is that the moment of creation occurred when the laws governing the universe were changed by the actions of the Absolute (an intelligence that dominates the universe). The universe then proceeded to develop according to the Ray of Creation, which we discuss in detail in the next chapter.

Objective science asserts that all levels of the Ray of Creation exist within each other. It does not suggest that the universe is expanding or that it is contracting. The presumed recession of the galaxies by astrophysicists is based upon the assumption that distant galaxies must be receding because of the Doppler shift of light (the redshift) from far galaxies. This is an extrapolation that we will discuss in more detail later. But first we need to discuss the merits of extrapolation.

The Problem of Extrapolation

In the objective science world view, extrapolation is not valid beyond a certain point. This is a simple consequence of the Law of Seven as described in the progress of an octave. An octave progresses from *do* to *re* and the same rate of progression persists from *re* to *mi*. Between *mi* and *fa* lies an interval which deflects the octave. From *fa* to *sol*, we get a rate of progression that persists from *sol* to *la*. However the progression from *la* to *si* is disharmonized. Between *si* and *do* there is a second interval which deflects the octave. With this kind of disruption in the progress of an octave, extrapolation is simply not valid. Things do not proceed in a straight line.

Given this fundamental law, backward extrapolation from the present to 13.7 billion years into the past has no validity. Whatever the circumstance, when we encounter any extensive extrapolation in the theories of contemporary science, we would be wise to consider the octave.

We can also look at this another way. From the perspective of objective science everything is alive and hence capable of action within the context of its existence. To suggest that the behavior of galaxies, suns and planets is mechanically predictable presumes that these living entities always repeat the same action without choice. In contrast, objective science proposes the existence of intelligence at every level, with at least some ability to choose.

Space

Our usual understanding of space is that it has three dimensions: length, breadth and height, and that these dimensions extend in every direction without end. We tend to think the universe exists in this infinite space and maybe it too extends infinitely. This is the view that Isaac Newton held of space.

The Lambda CMD model takes the view that the universe is expanding and that the only space which exists is the space within that expansion. Space is bounded and there is no space beyond it. The universe is thus like an expanding balloon with nothing, not even space, outside the balloon. From a philosophical perspective the distinction between these two views is almost meaningless. Whether there is or is not space outside that theoretical "balloon" can neither be proved nor disproved. So it does not matter. Whether the universe is infinite or finite within infinite space does not matter either.

The Lambda CMD model is deduced from Einstein's theory of relativity. In this theory, Einstein treats spacetime as a single four-dimensional framework, with time as the fourth dimension. While the three dimensions of length are not necessarily related to the contents of the four dimensional space, this is not so in respect of time. The dimension of time was deemed to relate directly to the force of gravity. Four dimensional spacetime could thus be curved, with time distorted by gravitational force and hence responsible for the curvature.

Objective science does not align with this view of time, as will be discussed later. Here it is sufficient to note that, from the perspective of objective science, the nature of time is determined by the cosmos within which time is being considered.

There is a conceptual question that can be posed in respect of space, which is as follows:

If a particle is moving at a constant speed in space in a given direction with respect to the three dimensions of space and suddenly the rest of the universe disappears, is the particle still moving?

This question cannot be resolved by experiment in the real world, but it is important because it is fundamental. Newton took the view that the particle would indeed still be moving, implying that empty space itself provides a frame of reference. A further question that can be asked is:

Gurdjieff's Hydrogens: The Ray of Creation

When the universe is removed, does the particle have any mass?

It could not have mass by virtue of gravity since there is nothing else in this hypothetical environment that could generate a gravitational force. So the question is whether it would have mass by virtue of inertia and this is not easily answered.

Inertia (the resistance to a change in motion) could only be detected by the intervention of some force on the particle. But the conditions of this "thought experiment" eliminates that possibility. If space were utterly empty, then there would be nothing to provide a resistance to change, and so the conclusion that it has no inertia and hence no mass, seems valid. However there is another possibility that can be considered—that space can never be empty, but is filled with "aether." If that is the case then the particle could have mass by virtue of the aether. To pursue this any further we need to discuss what we mean by "aether."

The Aether

The existence of aether was suggested by the Ancient Greeks and generally believed to be an aspect of the universe for many centuries, although there were few assertions about its properties and how it influences anything. The aether became important with the study of the wave characteristics of light. Vibrations of sound in the air, or ripples on the surface of water, exhibit a wave characteristic through the material mediums of air and water respectively. They are phenomena that are conducted by fluids (air and water). Because light also exhibited this wave characteristic, even in a vacuum, it was assumed that light waves propagated through a (fluid) medium. That medium was presumed to be the aether.

The ground breaking work of Clark Maxwell in providing a theory, complete with practical equations, to describe the propagation of light and magnetism assumed the existence of an aether through which electromagnetic radiation propagated.

Incidentally, another reason for presuming the existence of aether was the force of gravity. It was clear that masses attracted each other, but through what medium did such an attraction take place and how? The aether provided a possible answer to that too.

Objective science assumes the existence of an aether, as Gurdjieff states somewhat obtusely in *The Tales*. This provides us with good reason to be interested in contemporary science's experiments to determine the existence or otherwise of the aether.

The Apes of Objective Science

The assumption of an aether was never doubted until the famous Michelson-Morley experiment, which attempted to measure the aether but failed to find strong evidence for it. This warrants some comment. First of all there are two possible theories as regards aether. The first is that the whole of space is filled homogeneously with aether. The second also has the whole space filled with aether, but the aether is "entrained" by the activity of the material it surrounds. The second idea proposes that the aether is a fluid of some kind. So the entrainment of aether by the Earth is analogous to the "entrainment" of water by a ball under water, which is both spinning and orbiting some central point. The water in the vicinity of the ball will tend to be dragged with the ball, but water at a distance will not be affected.

The Michelson-Morley experiment was attempting to discover "unentrained aether." It did so by trying to find a difference in speed between two rays of light that travelled at right angles to each other, using an interferometer. We will not go into the fine details of the experiment here, they are well documented elsewhere. The experimenters expected that their results would reflect the approximately 30 km/sec speed of the Earth around the Sun. It did not.

Nowadays if you read about this experiment, the narrative describing it will often insist that the experiment produced a null result, proving that the aether did not exist. That was not the case. Michelson expected that because of the Earth's 30 km/sec motion, his interferometer experiment would yield a result showing a fringe shift equal to 0.04 fringes, but the highest value measured in a number of separate experiments showed a deviation from zero of only 0.018 fringes, and other measurements were much less. His conclusion was that the hypothesis of a stationary aether was not confirmed.

In reality he could never have obtained the result he expected even if there were a stationary aether, because, while the Earth moves in orbit at 30 km/sec, the whole of the solar system is also moving and it moves at much higher speed than that.

Another scientist, Dayton Miller, conducted far more detailed experiments over a period of almost 30 years, also using an interferometer, but with a far more precise set up. One distinct difference between the two experiments was that Miller's experiments were conducted on Mt Wilson, at an elevation of 6000 ft. He was not investigating the idea of stationary aether, but of Earth-entrained aether. The expectation was that the aether would move at a faster speed at such an elevation than at sea-level.

His experiment yielded consistently positive results, which varied according to time of day and season. Analyzing his experimental data, he eventually concluded that there was an entrained aether. He also concluded that the Earth (and the whole solar system) was moving at a speed of 208 km/sec. towards an apex in the Southern Celestial Hemisphere.

After Miller's death, it was discovered that the solar system is indeed moving—it is gradually orbiting the Milky Way and the speed of its motion is approximately 230 km/sec, not so different to the 208 km/sec figure Miller suggested. (And, incidentally, the Milky Way is also moving.)

Unfortunately for Miller, and his place in contemporary science, by the time he was publishing his results, the world of physics had become enamored of Einstein's relativity and so his results were ignored. Later, after he died, his experimental results were "pecked to death" by critical review. Nobody made any attempt to replicate his work, it was simply dismissed. The fact that Miller had predicted (within 10%) the speed of the solar system was also ignored.

Other experiments in the last hundred years which seem to suggest the existence of an aether, including those by M G Sagnac, M Allais, E Silvertooth, R DeWitte, and Y Galaev have tended to be ignored or dismissed, or simply "explained" away. No matter how much modern science would like to deny it, there are experiments that support the existence of aether.

As we shall soon see, the Standard Model of Cosmology is in considerable disarray. Until the model ceases to be based on Einstein's general relativity, it will never accommodate aether. Einstein himself declared that proof of the existence of aether would invalidate relativity.

The Recession of The Galaxies

There is a "law" derived from general relativity equations by Georges Lemaître which defines the expansion of the universe. The estimated value for the rate of expansion is called the Hubble constant and the "law" is called Hubble's Law. It states that:

- Objects observed in deep space (extragalactic space) show a Doppler shift in the light they emit. This can be interpreted as relative velocity away from Earth.
- The Doppler shift measured velocity of various galaxies receding from the Earth is approximately proportional to their distance from the Earth (for galaxies up to a few hundred megaparsecs).

The reason that Edwin Hubble's name is attached to both the constant and the law (and also the famous orbiting telescope) is that Hubble originally provided "evidence" for this expansion of the Universe using measurements of Doppler redshift from a collection of galaxies.

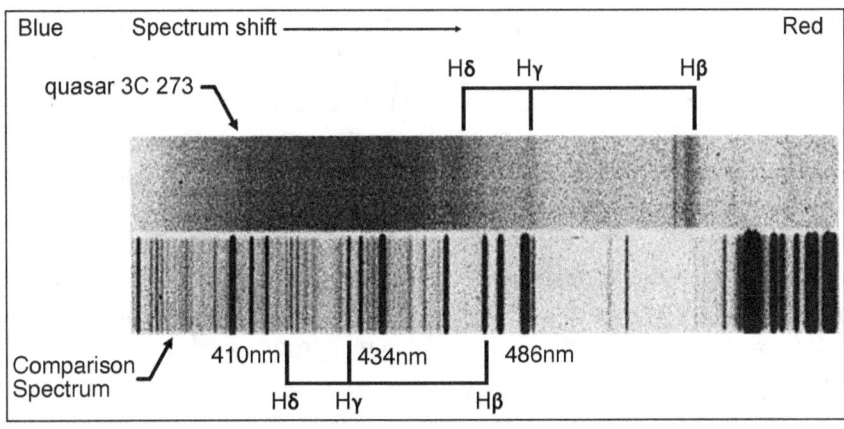

Figure 1. Redshift of Quasar 3C 273

It is important to understand what redshift means. You can recognize particular elements (hydrogen, helium, etc.) from their spectral fingerprint. An element emits light only on certain well known wave lengths. If you detect the spectrum of hydrogen from any source near or far, you know it is hydrogen from the pattern of wavelengths of light detected. Examining such spectra from a set of 23 galaxies, Hubble noticed that their pattern was as expected, but their wavelengths had shifted towards the red end of the spectrum—hence the term "redshift."

A redshift comparison of two spectra is illustrated in *Figure 1* above. The lower spectrum is an interferometer recording of a light source on Earth and the upper one shows the interferometer recording of light from the quasar 3C 273. The blue end of the spectrum is on the left and the red end on the right. In both spectra the location of specific wavelengths, $H\delta$, $H\gamma$ and $H\beta$ are marked and, as can clearly be seen, they are shifted towards the red end of the spectrum in the reading from the quasar.

An explanation of such shifts is that the source of light is receding and the shift is a Doppler effect. This would mean that the quasar's motion away from us slightly stretches out the wavelengths of the light it emits as it recedes, which reveals the speed at which it is receding. Galaxies that are fainter are probably further away, so if they register higher redshifts you have evidence that these more distant galaxies are probably receding at a greater speed.

Hubble's original graph of redshifts for 23 galaxies certainly seemed to indicate such a pattern. However, there were errors of extrapolation in the way that his graph had been constructed, and when such errors were removed, the presumed pattern disappeared. Hubble, to his credit, eventually admitted the error and denied that his work had demonstrated the redshift Doppler effect.

His admission came too late. The redshift of galaxies was almost instantly accepted as a genuine phenomenon demonstrating an expanding universe. General relativity suggested expansion and the general consensus was that Hubble had provided the necessary evidence. This laid the foundation for the Standard Model of Cosmology (The Big Bang Theory). If the distant galaxies are speeding away from each other, then you can extrapolate back in time to points where they were much closer together. Go back far enough—say 13.8 billion years—and then maybe they all emerged from a single point.

The situation subsequently became perplexing when astrophysicists began calculating a particular number generally referred to by the letter z. This "z" denotes the ratio between the recessionary velocity of a distant stellar object and the speed of light. It can be expressed simply by the equation $z = v/c$, where v is the recessionary velocity and c the speed of light. z classifies stellar objects according to their recessionary velocity as measured by Doppler redshift.

Unfortunately and inconveniently, some distant objects, particularly quasars, exhibited a value of z that was greater than 1—in other words the galaxy appeared to be speeding away from us at a speed greater than the speed of light. The quasar ULAS J1120+ 0641 currently holds the highest measured value of z, at about 7.1.

In theory, distant objects cannot recede at such speeds because Einstein's special theory of relativity insists that the speed of light places an absolute limit on velocity. As relativity lies at the foundation of The Standard Model of Cosmology, a clear contradiction had emerged. An adjustment to the Standard Model needed to be made—and it was.

The alternative explanation was this:

> *The expansion of the universe itself created new space-time in between our galaxy and those distant galaxies and quasars. As new space-time was created, it slightly stretched out the wave lengths of the light from those sources.*

The uncertainty surrounding Hubble's original work was dealt with by assuming that the expansion of space-time was not detectable in the local galaxy cluster (of which our galaxy is a component) because of local gravitational effects. All of this was acceptable because it could fit into a mathematical model that represented an expanding universe.

The Quasar Evidence

The word "quasar" is short for quasi-stellar. They are not stars as we normally understand them, for two reasons. First, they are extremely luminous—they emit vast amounts of light. How much? Some quasars are estimated to have a luminosity 100 times greater than that of our Milky Way, which is estimated to consist of anywhere between 100 billion and 400 billion stars. Secondly, the spectra of light from a quasar contains very broad emission lines, unlike any from known stars, indicating the presence of many more incandescent elements than are evident in typical stars.

The current consensus theory for quasars is that they are highly compact accretions of matter at the center of a galaxy that surround a massive black hole—the light being caused by mass from the quasar gradually vanishing into the black hole.

There is, however, a problem with the empirical evidence. Many quasars are found in visual proximity to galaxies, often in pairs—and the galaxies have very different redshift values to the quasars. When such associations between galaxies and quasars were first noticed, they were explained away as coincidental. The redshift values "indicated" that the quasars in question were simply billions of light years further away—they just appeared to be associated by chance.

Halton Arp, an empirical astrophysicist and avid collector of quasar images and data, has been embarrassing Big Bang theorists for many decades with evidence of this clear and obvious association. The reward for his honest efforts has been to have his research work suppressed by almost all the journals of astrophysics. Although he died in 2014, his work is available in a number of books including *Seeing Red: Redshifts, Cosmology and Academic Science*. This book presents a great deal of evidence for galactic-quasar associations and is justifiably critical of the academic establishment.

What Causes Redshift?

Because of Halton Arp's work, the Lambda CDM theory is dead in the water. Nevertheless, the astrophysics establishment has plugged it into an elaborate intellectual life support machine and declares it to be in perfect health, but it shows no signs of life—"its heart is not beating and it breatheth not."

From the perspective of objective science we need to pay it little attention. We were obliged to distrust it from its lack of empirical support and its over-investment in mathematical modeling. And we have no option but to reject it because of its devil-may-care extrapolations.

However, there are reasons why redshift should interest us. It is undoubtedly the case that redshift can be the result of a Doppler effect when a distant stellar object is genuinely moving away from us. Blue shift is also possible when a distant stellar object moves towards us. For example, the Andromeda galaxy appears to be moving toward our Milky Way galaxy because light coming to us from there exhibits a definite blueshift.

The problem with redshift is that the Doppler effect alone is inadequate to explain many empirical observations. The most blatant of these is the quasar data. Quasars, in many instances, are clearly associated with nearby galaxies, but their redshift exhibits a far greater displacement than the light from the associated galaxy. As the light from both passes through approximately the same regions of space to reach us, the causes of the redshift must, to some degree at least, be related to the source. It is possible that the redshift is caused by a quasar's gravitational forces or by local electromagnetic effects, but as we know very little about quasars, aside from the fact that they emit a great deal of light, all theories of the cause are speculative at best.

Such quasar light must also pass through a vast distance between its source and our solar system. We do not know for sure what occupies that space. The current theory is that it is close to being a vacuum, but contains a plasma of hydrogen and helium, plus various other particles, particularly neutrinos, and possibly traces of other elements (dust) that may or may not be in the form of plasma (plasma is a state of matter where the electrons are not bound to the atomic nucleus). Space is also dense with electromagnetic radiation passing through in every direction and may be awash with magnetic fields. We do not know for sure, because we cannot yet take accurate measurements, but the above description seems likely to be accurate given current evidence.

The plasma density in intergalactic space is currently estimated to be one hydrogen atom, i.e. one proton and one corresponding electron, per cubic meter. But estimates vary, with the space near the center of a galaxy estimated to be denser—as much as 1000 protons and electrons per cubic meter. Traveling through space with such a minimal density, light could pass from a source millions of light years from Earth with little chance of encountering a single proton.

If redshift is not caused by a Doppler effect, then the light must lose energy somehow between its origin and its destination. The wave length of the light has increased and its frequency has been reduced. It is possible then that the loss of energy occurs during its journey. Theories that propose how such a loss of energy could happen are generally referred to as "tired light" theories and there are many. We will list them here without venturing to explain any of them:

> Thomson/Compton scattering, Rayleigh scattering, gravitational drag, atomic secondary emission, dispersive extinction, plasma redshift, redshift theorem, coherent Raman Effect on incoherent light, electronic secondary emission, Wolf effect, spectral transfer redshift, extinction Compton scattering by relativistic electrons, thermalization, gravitational interaction and eternal contracting universe.

This list was sourced from the book *The Static Universe**, which we recommend to those readers who wish to explore the competing cosmological theories of contemporary science. We include it here primarily because it clearly indicates that there are more than one or two astrophysicists who have not been mesmerized by the currently dominant Standard Model of Cosmology (The Lambda CDM theory).

To that list we could also add "the loss of energy due to movement through the aether," which was measured by Dayton Miller during the early part of the 20th century in an experiment that no-one has ever attempted to replicate, but was "pecked to death."

Darkness

A schoolboy prank that used to do the rounds in the 1960s worked like this:

You pick a victim whom you were reasonably certain was unfamiliar with the prank, and ask them: *"What is the first sign of madness?"*
They would respond predictably with: *"I don't know."*
You then say: *"Hairs on the palm of your hand."*
The victim would invariably and immediately take a look at the palms of their hands.
You would then say: *"Do you know what the second sign of madness is?"*
They would respond *"No."*
You would say *"looking for them."*
Caught by the prank they would usually smile and as they did so, you would continue: *"And you know what the third sign of madness is?—finding none."*

We could rephrase this a little for the benefit of many modern astrophysicists. The first sign of madness is: belief in the idea of dark matter. The second sign of madness is looking for it, and the third sign of madness is finding none.

---
* *The Static Universe* by Hilton Ratcliffe

Gurdjieff's Hydrogens: The Ray of Creation

The sad truth is that the motion of stars within galaxies cannot be explained by the force of gravity. The assumption is that you can estimate the mass of stars by measuring their brightness. The brightness of a star at the estimated galactic distance is taken to indicate its mass. However, if you model galaxies in this way, the outer stars appear to be orbiting the galactic center too quickly. You either abandon the idea that gravity is the only force involved, or you invent the existence of matter that is undetectable: dark matter.

The opponents of Copernicus and Galileo, who rejected the heliocentric model of the solar system and adamantly defended the once dominant Ptolemaic model, did not resort to a magic invisible fix to defend their cherished model. When the empirical evidence became increasingly unfavorable, they gradually acceded to the heliocentric model. And yet, from a mathematical perspective, there is no difference between the geocentric and heliocentric models. It merely reduces to the question of where you place the origin of the three dimensional framework marked out by the x, y and z axes.

The heliocentric model is preferable merely because it results in much simpler equations to define the orbits of the planets and their moons. The addition of the force of gravity to either mathematical model (along with accurate estimates of the mass of the planets and the sun) does not alter this. The heliocentric model is still far simpler, but both models can be made to work mathematically. The problem that astrophysics faced was that it could find no way to explain the outer orbits of suns (and their solar systems) around the center of a galaxy. Their model simply did not work. Kepler's laws of planetary motion, which Newton demonstrated as derivable from his laws of motion and gravitation, do not apply to the empirical data of the orbits of suns within a galaxy.

Once astrophysics was faced with such empirical evidence, it had to make a choice. It could either search for forces other than gravity that might be responsible for this inconvenient galactic behavior, or invent the existence of mass that was "currently undetectable." The fact that the consensus of physicists chose to adopt the "magic dark matter" proposition is a testament to their identification with the Standard Model of Cosmology.

Current estimates suggest that dark matter makes up about 27% of our universe. How is that figure arrived at? It's the ratio of "dark matter" to real matter that would need to exist to account for the orbit of suns within galaxies according to standard gravity models.

Just as you do not need to be a master carpenter to detect a wobbly table, you do not need to be an astrophysicist to realize that dark matter is an absurdity.

Nevertheless, if you search the Internet you will find abundant "proof" of dark matter's existence—"proof" of hairs on the palm of your hand. It is a testament to the power of mathematics that, given a little selective empirical evidence and a strategic choice of variables and constants, you can produce impressive models of things that cannot, do not and never will exist in reality.

As for dark matter, so for dark energy. Dark energy is a newer fantasy than dark matter. In 1998 two independent supernova projects which treated particular types of distant supernovae as standard candles produced results that could be interpreted as demonstrating that the universe was not only expanding but the expansion was accelerating. This was of course a shock to the world of astrophysics, because there was nothing in the Standard Model that could account for this inconvenient acceleration. It had to make a choice. It could either question all the assumptions of the Standard Model or invent the existence of energy that was "currently undetectable."

The fact that the consensus of physicists chose to adopt a "magic dark energy" solution pretty much demonstrates that the Standard Model of Cosmology is no longer a theory. It is an atheistic "article of faith."

Chronologically, the first of the dark ideas to emerge from the mind of modern physicists was the black hole. You would think, given the plethora of articles appearing over decades in science magazines and web sites that there could be no doubt whatsoever about the existence of black holes. After all, have not astrophysicists classified black holes according to whether they are stellar, supermassive or miniature and additionally whether they have retrograde rotation, prograde rotation or no rotation? And is it not the case that every month or so some article appears heralding the discovery in some sector of the sky of yet another black hole? Didn't they even produce a picture of one?

In reality a black hole is a theoretical construct based on mathematical modeling and extreme extrapolation. By extreme extrapolation, we mean that assumptions are made about what happens in conditions that are impossible to produce in a laboratory or observe in the sky. They are very much like hen's teeth.

The Cosmic Microwave Background Radiation

"Our vast universe expanded to its current vast size over 13.8 billion years after its initial emergence from a singularity." This is undoubtedly a very bold assertion, especially as the same physicists who assure us this is the case, also assure us that when a critical mass of matter exists within a given volume of space it will collapse into a black hole. Clearly, this collapse always takes place, except in the rare circumstance that a universe is being created.

It is mathematically challenging to model the creation of the universe from a singularity, especially given that there is no empirical evidence at all and never can be. Nevertheless, physicists have done their best with their "laws" and mathematical equations, with the help of a convenient assumption here and there. If you throw in the idea of dark energy, which is currently estimated to account for about 70 percent of the whole universe, it's hard to imagine that you wouldn't be able to cook up such a model in some way.

Before there was any evidence of any background radiation, the Standard Model theorists had predicted what (perhaps) should be found:

1. It should be isotropic (i.e. detectable in every direction and always be the same).
2. It should exhibit black body radiation.
3. The radiation should have a longer wavelength than light.
4. It should exhibit a temperature between $28°$ and $50°K$.
5. It should not be explicable as coming from any other source.

Clearly, if we are looking back into the origin of a universe that began with a big expansion, then the most distant thing we should be able to detect is the initial expansion, which logically would have to be the same in every direction. The expansion model suggests that the primordial matter that we might be looking at in this way should have cooled down and be inert. It should thus exhibit black body radiation. It would have to have a lower wavelength than visible light otherwise we would detect it as light or very energetic electromagnetic radiation, and anyway given the temperature estimate ($28°$ and $50°K$) it is only likely to be detectable as thermal radiation.

When Arno Penzias and Bob Wilson were working together at Bell Labs on a telecommunications satellite project in 1965, they kept encountering interference to signals in the microwave band. Naturally they tried to eliminate the interference in various ways, but they soon discovered that they couldn't, because the interference came from all directions and its level was always roughly the same. Their accidental discovery had them scratching their heads to determine a cause, but that problem was quickly snatched from their hands. Standard Model physicists were soon swarming all over this newly discovered radiation like ants on an ant hill. They had little doubt as to what the cause was.

This microwave interference was soon named CMBR, declaring by its nomenclature that it came from the "very edge of the universe." The fact that this radiation was definitely not isotropic and its temperature proved to be a mere 2.7°K, well below the predicted value, were not allowed to intrude on this "clear validation of the Big Bang."

There has been a great deal of study of the CMBR. Two different satellites have been launched at great expense and they have gathered masses of data in an effort to measure the CMBR in every possible direction. Scientists have tortured the data and manipulated their models every which way but loose in an effort to explain why the empirical evidence does not show this radiation to be isotropic.

Rather than being concerned by their heroic effort to marry the data with the theories—an arranged marriage of epic proportions—let's consider one simple question.

When scientists calibrate this radiation, whether doing so from Earth or from a satellite, how do they know how far away the source of the radiation is?

The answer to this is that they don't. Science has no means for distinguishing between electromagnetic radiation of a given wavelength that has its source a few miles down the road and round the corner, and radiation that has its source 10 billion light years away. And this matters. There clearly is a ubiquitous microwave radiation, but its source could be the local space between the planets of our solar system, which is also ubiquitous from our viewpoint. Or it could be the space between our viewpoint and the nearest stars, or it could be the space between our viewpoint and the rest of the Milky Way, or it could be intergalactic space, or it could indeed be the edge of the universe (if there is one).

The only clue available to determine its true origin is the fact that it is not isotropic even though, according to Big Bang theory, it should be. Astrophysicists explain away this fundamental disparity by claiming that the electromagnetic radiation from the CMBR is distorted in various ways in its passage to our system by interstellar dust or gravity wells or whatever.

Distances

You might believe that when astrophysicists assign distances to various objects in the universe they have an accurate means of measuring such distances. This is not exactly the case.

Currently the best technique for estimating distance of an astronomical object is through parallax. In half a year the Earth goes from one side of the Sun to another, points that are roughly 186 million miles apart. Consider a very distant galaxy that appears stationary and a star that is relatively close to our solar system in almost the same place in the sky. As the Earth moves, the apparent distance between the two objects will vary slightly due to stellar parallax. By measuring the variance, the distance of the far nearer star can be estimated with reasonable accuracy.

Sadly, the best instruments we have, space-based telescopes, are not capable of measuring parallax angles of less than 0.001 arcsec, which limits this technique to about 1000 parsecs (3260 light years). The Milky Way is about 30,000 parsecs in diameter, hence this technique is only useful for a small population of stars in our own galaxy.

Beyond such distances, in trying to measure the distance to other galaxies, astrophysicists seek out "standard candles." In theory, if we know the actual brightness of a distant object in another galaxy then we can estimate its distance from us according to how bright it appears to us. Astrophysicists currently use Cepheid variable stars as one means of making distance estimates. These stars are very bright, they pulsate in a predictable way, and conveniently, the Cepheid star's period (its frequency of pulsation) is directly related to its luminosity.

The problem with this is threefold:

1. No Cepheid star is near enough to provide a baseline that could be confirmed by the parallax method.
2. The inverse square law used to determine the distance presumes that there is nothing present in interstellar space that might affect the observed luminosity.
3. Redshift distortion, which we have already discussed.

The point is that while you can find purported "maps of the universe" in books and on the Web, they only represent "the current best guess."

The Birth of The Elements

Where did the elements come from? The scientific narrative is roughly as follows:

> After the Big Bang, the universe expanded rapidly (exponentially) during "inflation," as a kind of soup of fundamental particles, gradually cooling as it grew. At first there was radiation and then quarks combined together to form protons and neutrons. When 3 minutes old, it became cool enough for the protons and neutrons to combine into nuclei. The four lightest elements, hydrogen, helium, lithium, and beryllium then formed, but only as

nuclei, not as fully fledged atoms with orbiting electrons. After about 300,000 years atoms formed. And that was phase one. Having formed, these elements then gathered together in clouds, which condensed to become stars, which organized themselves into galaxies.

Once a star had formed, a continuous process of nuclear fusion was presumed to begin in the star's interior giving rise to its output of electromagnetic radiation. Thus heavier elements than the initial four, all the way down to iron, could form by nuclear fusion in stars that were large enough. In stars that were very large, supergiants, the theory is that the center of the star would eventually become dominated by iron atoms, and that this concentration of iron would eventually lead to a massive explosion. The nuclear fusion process to create elements heavier than iron absorbs rather than emits energy. It is presumed that up to a point there is a balance between the energy streaming out from the star and the star's own gravity. This now collapses, the force of gravity condenses the mass of the star into a smaller sphere and the sphere explodes.

In theory, the core of the supergiant generates gamma rays that are powerful enough to break apart the iron atoms and a vast amount of energy is released in what becomes a Type II supernova. The shock wave that tears the star apart is supposed to be hot and dense enough to allow the formation of all the elements heavier than iron and these are scattered around the neighborhood by the vast explosion. These elements then go to participate in the formation of planets

The purpose of this myth is to explain the existence of elements heavier than iron. There are far more credible theories. Nevertheless this myth and the observed abundance of particular elements in the firmament is usually deemed to offer support for Big Bang, although it does not offer proof of any cosmological model. The universal abundance of most elements was predicted correctly by Fred Hoyle decades ago on the assumption of a steady state universe.*

Big Bang theorists co-opted the data on observed elements and focused on the abundance of light elements. However, their mathematical models for predicting the abundance of each such element necessitated an adjustable parameter. There is no indication of what this adjustable parameter corresponds to, other than it makes the model work.

---

* https://en.wikipedia.org/wiki/Steady-state_model

The Evolution of Galaxies

The final "scientific" pillar that is supposed to prove the correctness of the Big Bang model is the evolution of galaxies. Because galaxies change very slowly we are unable to observe their evolution directly, but it is clearly the case from observation that very distant galaxies (where what we see through telescopes occurred 10 billion or more years ago) look different to closer galaxies (where what we observe happened just a few billion years ago). Astrophysicists have thus been able to model galactic evolution and their view is that galaxies form roughly as follows:

- They begin with smaller clusters of stars.
- These clusters coalesce (or collide) to form larger disk-shaped clusters.
- The rotation speed appears to increase.
- Stars form rapidly inside the disk and a spiral structure emerges.
- Galactic mergers occur, possibly leading to the creation of elliptical galaxies.

When small galaxies collide or coalesce there is no indication that any of the stars within those galaxies collide with each other directly. Big Bang theorists naturally assume a prior stage to the first step described here in which small clusters of stars are formed from gas clouds.

Galactic evolution is the best evidence that there was a beginning of some kind to the universe. No mature galaxies have been observed at very far distances. It is also worth noting that quasars appear to be most numerous in very distant galaxies.

The Big Bang Objections

There are those who scoff at the various myths that ancient peoples recorded and gave credence to (in some way) as explanations of the creation of the universe. They are many and various, from the familiar Biblical creation of the Earth in seven days to the very philosophical formulation in the Tao Te Ching, which states:

> *There was something featureless yet complete,*
> *Born before heaven and earth;*
> *Silent—amorphous—it stood alone and unchanging.*
> *We may regard it as the mother of heaven and earth.*
> *Not knowing its name, I style it the 'Way.'"*

None of these creation myths seem as outrageous and intellectually bankrupt as the Big Bang myth. It claims that, at the beginning of time, something occu-

The Apes of Objective Science

pying a space as small or smaller than the Planck volume* expanded over approximately 13.8 billion years to become the current universe. The Plank volume is considerably smaller than a hydrogen atom. Indeed, a hydrogen atom is roughly 10^{73} times larger than the Planck volume.

Think about that. The current universe is estimated to contain roughly 10^{80} atoms. And, in current theory, it also contains a vast amount of invisible dark matter and dark energy, made up of no-one knows what. There's a lot more of this dark stuff than the atoms we know and love. All in all, that's quite a magic trick.

Our great great grandchildren will laugh at us for taking anything so absurd seriously. However, if you aspire to be an astrophysicist at the moment, you'd better take it seriously, otherwise you're likely to be in want of a job.

Here below, for the record, are a list of scientific objections to this absurd theory. Although very long it is by no means exhaustive:

- The static universe models fit observational data far better than any expanding universe models. (Of course the fit is not perfect, but it requires far fewer "adjustable parameters.")
- The element abundance predictions based on the Big Bang is awash with adjustable parameters. They are not predictions within the usual meaning of the word.
- The microwave "background" (CMBR) is clearly not isotropic unless you add many adjustable parameters. There are a variety of more credible explanations, one of which is: the CMBR is simply the temperature of space ($2.8°K$), the minimum temperature that any body in space would cool to if only warmed by distant starlight. No adjustable parameters are required for this explanation.
- The predictions of the CMBR background temperature based on the Big Bang do not indicate a temperature anywhere close to $2.8°K$ without the intervention of an adjustable parameter.
- The Big Bang time-scales don't work. There are too many large galaxies to credibly form in the supposed time of 13.7 billion years (unless you add in some adjustable parameters).
- The proposed age of the universe is questionable even if you accept the ridiculous extrapolation. Using the same time measurement approach you can find globular clusters (early galaxies) that appear to be older.

---

* https://en.wikipedia.org/wiki/Planck_units

There's even a star, known as the Methuselah* star, in our own galaxy which has been dated as 14.5 billion years old.

- The most distant galaxies in the Hubble Deep Field show evidence of too large an amount of metals in their composition. The Big Bang theory requires the stars, quasars and galaxies in the early universe to be mostly metal-free. A great number of supernovae would be needed to build up the metal content found in stars. Those distant galaxies have more metal than they should have. This challenges the calculated age of the universe in another way.

- The Big Bang asserts that exactly equal amounts of matter and antimatter were created in the initial expansion that began the universe. Antimatter is identical to matter but is oppositely charged. So an antimatter electron (a positron) spins around an antimatter proton in an antimatter hydrogen atom. Antimatter has been created in laboratory conditions—it is not a theoretical fiction. There is no trace whatsoever of the antimatter that was supposedly created in the Big Bang. If it was destroyed by encounters with matter, then all the matter in the universe would have been destroyed, leading to nothing.

- Quasar data and observations are problematic for Big Bang, by virtue of redshift measurements. When the z redshift value is 7 for example, it implies a distance of 13.172 billion miles, making the quasar older than the galaxy it is associated with. As the whole Big Bang theory rides on the back of redshift assumptions, redshift anomalies like this undermine the theory.

- Dark matter is best thought of as the most adjustable of all adjustable parameters. We cannot detect it, so we can assume it is anywhere we want it to be—for example, to explain the motion of spiral galaxies, which gravity alone cannot explain.

- Dark energy is yet another adjustable parameter—for which there is no evidence. It has been invented in yet another effort to breathe life into a dead theory.

- The Big Bang violates the first law of thermodynamics. This quite reasonable law maintains that energy can neither be created nor destroyed. The Big Bang requires that new space filled with "zero-point energy" is continually created between the galaxies.

---

* https://www.space.com/20112-oldest-known-star-universe.html

The Apes of Objective Science

- The whole narrative of the Big Bang violates the third law of thermodynamics, which can be roughly stated as: in a closed system entropy increases. According to this law, the initial gas clouds that created galaxies would not form as the universe expanded, they would disperse.

The most damning objection to this theory is none of the above. It is that the Big Bang theory has predicted almost no new observations correctly. When a theory is close to the truth the tendency is for new observations to confirm the assertions of the theory. With Big Bang quite the opposite has happened. New astronomic observations result in astrophysicists having to perform painful intellectual contortions.

The theory is based on fantastical extrapolation, mathematical modeling and nothing else. At every turn, astrophysics seeks to bend reality to its models and when it cannot, it creates new adjustable parameters to bend its models towards reality.

The Quantum Mechanical Morass

Einstein spent a substantial amount of time trying to harmonize his theories of relativity with the experimental results and theories of quantum mechanics, in search of a theory of everything. Physics believes there to be four fundamental forces acting at various scales in the universe. These are: gravitation, the electromagnetic force, the strong nuclear force and the weak nuclear force. If it were possible to find some equation which related all four of these forces together in some proven way, it would probably be possible to unite astrophysics (the physics of objects on a cosmic scale) and quantum mechanics (the physics of objects on a very small scale).

From our perspective, having investigated the Standard Model of astrophysics and found it wanting, it makes sense for us to take a look at quantum mechanics in the hope of finding a more useful basis for examining the phenomena of our universe. As with astrophysics, quantum mechanics has a Standard Model which claims to provide a map of all subatomic particles. This is illustrated in *Figure 2* and includes what physicists currently believe to be all the basic particles from which an atom is formed.

If you were taught at school that atoms are composed of protons, neutrons and electrons, and you have not kept pace with the development of physics, you may be wondering why you do not see either a proton or a neutron in this diagram. The explanation is that both are deemed to be composed of other particles bound together. A proton is composed of three quarks, two up quarks

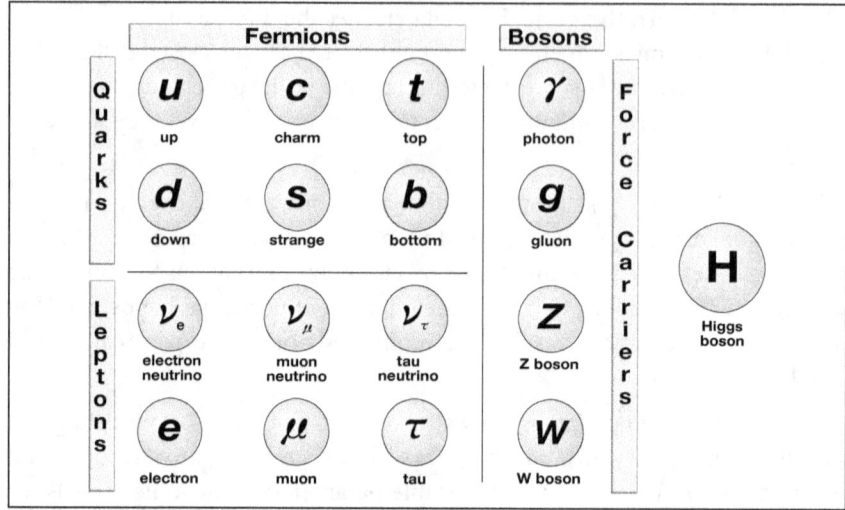

Figure 2. Quantum Mechanics' Standard Model

and one down quark, while a neutron is composed of one up quark and two down quarks.

If you are wondering why other particles you may have heard of: positrons, mesons, tachyons, and so on, are not in this table, it is either because they are antiparticles (like the positron), or they are compound particles composed of particles in the Standard Model (like various types of meson), or they are theoretical and yet to be verified (like the tachyon).

In theory, for every particle with an electric charge there is an antiparticle with an opposite charge. Antiparticles do not last long, since when they meet their corresponding particle they are both annihilated, producing energy in the form of photons. For reasons unexplained there is, in theory, a large imbalance between particles and antiparticles in the universe. Experiments suggest that antiparticles are formed in beta decay, a form of radioactive decay, and also by the interaction of cosmic rays (very high energy radiation) when they encounter the Earth's atmosphere. In theory there is a collection of antiparticles that mirrors the Standard Model of particles shown above, but if so this has not been demonstrated.

The twelve fermions in the diagram (six quarks and six leptons) are fundamental particles. Theory has it that these particles cannot be split into smaller particles. Composite particles like the proton are also fermions. The simple distinction between quarks and leptons is that quarks combine to make the compound particles (protons, neutrons, mesons, etc.) while leptons do not; they are particles that can be produced by nuclear reactions.

The four bosons shown in the diagram above are force carriers—particles that give rise to forces between other particles. The photon carries the electromagnetic force and manifests as electromagnetic radiation, such as light, heat, X-rays, etc. The gluon carries (or mediates) the strong nuclear force between quarks, "gluing" them together; hence the name. The W and Z bosons carry the weak nuclear force, providing an explanation for nuclear fission and fusion reactions.

The Higgs Boson is slightly different in that it is deemed to be the particle that confers mass on all other particles. In theory, without such a particle, no other particle would have mass. To better explain the Higgs Boson, we need to also describe the Higgs field. This is deemed to be an energy field that permeates the universe. This energy field continuously interacts with particles, via the Higgs Boson, which carries mass in the form of energy, and thus particles have mass.

Any quantum physicist reading the above description would (and should) regard it as far too simplistic and bereft of many important details. A comprehensive description would require several books and a deep dive into the theoretical characteristics of all these particles (electric charge, spin, mass, color, quantum states, etc). However, the intention here is not to try to explain this Standard Model, just to provide a rough description of it.

It is worth mentioning that the Standard Model of quantum mechanics has a much better (but not perfect) record of predicting and explaining experimental outcomes than does the astrophysics Standard Model.

The Wave/Particle Problem

The question of whether light consists of waves or particles goes back to Isaac Newton, who was convinced that light was made up of particles. His theory was opposed by Christiaan Huygens who proposed that light consisted of waves. This difference of opinion has never been completely resolved and is still under scientific investigation.

In 1801, Thomas Young performed the first double slit experiment which placed a light source in front of a plate in which there were two thin slits, behind which there was a screen. When this is done, an interference pattern (bright and dark bands) can be seen on the screen as illustrated in *Figure 3*.

If light consists of waves then as the light passes through the two slits, waves of light fan out from each slit. The two sources of light interfere with one another causing the interference pattern. Thus Young concluded that light consisted of waves, not particles.

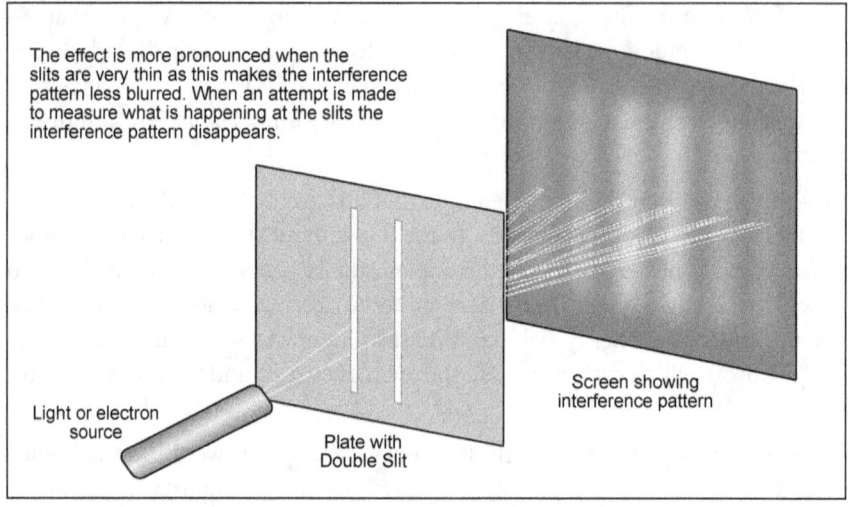

Figure 3. The Double Slit Experiment

Nevertheless, as more precise versions of this experiment were carried out, eventually to the point where individual photons of light were sent one-by-one, the light was always found to be absorbed at the screen at discrete points, implying that the light consisted of individual particles. The interference pattern appeared from the varying density of the points of arrival of the many photons.

To make matters more complex, it seems as though individual photons travel as though they are interfering with other (non-existent) photons. Further complications result from the problem that any experimental design which tries to determine which slit the photon passes through, eliminates the interference pattern.

The experiment has been done with electrons, and entities much larger than electrons. Currently, the largest entity used in a double-slit experiment that produced the usual result was a molecule comprised of 810 atoms. The important point about this is that this wave-particle problem is not just confined to subatomic particles.

Philosophical Discord

Quantum theory was born in 1900 when German physicist Max Planck published his study of the effect of radiation on a "blackbody" substance. He had demonstrated that energy, in some situations, exhibited the characteristics of physical matter. This conflicted with classical physics, which regarded energy as a continuous wave-like phenomenon, independent of the characteristics of physical matter. Planck's theory maintained that radiant energy was made up of particle-like components, known as "quanta." His theory helped resolve previously unexplained phenomena, such as the behavior of heat in solids and the nature of light absorption at an atomic level.

So at its base, quantum mechanics regards energy as both matter and a wave, depending on particular variables. So in the Standard Model one should not think of there being 17 particles, but 17 wave/particle dualities (quarks, leptons and bosons). Quantum mechanics consequently took a probabilistic view of the world, while classical mechanics had always taken a deterministic view of the world, where objects had precise properties, everything could be measured and nothing was left to chance.

Niels Bohr and Werner Heisenberg maintained (in what is called the Copenhagen interpretation) that physical systems generally do not have definite properties prior to being measured, and quantum mechanics could only predict the probabilities that measurements will produce certain results. The act of measurement affects the system, causing the set of probabilities to reduce to only one of the possible values immediately after the measurement. This didn't sit well with Einstein who responded to this idea with the statement, "God does not play dice."

The practical point was that quantum mechanics needed to use probability theory and copious amounts of statistics in order to estimate possible outcomes of events at subatomic levels, and it employed such mathematics with considerable success.

However, common sense suggests that the Copenhagen interpretation is absurd. Simply consider the situation of Mr. Predictable. If it is not raining, Mr. Predictable goes for a jog in the morning, but if it is raining, he does not. Let us also add the fact that it rains in his area 20% of the time in the morning, when he would normally go jogging. If we do not know why Mr. Predictable decides to jog or not to jog, then we will need to observe him to know whether he is jogging at the usual time and why. However, it is absurd to suggest that Mr. Predictable and his environment do not have definite properties until we make such observations.

The Abuse of Probability and Statistics

Statistics is a useful branch of mathematics. In particular, when we have a situation where we cannot gather sufficient data to predict its outcome, we can use statistical techniques to project possible outcomes. Consider the well proven business of life insurance. We do not have any reliable way of knowing when a specific individual will die. Even in abnormal situations when a particular individual has contracted a possibly terminal disease, all we can do is estimate when death is likely to occur. It is not possible to know and, very occasionally, the disease may not actually cause death.

However if you take a large enough sample of people who suffered exactly the same disease in the past, you can use statistical techniques to predict the likely outcome for any individual and, most of the time, you will be reasonably accurate.

Now, consider the very predictable situation of two balls on a pool table. One ball is stationary and the other has been accurately struck with a cue and is moving exactly towards the stationary ball. It will hit the ball, unless something highly unusual happens, such as an earthquake or, perhaps, someone unexpectedly picks up the stationary ball. We consider such situations to be deterministic: all the possible influences are known and the outcome is thus highly predictable.

Our lives are, to a great extent, deterministic. When we send a letter, we expect it to be delivered, because our experience is that letters get delivered. We expect machines to work as intended. We expect the banking system to work. We expect telephones to work, and so on. Nevertheless the future for most people is uncertain, because we do not know many of the influences that will act. In some situations, statistics can be used to make useful predictions, but such predictions only apply to populations. For individuals, all we can do is assign probability.

As regards quantum mechanics, the situation is no different. Even though it is often thought of as depicting a "weird" subatomic world, the reality is that physicists have been able to gather so little data that they are forced to work with statistical probability.

Fundamental Forces

Currently physics believes there to be four fundamental forces, two of which we recognize from our own experience. They are:

- **The force of gravity.** We may not know exactly how it works, but we experience gravity by virtue of our own weight and see it in action when things fall to the ground. At school we were taught that the force of gravity obeys an inverse square law, the gravitational forces between two objects weaken as the square of their distance increases. What we were probably not taught is that gravity is a very weak fundamental force even though it is supposedly able to cause the formation of super-massive black holes.

- **The electromagnetic force.** This is the force that we witness via electricity and magnetism, although we may not be sure how it works. Opposite poles of a magnet attract each other and similar poles repel; something that we have no doubt observed. Electricity flows from high potential to low potential. We may be less familiar with the fact that the electromagnetic force also obeys an inverse square law.

Both these forces are deemed to be infinite in range, although their influence becomes minuscule very quickly as distances increase. What you may not be aware of is that the electromagnetic force is vastly stronger than gravitational force. How much stronger?

Roughly 4.4×10^{37} stronger—that's a very large figure.

The other two fundamental forces are deemed to exist in the nucleus of an atom. They are:

- **The strong force.** This is the force that is believed to hold the nucleus of an atom together. It is estimated to be 137 times stronger than the electromagnetic force, but is estimated to have a very short range of 10^{-15} meters which is roughly the diameter of a medium sized nucleus. Thus it has no effect outside of an atomic nucleus.
- **The weak force.** This is the force believed to be responsible for nuclear decay and hence radioactivity. It is estimated to have a strength of only 10^{-6} of the strong force and its range is roughly one tenth the diameter of a proton.

These two forces are entirely theoretical. There was a need to explain why a collection of protons in a nucleus did not simply repel each other and the nucleus disintegrate. Similarly there needed to be an explanation for nuclear decay, so another force was theorized. So these two theoretical forces were invented accordingly.

The bosons of the Standard Model "explain" how these forces manifest at the atomic level. The photon is the carrier of the electromagnetic force, the gluon carries the strong force and the W and Z bosons carry the weak force.

What is missing from the model is a particle to carry the force of gravity. There is a theoretical particle called the graviton which is supposed to carry that force, but there is no direct evidence for its existence at the moment.

The Large and The Small

The practical problems of astrophysics and quantum mechanics are at opposite ends of a stick. Astrophysics is trying to explain the whole universe but it is handicapped by the fact that many of the objects it wishes to gather data on are far away. Distant galaxies cannot be easily analyzed because they move so slowly and very little data can be gathered from them—just the electromagnetic radiation that they throw off. But at least they can be seen. The best data we can get about a star comes from the Sun, but we have just one example. The best data we can acquire about a galaxy comes from the Milky Way, but again we have only one example and it measures somewhere between 100,000 and 200,000 light-years in diameter.

At the atomic level the problem of observation is far worse. With optical microscopes the limitation to what we can observe is about 50 nanometers in diameter, which is better than most experts once believed was possible. That level is remarkable but it doesn't take you down to the atomic level. Electron microscopes are thousands of times better, because the wavelength of an electron can be 100,000 times shorter than the wavelengths of visible light. They can provide pictures of atoms!

However an electron microscope doesn't passively record an image, it focuses a beam of high voltage electrons on the target using a magnetic lens and deduces an image from the interaction between the target and the stream of electrons. Other forms of advanced microscopy (such as photoionization microscopy) suffer from the same problem; the microscope interferes with what you are trying to observe.

The other problem is that, in the extremely small environment under observation, everything is in motion to some degree and sometimes moves very rapidly. Any picture one attempts to take is inevitably blurred and distorted in some way, and there is no means of correcting the distortion, because it isn't possible to compare the picture to the real thing. The use of such observing devices requires the application of deduction and extrapolation.

When one investigates how the Large Hadron Collider gathers its data one encounters the same situation. For example, protons can be accelerated to a high speed to collide with the nuclei of lead atoms. Various data collection devices record the event and the results are analyzed using deduction and extrapolation and a good deal of software to try to model what happened. Just consider the (surprising) estimate that both the W boson and the Z boson have a mean lifetime of 10^{-25} seconds. That isn't something that can be directly observed and measured. Any experimental result can only be arrived at by extrapolation from collected data.

There is a great deal of uncertainty in LHC experiments. The physicists have no doubt done their best to eliminate all possible defects in the 17-mile-in-circumference experimental apparatus, but it is trying to identify such small distances and time scales that any small anomaly could interfere with results. That doesn't just apply to the 17 mile ring, but also all collectors and all preparation devices and so on.

To add to the uncertainty, this is not dealing specifically with specific particles in specific locations, but with probabilities of particle locations. A single software error could lead to the apparent discovery of something new, when in fact it is not. Plus there's the problem of whether the LHC is truly a sealed system. How can we know whether some particular result wasn't influenced by a passing neutrino or two. And in strict violation of the usual scientific methodology, there is no possibility of using another identical LHC to test any supposed results.

We are not claiming here that quantum mechanics is wrong, only that there are good reasons to be skeptical. Its most disturbing aspect is the number of "adjustable parameters" it requires. The Standard Model, for example, needs 20 such parameters. In modern physics, adjustable parameters ride to the rescue of doubtful theories. The simplest supersymmetric extension of the Standard Model has no less than 105 additional parameters.

Another good reason to be skeptical of quantum mechanics is provided by pilot waves...

Pilot Waves

Pilot Wave Theory was created by Louis de Broglie in 1927, and later gave rise to the De Broglie-Bohm causal interpretation of quantum mechanics. This provides an alternative explanation for the double slit experiment. In 2004 physicists Yves Couder and Emmanuel Fort used pilot wave theory to reproduce many of the quantum effects.

Rapidly vibrating an oil bath, they were able to bounce silicon droplets on the surface, which walked along the surface producing waves through the oil as they moved. They were able to demonstrate single-particle diffraction, tunneling, quantized orbits, and orbital level splitting. Such systems are now known as hydrodynamic quantum analogs.

This could be an explanation for quantum behavior since it uses the same mathematics as other interpretations and thus is supported by the same evidence. In pilot wave theory the particle and the wave are separate but related phenomena. The particle induces a pilot wave in a medium and that determines how the particle moves. The position and momentum of a particle are considered to be the hidden variables and the observer doesn't know the precise value of these variables, and cannot know them precisely because any measurement disturbs them.

This is analogous to the Heisenberg uncertainty principle.

If we consider an all-pervading aether to be the medium for the wave then we have, conceptually at least, an alternative theory. And as we are in the orbit of quantum mechanics, we could postulate that the Higgs field and the aether are in fact the same thing. Thus the particles can be viewed as moving through the medium of the Higgs field in tandem with a wave in that field.

This brings us face to face with the problem that the only wave phenomenon that is believed to occur without the participation of some medium is light (the photon)—since photons are deemed to be massless because they travel at the speed of light.

Mach and Weber

Aware of the current malaise in physics, both at the macro and micro level, some physicists (a relatively small group) have gone back to basics and have been reevaluating the theories of Ernst Mach (1836 - 1916) concerning the origin of mass, and the theories of Wilhelm Weber (1804 - 1891) concerning electromagnetism and gravitation.

Going back to basics means exactly that. There are two "thought experiments" in physics that are fundamental. The first concerns the motion of an object in a straight line. The question is: from what does an object get its mass?

According to Newton's first law of motion, an object that is at rest will stay at rest unless a force acts upon it, and if it is in motion, it will stay in motion, moving at a constant velocity unless a force acts upon it. For the situation where the object is in motion, this law establishes a frame of reference for its motion, as illustrated in *Figure 4*.

The thought experiment is this:

The Apes of Objective Science

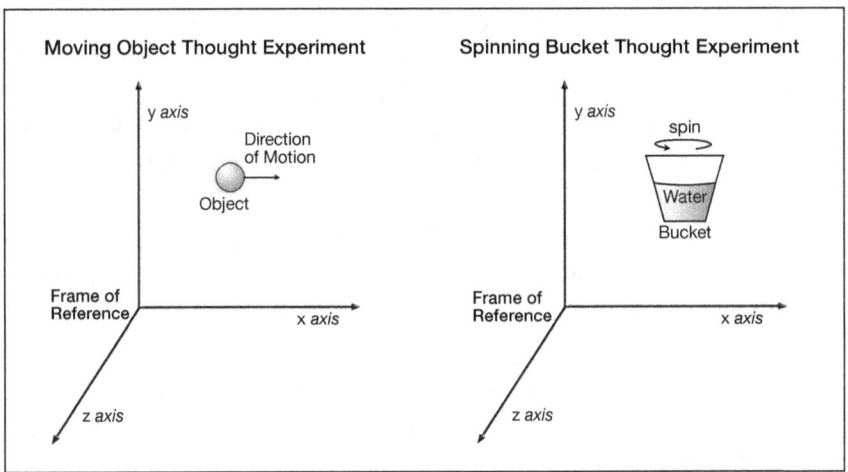

Figure 4. *The Thought Experiments*

If you have such an object in motion and immediately the whole of the universe except that object disappears, does the object still have mass and is it still in motion?

As this is a conceptual experiment that can never be carried out, the preferred answer can only be an opinion. The answer "no" implies that the object acquires its mass and its motion from the rest of the universe, so if that were removed it would have no mass and exhibit no motion. So mass and motion are only in relation to something else. The answer "yes" implies that the mass and motion are intrinsic to the object or in relation to empty space.

Newtonian and Einsteinian physics take the view that the answer is "yes." Mach based his physics on the assumption that the answer is "no."

A second and equally important experiment, proposed by Newton, concerns a bucket half-full of water which is spinning. When you have such a bucket, not only does it have angular momentum (the momentum of its spin), but the surface of the water, which would be flat if the bucket were not spinning, will be concave, as illustrated in *Figure 4*.

Again the question can be asked:

What happens if you remove the rest of the universe? Is the bucket still spinning and will the surface of the water still be curved?

The Newtonian and Einsteinian approaches assume that the answer is "yes," while the Machian approach assumes that the answer is "no."

Gurdjieff's Hydrogens: The Ray of Creation

Consider the idea of removing the whole of the universe, you have no basis for creating a frame of reference, since empty space is empty and infinite in every direction. Ernst Mach maintained that the universe in its totality conferred mass and spin on every object that was part of it.* He could as easily have depended on the existence of a mediating aether that conferred mass and spin.

The attraction of Weber's electromagnetic and gravitational theories** are that he proposed a credible model of the atomic nucleus that needed neither the strong nor weak forces to explain atomic behavior.

Our primary reason for drawing attention to Mach and Weber is not a deeply held conviction that their theories are correct, we are simply indicating that modern physics may soon turn in a new direction, because the old direction looks like a dead end.

What is a Field?

When you study physics you often encounter the word "field," and you rarely encounter a satisfying explanation as to what such a thing actually is. The term "field" first cropped up in the study of magnetism. The magnetic field was deemed to be "something" that conducted the magnetic force in the region of a magnet. After that other fields were postulated: electric fields, gravitational fields, and most recently, the Higgs field of quantum mechanics.

Nevertheless the word "field" is not well-defined and hence can easily confuse. Here are a range of possible definitions:

- A field is a region of space.
- A field is a state of a region of space.
- A field is a real physical entity filling a region of space.
- A field is a medium within a region of space that enables the propagation of a force.
- A field is a mathematical function that returns a value for every point within a region of space.

We can reduce these possibilities, if we decide, as we should from first principles, that, while a field can occupy a region of space, the space itself cannot have any properties. Thus the properties exhibited by a region of space must exist as a result of something which occupies the space. This eliminates the first definition. And we may want to head in that direction anyway, since

---

*The Science of Mechanics: A Critical and Historical Exposition of Its Principles by Ernst Mach
**Weber's Electrodynamics by A.K. Assis

the current best theory we have of particle physics insists that, at the very least, there is a Higgs field that fills all space.

The final definition—the mathematical one—is nicely precise, and if we adopt that one, while insisting that the region of space referred to contains "something" that causes the function to produce the value it does, we can collapse all these definitions into one. This means that "state," "physical entity" and "medium" are simply terms that refer to a "something." It does not at all imply that it is necessarily the same "something" in respect of magnetism, electric fields, electromagnetism, gravity and subatomic particles. However, in every case, the possibility of an aether is not excluded.

Gurdjieff's Hydrogens: The Ray of Creation

Chapter 2

THE THRESHOLD OF OBJECTIVE SCIENCE

"To know means to know all. Not to know all means not to know. In order to know all, it is necessary to know only a little. But, in order to know this little, it is first necessary to know pretty damn much."

~ *Gurdjieff*

The basic tenets of objective science do not align with those of contemporary science. This needs to be acknowledged and accepted without compromise by those who wish to study and apply objective science. We are suggestible. It is a human failing. Consequently, we are inclined, by habit, to accept frequently repeated assertions from any source we regard as credible, including contemporary science.

We often do not attempt to critically analyze such assertions, and we may even try to "fit them" into whatever world view we have constructed in applying objective science—assuming we have indeed constructed one. The possibility of being deflected by such opinions warrants the adoption of a meticulous approach to analyzing information we encounter from that source. Before we take another step, let us draw a line, with the following statement:

There is no possible basis for debate between those who study objective science and adherents of contemporary science—nor is there reason to debate.

Objective science comprises a theory of everything, from the Absolute to the smallest particle that exists. The objective scientist does not, at the outset, know the truth of this broad theory and will only be able to verify it individually. If one man proves it true to his own satisfaction, no matter who he tells or how he describes it, he cannot prove it true to another. Objective science works from theory towards confirmation on an individual level.

Contemporary science invents hypotheses and seeks their confirmation through a process of experiments that are peer-reviewed and repeatable (in theory) by anyone.

No common basis exists between these two approaches to science, not even a partial one, to provide a foundation for debate. Without a common basis from which to venture forth, productive discussion and genuine agreement is impossible.

Nevertheless, it may prove useful for the objective scientist to interact with contemporary science representatives to learn why they adhere to a particular theory and discover what data they consider important. Beyond such data gathering, there is unlikely to be any useful outcome. Similarly, contemporary scientists may be interested to know about ideas from objective science even if they reject its approach out of hand.

The objective scientist has no reason to proselytize. It would be fruitless.

Objective science considers the universe to be alive at every level.

A fundamental disagreement between the two approaches to science is that contemporary science only regards life as a characteristic of what it defines to be living biological organisms. The dispute here concerns the very definition of life. To the objective scientist, every cosmic unit is alive in some way. At the large scale, planets, suns, and galaxies are viewed as life-forms. At a smaller scale, molecules are viewed as life-forms.

Objective science considers man's inner world to operate as a system in the same manner as the universe as a whole.

The objective scientist considers the existence of two separate but related worlds: his inner world and the outer world he experiences and with which he interacts. He views his inner world, including his body, as a cosmos. Thus, he views the inner world and the outer world as subject to the same laws, although acting at different levels.

This perspective is enshrined in the words "As above, so below."

Contemporary science does not regard the inner world, insofar as it acknowledges its existence, as having any importance to science. The term "cosmos" is used by contemporary science only to describe the universe as a whole—as a complete system. There is disagreement about the meaning of this word.

The Threshold of Objective Science

An important purpose of objective science is to assist the individual evolution of human beings.

The purpose of objective science is to assist individuals in their personal evolution and particularly the evolution of their understanding. It is a collection of information that can, in theory, be transformed into knowledge and then into understanding.

Contemporary science has no purpose of this kind and does not (currently) acknowledge the possibility of personal evolution. It seeks to arrive at the truth—it has that in common with objective science—but it seeks to do so only by empirical activity beyond the inner world of man and thus regards as true only those things that can be established collectively in that way.

The theories of objective science are immutable.

The theories of contemporary science change over time. It is usually presumed that such change indicates progress and that human knowledge is increasing accordingly. Individual scientists seek to add to that "progress."

The theories and models of objective science are given. Gurdjieff never suggested that they could be improved or extended. They are said to be "the product of higher mind" given in a form suited to "seekers after truth" in our era.

The objective scientist does not blindly believe the assertions of objective science.

Gurdjieff said that objective science can only be thoroughly understood by higher mind. Our activity in the study of objective science runs in parallel with work on being—work on ourselves to develop the capacity to understand objective science. We are advised to adopt an attitude of skepticism as we investigate and evaluate objective science's assertions.

It will be challenging to establish the truth of something if we set out believing it to be true. To establish the truth, we need to observe it clearly, and a belief in the "truth" of something may make it more rather than less difficult to see. Many of the assertions of objective science challenge our ability to comprehend. At times we need to carry out personal experiments in our efforts to comprehend. As such, we confirm, refute or reformulate what we have been given.

Objective science avoids specialization.

Contemporary science categorizes itself into topics, which result in individual specializations. For example, natural science, which covers the description and understanding of natural phenomena, is divided into life science and physical science, essentially a division between the animate and the inanimate.

Life science is also called biology and is further divided into ecology, zoology, botany, microbiology, and biochemistry. Physical science is broken up into astronomy (or astrophysics), earth science, meteorology, geology, oceanography, physics, and chemistry. The social sciences are similarly divided, and the tools of science, termed the "formal sciences," are classified under mathematics, logic, statistics, systems theory, decision theory, and theoretical computer science.

This division can lead to a kind of "compartmentalism" where information and associated theories are gathered together and studied but seek only for consistency and completeness within their own compartment and ignore inconsistency between compartments.

In objective science, there are no compartments. Knowledge constitutes a unity. The objective scientist's initial goal is "to know ever more and more concerning the laws of World-creation and World-maintenance." The objective scientist's ultimate goal is to completely connect all aspects of knowledge together to form a unity within their being.

The data of contemporary science may be useful to the objective scientist.

Many of the questions that concern contemporary science, such as: "What is life?" "What is the origin of man?" "How was the universe created?" "How do specific phenomena occur?" are also questions that objective science seeks to address. Contemporary science performs many useful experiments to investigate such questions.

The objective scientist may find many of these experiments worthy of study as they can provide useful data for our understanding of the external world. The only caveat is that contemporary science theories associated with the data should be treated with skepticism.

Sadly, it is also the case that some contemporary scientists are not beyond fabricating data and hiding or destroying inconvenient data.

Fundamental Concepts of Objective Science

Some concepts of objective science are fundamental to any attempt to describe it. Thus, although we will revisit these concepts later to explore them in greater depth, it makes sense to declare them now and describe them briefly.

They are:

Cosmos

The most fundamental concept of objective science is the idea of a cosmos. A cosmos is an ordered system. The word is Greek in origin (*kosmos*) and was used by Pythagoras to describe the universe. In the Work, the terms "cosmos" and "cosmic unit" describe living beings. The term frequently occurs in *The Tales*. A galaxy is a cosmos, as is a star or a planet or a moon. These are large cosmic units. Human beings and animals are smaller cosmic units, and individual cells smaller still. Atoms, as contemporary science defines them, might be cosmic units.

Life

Objective science proposes that a cosmic unit is alive. In general, objective science considers something to be alive if:

- It has an individual existence.
- It has a lifetime.
- It eats, breathes, and senses.

Cosmic units feed on one another or substances created by other cosmic units. This idea is expressed in *The Tales* by the word "Trogoautoegocrat." The word is Greek in its component morphemes, as follows: "trogo" is from *trogein* meaning "to gnaw, nibble, munch"; "auto" is from *autos* meaning "self"; "ego" is "I" and "crat" is from *kratos* meaning "strength, power, rule." Taken together, "I govern by consuming myself."

Each cosmic unit consumes substances and discards residues as an inevitable activity of maintaining its existence. Each cosmic unit has three foods, which at the human level are normal food, air, and impressions. Should a cosmic unit cease to take in one of its three foods, it dies and ceases to be a cosmic unit. Its various parts lose their coherence, and its individuality, such as it is, is destroyed. It becomes a collection of substances that themselves, in various ways according to their properties, become food for other cosmic units.

We can see this from one perspective when we consider our own existence. No matter what our eating habits, everything we eat was once part of some other cosmic unit, whether it was plant, animal, or chemical compound (such as salt). This is true of other life-forms at the planetary level, whether animal, plant, or individual cell. However, for very large beings (planets, for example) and very small beings (molecules, for example), we cannot easily identify their three foods or how they digest such foods.

We cannot easily comprehend this, partly because it is not clear how a planet or a molecule feeds. However, if such things are alive, then feed they must.

Reproduction

Life-forms reproduce. In Nature, this is primarily sexual among multicellular life-forms, although many plants and fungi reproduce asexually. There are indications of reproduction at the macro scale (planets, suns, and galaxies), which we shall discuss later.

Because cosmoses have a lifetime, new cosmoses must arise at every level to replace the dead. The Trogoautoegocratic process requires this balance—otherwise it would inevitably fail.

Substances

If something is not alive, in the sense of being a cosmos, then we can think of it as a substance. Thus a substance is either within a cosmic unit or outside of it. Consider, for example, the dead branch of a tree that has fallen to the ground. It was once part of the tree (a cosmic unit) but is no longer so—it is now just wood. As time passes, it will be consumed by other cosmoses: termites, carpenter ants, powderpost beetles, fungi, and so on.

Many creatures feed on other cosmoses directly—predators feeding on prey. When they have killed their prey, they feed on its substances.

Mammals feed their young on milk—prepared food consisting mainly of nutrients, which are substances, not cosmoses. Trees and plants produce fruit, preparing food for other species, usually helping their own reproductive strategy. A fruit is not a cosmic unit, but contains the seed of a cosmic unit.

Atoms

Objective science's definition of an atom is distinctly different to that of contemporary science. It insists that an atom of a substance is "the smallest quantity of the substance which retains all its properties, including its psychological properties." By this definition, an atom of ice is different from an atom of water, even though, from a chemical perspective, we would classify them as the same substance. Also, by this definition, atoms are alive—if they were not, they could not have psychological qualities. As such, the atoms of objective science are cosmoses, living beings. The atoms of contemporary science are not necessarily cosmoses. Nitrogen, for example, only occurs naturally as a molecule, N_2, and never as an atom N. Hence an atom of nitrogen is not a cosmos, whereas a molecule of nitrogen could be.

Cosmoses are compositions of smaller cosmoses. While a cosmos may appear to be a single living organism, it is formed from aggregations of other life-forms with far shorter life spans. Human beings are an aggregation of diverse cells, including many life-forms (bacteria) that are not of our life pattern (our DNA) but are symbiotes or parasites.

Cells themselves are aggregations of diverse molecules, which may be cosmoses within their own context. And those molecules are composed of atoms and subatomic particles that may also be cosmoses.

We can also go up in scale. An individual man is an insignificant particle in the life of humanity, which may also be a cosmic unit in its own right. Humanity is merely one of the many life-forms that constitute Great Nature, which is itself vibrant and alive.

Material

A fundamental axiom of objective science is that everything is material—not just the universe and everything within, but even the Absolute is material. This assertion denies the idea that anything can lack weight or materiality. Light is material. If there are ghosts or spirits, they are material in some way. If there are "unseen forces" that affect our reality, they too are material, even if they are not visible. There is no exception.

Everything that exists also vibrates. The higher the vibration, the greater its energy and the more rarefied its material. In this way, the density of something material has an inverse relationship to its level of vibration. Ice is denser and has a lower vibration than water, which is denser and has a lower vibration than steam.

Viewing the universe from this perspective, we can think of it as being awash with material. There are points of concentration or points of stability where denser material gathers and is transformed. Such points of stability appear to be rare. Moons are small aggregations of dense material that spin around planets. Planets are larger concentrations of dense material that revolve around suns. Suns are larger concentrations of material with higher frequencies of vibration than planets or moons.

While the space between planets and the space between solar systems may appear to be "empty space," this is not the case. It is filled with various materials with very high frequencies of vibration and low materiality: plasma, streams of particles, and energy in the form of magnetism and electromagnetism.

Gurdjieff's Hydrogens: The Ray of Creation

Aether

It is a fundamental axiom of objective science that a medium, aether, permeates the whole universe, acting as a substrate that supports its materiality. Space is not a vacuum.

The Ray of Creation

We stand on the Earth, and, looking into the sky at night, we observe the Moon be in close proximity to the Earth and orbiting it. The idea of the Moon as a satellite of Earth is not difficult to accept. With an optical telescope, we can easily confirm that other planets also have moons.

We can see that other planets orbit the Sun. So we can think of the Earth as an individual planet on the one hand and a member of the family of Planets within the solar system. The Sun dominates the solar system. In the night sky, we observe stars, and it seems reasonable to assume that these are also suns that may have families of planets. Recent astronomical observations confirm this.

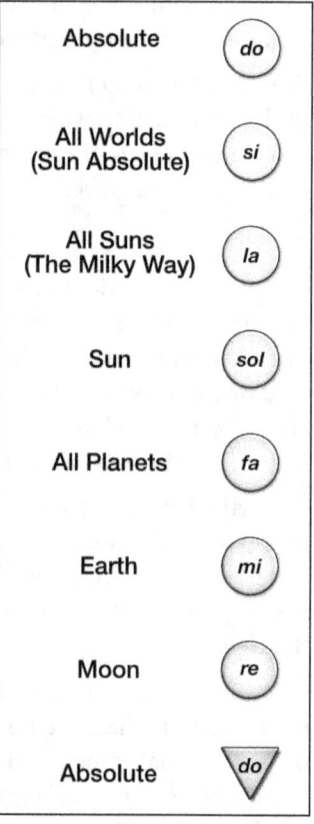

Figure 5. *The Ray of Creation*

In the night sky, we can also observe the vast arc of the Milky Way, an extensive family of stars. Current estimates suggest it is composed of hundreds of billions of stars. We refer to large clusters of stars as galaxies, and modern telescopes confirm that there are vast numbers of these galaxies. So if we take the universe as one entity, the collection of all galaxies, it is reasonable to think of it as comprised of "All Worlds." Referred to as the Sun Absolute, it defines the space of the universe and contains it. It can also be thought of as the abode of the Absolute.

The only aspect of the Ray of Creation illustrated in *Figure 5*, that is theoretical to us is the existence of a supreme intelligence, the Absolute, which pervades all worlds and exists at both ends of the Ray. Each level of the Ray is located within the levels above it. Thus the Sun Absolute (the universe) lies within the Absolute. The totality of the universe is invisible to man.

The Threshold of Objective Science

Within the universe there are billions if not trillions of galaxies. One galaxy—our galaxy—appears to be visible to man as it stretches out across the night sky, but in fact, it is not. We do not see it all, and even if we did, we would see only its vast population of stars. With the aid of telescopes, we can see that whole galaxies usually appear to be disk-shaped. We can even deduce from astronomical evidence that this is true of our galaxy. But we do not see our galaxy like that—we see it as a cloud of stars.

The Sun is just one of these stars. The body of the Sun is clearly visible to us, and we sense its light and heat. Indeed we are within its atmosphere, but we do not see the extent of that atmosphere. Within the Sun's atmosphere exist the family of Planets. Our Earth is a member of that family, and it has a dependent satellite, the Moon, which lies within its influence. We might be inclined to think that there is nothing below that, but the Ray of Creation asserts that there is—the Absolute as the central foundation of everything that lies above it. The Absolute at the end of the Ray of Creation sits at the center of every moon, planet and star.

Figure 5 illustrates the Ray of Creation from our perspective—it is our subjective Ray of Creation. There are many other such Rays for other planets and Moons. It is essential to understand that each level of the Ray of Creation possesses a higher level of reason than the level below. At every level, the Ray is alive and represents a living being or living beings. Thus the Moon is the least intelligent such being in our Ray. The family of Planets represents a higher intelligence than our Earth, the Sun is more intelligent still, and so on.

In *In Search of the Miraculous*, Gurdjieff states* that as living beings, planets and moons are able to live for a specific time, have specific possibilities of self-development, and may transition to higher levels of being. Directly opposing the view of science, he depicts the Moon not as a "dead planet" but as a planet in embryo, far from achieving the degree of intelligence possessed by the Earth, but growing towards that.

If the Moon eventually achieves Earth's level, it may give birth to a moon of its own. In that case, the Earth will have achieved a higher level and will become a sun for the evolved Moon and its new satellite. At some time in the past, the Sun was merely a planet and the Earth was a moon for it. Earlier still, the Sun was a moon.

However, the evolutionary development of a planet to a sun and a moon to a planet is not guaranteed. If our Sun died, it would bring death to all its planets

---

* *In Search of The Miraculous* by P D Ouspensky, p25

and moons. The same applies to man. If a man dies, each cell of his body will also die. He will be broken apart as his physical body is gradually deconstructed to become food for other local life-forms. So if a solar system dies, then it too will serve as food for something else within the Ray.

Before we can dive more deeply into the intricacies of the Ray of Creation, we need to discuss both the Law of Three and the Law of Seven.

The Law of Three

Objective science asserts that all interactions, all phenomena at every level from the atomic to the macrocosmic level, occur because of one law, the Law of Three. The Law of Three acts when three different and opposing forces come together.*

One force acting on its own or two forces together never produce a phenomenon. It is only with the help of a third force distinct from the other two that a phenomenon can occur at any level. The three forces are called active, passive, and neutralizing.

The Law of Three is not entirely a new idea; it is possible to find traces of it in many ancient traditions and systems, although one rarely encounters any coherent explanation. The first force is referred to as "active," the second "passive" and the third "neutralizing." These labels may be misleading as all the three forces are active—in the sense that they act. They appear active (or affirming or pushing) and passive (or denying or resisting), and neutralizing (or reconciling or equilibrating) in relation to each other when an event occurs.

The first two forces are not particularly difficult to understand. Whenever anything is attempted, there is some resistance to overcome. However, the third force is often difficult to see, either because it seems to blend into the medium within which an event occurs or because it is wrongly viewed as part of the result. In some situations, for example, when it is known that a chemical reaction requires a catalyst, the third force is known, but in other situations, it is not seen at all.

Situations can be confusing sometimes. For example, sometimes a chain of events appears to be a single event because it happens very quickly. Sometimes some substance may appear to be involved in an event when in fact it is not, and so on. *Figure 6* illustrates the Law of Three in detail. We describe it in a list of bullet points.

---

* *In Search of The Miraculous* by P D Ouspensky, p77

The Threshold of Objective Science

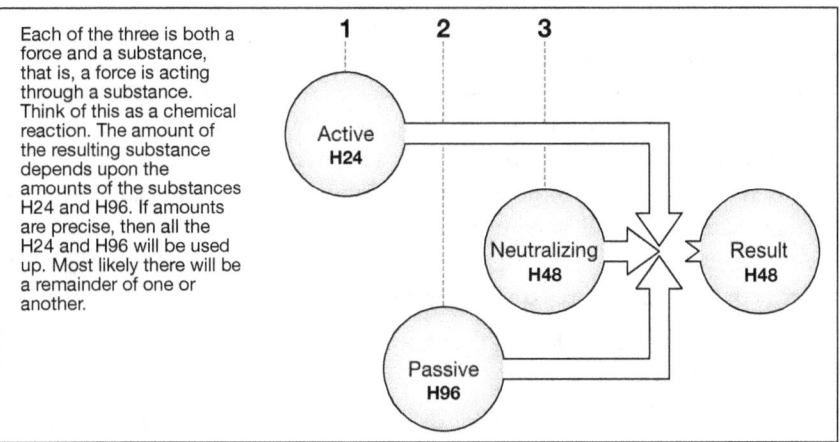

Figure 6. The Law of Three

- All events, at large scale or small scale, happen in accordance with the Law of Three.

- There are aways three substances involved through which the three forces act.

- In the diagram, the three substances that interact are labelled H24, H48 and H96. In the next chapter we explain the derivation of this classification of substance (H1, H3, H6, H12, H24, H48, H96, H192, etc.) using the Law of Three and the Law of Seven.

- In all interactions, only substances from three adjacent categories, such as H24, H48, and H96, can interact. The middle one always carries the neutralizing force.

- The substance that carries the neutralizing force is never changed by the event in any way. Because of this fact, it is possible to identify substances in particular contexts that often act as the third, neutralizing force. Water is often a neutralizing force for example, as are enzymes in digestive processes.

- The result produced by the event is always in the same category of substance (the same *Hydrogen*) as the neutralizing force.

- The interaction is best envisaged as a chemical interaction where the three substances involved are, for some reason, brought together. It completes when one of the substances is exhausted so that three substances are no longer together in the same place.

Gurdjieff's Hydrogens: The Ray of Creation

Man can study the Law of Three in his inner world, but to do so requires knowledge. Indeed knowledge is likely to manifest as the third force. Consider a man who wishes to know himself. His desire will be active force, but it will encounter resistance in his passive habits. Specifically, he identifies with everything that happens, and thus his attempts to observe himself are fruitless. The two forces simply spin around each other. But one day the man is taught about dividing attention. He has new knowledge and this causes an event to occur. He sees something about himself. The third force was knowledge, and its outcome is the man acquiring knowledge about himself.

In *The Tales* Gurdjieff refers to the Law of Three as Triamazikamno, and describes it in the following way:*

> *"'A new arising from the previously arisen through the "Harnelmiatznel," the process of which is actualized thus: the higher blends with the lower in order to actualize the middle and thus becomes either higher for the preceding lower, or lower for the succeeding higher; and as I already told you, this Sacred-Triamazikamno consists of three independent forces...*

The term "Harnelmiatznel" means a "mixing together." The Law of Three always involves the mixing together of two substances, one of which has a higher level of vibration and one a lower level of vibration, with the result being a substance with a level of vibration in between the two.

He also points out that the Law of Three inevitably manifests as chains of events when he writes *"and thus becomes either higher for the preceding lower, or lower for the succeeding higher."* The resulting substance will either participate in a subsequent event with a substance whose vibration level is lower than the first or second force in the interaction that produced it, or with one whose level of vibration is higher.

The Law of Three indicates that everything is in motion, either ascending or descending.

The Law of Seven

While the Law of Three governs how things happen at a specific time and place, the Law of Seven governs how events unfold over time. The Ray of Creation illustrates the Law of Seven as a creative octave: *do-si-la-sol-fa-mi-re-do*. We call this a creative octave because it commences with an active *do* and proceeds downwards through all the notes until it completes—thus, it is involutionary.

---

* *The Tales*, Chapter XXXIX, The Holy Planet "Purgatory", p751

The Threshold of Objective Science

In *The Tales*, Gurdjieff formulates the Law of Seven with the following words:*

'The-line-of-the-flow-of-forces-constantly-deflecting-according-to-law-and-uniting-again-at-its-ends.'

He then states:

"This sacred primordial cosmic law has seven deflections or, as it is still otherwise said, seven 'centers of gravity' and the distance between each two of these deflections or 'centers of gravity' is called a 'Stopinder-of-the-sacred-Heptaparaparshinokh.'

Gurdjieff uses the word "stopinder" to indicate the distance between the rates of vibration, from one note (or pitch) to another. Each note or 'center of gravity' deflects the line of the flow of forces. If we look again at the Ray of Creation, in *Figure 5*, it is represented as if it were a line descending from the Absolute. However, it could also be represented as a circle, uniting *do* at the top, to *do* at the bottom.

The path between two adjacent notes, *do-si* for example, is a stopinder and forms an inner octave. In *Figure 7*, we depict four octaves, all of which commence with the top *do*. In the diagram, the outer octave represents the Ray of Creation. So, each top *do* in the inner octaves vibrates with exactly the same frequency as the *do* of the Ray of Creation. The initial note of each inner octave is *do* and the final *do* of each inner octave is *si* of the adjacent octave. This repeated fractioning of stopinders into inner octaves proceeds seven times.

The notes in each octave and all octaves represent frequencies of vibration. Any specific material, for example, air molecules, can vibrate across a range of frequencies. Heated air vibrates at a higher rate than cool air, but it is the same substance. The same could be said

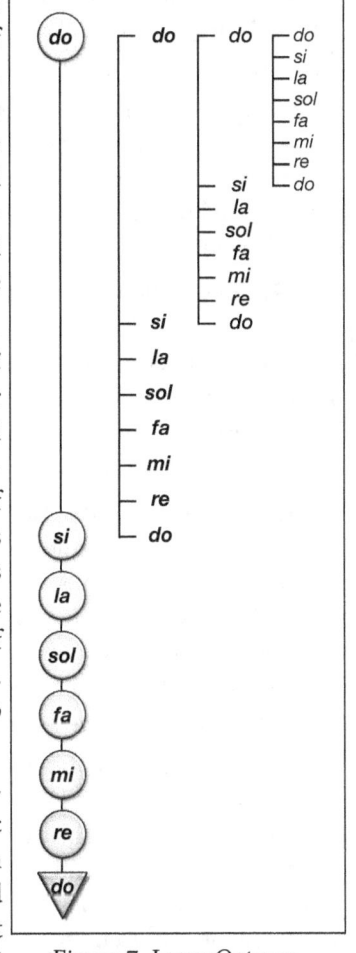

Figure 7. Inner Octaves

---

* The Tales, Chapter XXXIX, The Holy Planet "Purgatory", p750

Octave	do	re	mi	fa	sol	la	si	do
Stopinder		3	3	2	4	4	5	3
Example	24	27	30	32	36	40	45	48
Ratio	1	9/8	5/4	4/3	3/2	5/3	15/8	2
Increase		9/8	10/9	16/15	9/8	10/9	9/8	16/15

Figure 8. Octaves, Notes and Stopinders

of the substance of electromagnetic radiation. When it manifests as radio waves, the frequency can vary from low to high.

While we might think of electromagnetic radiation as a single phenomenon (as energy of varying wavelengths), from the perspective of objective science, it is a variety of substances, many of which have distinct properties. It can vibrate molecules (as it does in a microwave oven), transmit heat directly (as with infrared radiation), manifest as light, and even penetrate matter as X-rays and gamma rays do.

The Ray of Creation is the unique octave that creates the universe we experience. Its high and low *do*s, meet and unite in a way that may be difficult to imagine. All other octaves are inner octaves at some level that echo this great octave. They are distinct in one specific way: the higher and lower *do*s do not unite, although the lower *do* resonates with the higher *do*.

Figure 8 illustrates an octave numerically. The number 24 denotes the rate of vibration of the lower *do*. The higher *do* is thus double that number, 48. Beneath those example numbers are ratios of the note's vibration compared to that of the lower *do*. Thus *re* is 9/8, *mi* is 5/4, *fa* is 4/3 and so on. In the line labeled "increase" we show the ratio of vibration of the note to the previous note. So *re* is 9/8 of the vibration of *do*, *mi* is 10/9 of the vibration of *re*, and so on. If we assign numbers to each stopinder (each gap between two notes), we get *do-re* = 3, *re-mi* = 3, *mi-fa* =2, and so on.

The theory is that all octaves have this structure. Every activity at every level from the whole universe down to the subatomic particle, proceeds either as an ascending or descending octave. The universe is perpetually in motion. At every point of activity in the universe, substances react with other substances in accordance with the Law of Three, and each such action is a note in either an ascending or descending octave. Within the universe, there are cosmoses at every level, and these cosmoses absorb material in three ways, transforming it internally, causing some of that material to ascend to a higher level and some

The Threshold of Objective Science

to descend. In either case, the material then serves some purpose, either within the cosmos or, if it is discarded by that cosmos, for the benefit of some other cosmos. This is what Gurdjieff calls the Trogoautoegocrat, a universe that operates by reciprocal feeding.

If there were no movement back up from the lower levels of the Ray to higher levels, the universe would simply run down. All of the higher substances would gradually descend in an involuntary manner until they were all but lifeless. But this does not happen. Evolutionary octaves run in the opposite direction, enabling lower substances to climb to a higher level.

The Ray of Creation is an involutionary octave from the high *do*, complemented by an evolutionary octave, ascending from the lower *do*. There is a balance because there has to be. Conceptually, if a substance ascends in one place, the same amount of the same substance will inevitably descend in some other place. Of course, it is more involved than that, but the principle applies.

The Creative or Involutionary Octave

Even if we can envisage the Absolute's creation of the universe as an octave, it may be difficult to see how this octave applies in our lives. It will help if we consider its structure.

It begins with the high *do*, but immediately it encounters an interval that it must overcome to proceed further. If it passes the first interval, it can proceed through the notes *si*, *la*, *sol*, and *fa*, where it encounters a second interval that it must overcome to proceed further. If it passes the interval between *fa* and *mi*, it can proceed downwards until it completes the note. If it does not, then it will stagnate at the note *fa*.

As an example, think of designing and then building a house. It begins with someone having the intention (the will) to bring a house into existence. It proceeds through design activity (where to build, how many rooms, purpose of each room, etc.) to become a series of architectural drawings. It then goes through the building process, until eventually, it is complete and can be occupied.

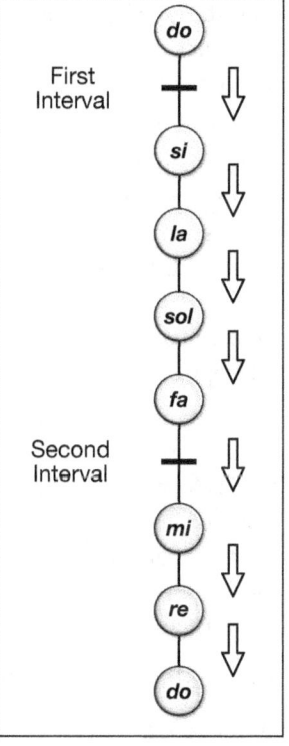

Figure 9. Involution

69

Gurdjieff's Hydrogens: The Ray of Creation

We begin with an intention at the level of thought, and we descend to a point where the intention has become a solid physical reality. The physical reality is far more constrained than the idea that conceived it. As the house goes from concept to physical house, it becomes increasingly constrained until, when completed, it is not subject to the laws that constrain design but to the laws that govern rock and wood.

Now consider a creative octave that fails. I have an idea for a science fiction novel. It is born of some comment I heard that people do senseless things. I dream up the idea that a small group of aliens has landed on planet Earth with a remarkable machine that allows them to project ideas into a person's mind, which they cannot help but act on. Developing this idea, I imagine that the aliens arrive on Earth about 1910 and decide it would be fun for humanity to have a massive World War. The idea intrigues me, and I convince myself that this novel would be a best seller and it would earn fabulous amounts of money.

But sadly, I am no writer. I have never been able to focus on writing for very long, and it takes a long time for me to write even a few paragraphs. So I do not even start, or I start but cannot even complete the first page. The truth is that I am stuck at the note *do*, and this project will go no further. As the weeks pass, the force of my ambition to become a writer fades, and I rarely even think of writing another word. The *do-si* interval stopped me.

Now imagine instead that I am a practiced writer of fiction. I commence writing and get most of the way through the book. I send a specimen chapter and a synopsis to several agents and publishers. However, none of them are impressed. Some suggest changes to make, but I think their suggestions are terrible. I cannot get published, and I put the project to one side and start something else. The manuscript gathers dust. In this example, the *fa-mi* interval stopped me.

If you honestly examine your life, you will discover examples of unstarted projects and uncompleted projects. The involutionary octave is creative, so it applies to all attempts to create something, whether to make a movie, compose music or invent a new mouse-trap. It begins with something high (most likely an idea married to an intention), and it completes with the production of something concrete (a house, a painting, a book). A relatively high *Hydrogen* gives rise to a lower one.

The Evolutionary or Receptive Octave

The evolutionary octave ascends, proceeding in the order *do-re-mi-fa-sol-la-si-do*. Think in terms of such an octave applying to the restoration of something or the genuine evolution of something. An example of restoration is: healing someone who is ill, where the evolutionary octave applies to the patient who goes from illness to health. For evolution, we can think of the evolution of life on Earth, ascending from bacterial life through plants, then animals, then man. Or we can think of acquiring a skill.

The first interval in an ascending octave is at *mi-fa*. Here the initial impulse at *do* loses momentum to the point where it can go no further. A person who chooses to work on themselves is initiating an evolutionary octave. The octave begins with a normal human being and, if it completes, ends with the creation of higher being bodies.

It may commence with the discovery of Work ideas in a book, or on the Internet, or by talking to someone. But many people encounter these ideas and are not sufficiently motivated by them to go any further. They do not manage to traverse the *do-re* stopinder. They are stopped by one of the two intervals within it.

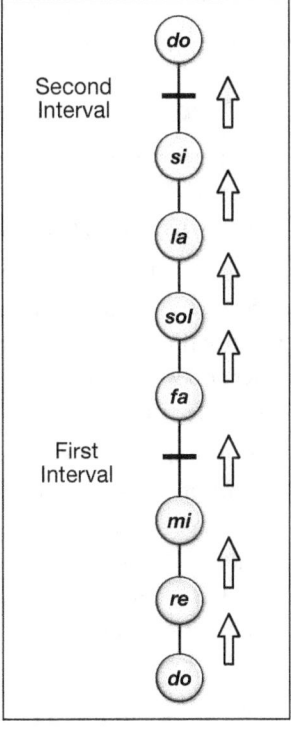

Figure 10. Evolution

In the octave of the Work, the *do* is "receiving the ideas." To pass to the next note, the individual must, at a very minimum, value the ideas and make efforts to understand them. The note *re* is "applying the ideas to oneself." Having recognized that some of the ideas may be accurate, the individual reflects on them and inwardly observes behaviors that correspond to how human behavior is described by the Work.

They thus pass through the note *re* to the note *mi*, where they begin to understand their personal limitations. They try "to do" and discover that they can achieve very little in respect of self-remembering and self-observation—so little that they may eventually cease to try. The likely outcome is that they begin to descend back through *re* to *do* from the note *mi*. At this point, they may give up and move on. Alternatively, they may seek out and join a Gurdjieff group.

It is in the nature of the octave that what does not ascend will eventually descend. Nothing stays in the same place indefinitely. It either combines with something higher and ascends or combines with something lower and

---

* *In Search of the Miraculous* by P D Ouspensky, p125

descends. The progress of an octave, whether ascending or descending, involves the Law of Three at every step.

The Modern Musical Scale

Before we proceed, we need to discuss the musical octave to dispel a possible source of confusion. It can arise from the fact that western musicians adopted and now use the equal temperament scale of music almost exclusively.

Consequently, the ratios of vibrations in the octave that Gurdjieff describes in *In Search of the Miraculous** and illustrated in *Figure 8* are not the same ratios as those normally employed by Western music. As shown in Table 1, Gurdjieff describes what is termed the "just scale," while Western music employs what is called the equal temperament scale.

Equal temperament came into common use in the 17th century. It had the distinct advantage of enabling a piece of music to be transposed from one key to another. To achieve this, it was necessary to slightly adjust every note and semitone in the octave using a mathematical scheme. A musical octave has eight full tones, from *do* to *do*, and five semitones between each pair of notes except *mi-fa* and *si-do*. Table 1 shows both the just scale and the equal temperament scale, along with the ratios of vibration that define each.

For each scale, the ratio between each note or semitone and the octave's fundamental *do* is shown, with the exact ratio in mathematical terms and a decimal value. This is set in the context of an octave that commences from Middle C, so we can show the exact frequency of each note for both scales. In the table's final two columns, we show the differences between the two scales numerically (between frequencies from Middle C upwards) and as percentages.

The main point to note is that the just scale and the equal-tempered scale do not agree exactly except the first and last *do* of the octave. For example, in the just scale, the note *sol*'s (the fifth) frequency is exactly 1.5 times higher than the lower *do*. In the equal temperament scale, it is 1.4983 higher.

As can be seen from the table, the tones and semitones of the equal temperament scale are determined mathematically according to the number 2 taken to ascending powers of one-twelfth. The equal temperament scale is best thought of as a "useful approximation" to the just scale.

It is perhaps surprising that this works as a musical scale that pleases the ear, but it does and has done for centuries. The just scale is (in practice) more

---
* *In Search of the Miraculous* by P D Ouspensky, p125

The Threshold of Objective Science

Name	Just Scale			ET Scale			Difference	
	Ratio	Dec.	Freq.	Ratio	Dec.	Freq.	#	%
Unison (C) *do*	1	1.0000	261.63	$2^{0/12}$	1.0000	261.63	0.00	0.00
Minor Second (C♯/D♭)	25/24	1.0417	272.54	$2^{1/12}$	1.0595	277.18	4.64	1.70
Major Second (D) *re*	9/8	1.1250	294.33	$2^{2/12}$	1.1225	293.66	-0.67	-0.23
Minor Third (D♯/E♭)	6/5	1.2000	313.96	$2^{3/12}$	1.1892	311.13	-2.84	-0.90
Major Third (E) *mi*	5/4	1.2500	327.03	$2^{4/12}$	1.2599	329.63	2.60	0.80
Fourth (F) *fa*	4/3	1.3333	348.83	$2^{5/12}$	1.3348	349.23	0.40	0.11
Dimin. Fifth (F♯/G♭)	45/32	1.4063	367.92	$2^{6/12}$	1.4142	369.99	2.07	0.56
Fifth (G) *sol*	3/2	1.5000	392.44	$2^{7/12}$	1.4983	392.00	-0.44	-0.11
Minor Sixth (G♯/A♭)	8/5	1.6000	418.60	$2^{8/12}$	1.5874	415.30	-3.30	-0.79
Major Sixth (A) *la*	5/3	1.6667	436.05	$2^{9/12}$	1.6818	440.00	3.94	0.90
Minor Seventh (A♯/B♭)	9/5	1.8000	470.93	$2^{10/12}$	1.7818	466.16	-4.77	-1.01
Major Seventh (B) *si*	15/8	1.8750	490.55	$2^{11/12}$	1.8878	493.88	3.33	0.68
Octave (C) *do*	2	2.0000	523.25	$2^{12/12}$	2.0000	523.25	0.00	0.00

Table 1. *The Just and Equally Tempered Scales Compared*

pleasing to the ear since it is constructed from simple vibrational ratios (3:2, 4:3, 5:4, 9:8, and so on). However, it creates many problems for the design of musical instruments, particularly fretted instruments. The fret locations beneath each string need to be different, so the fret can't cross the instrument's neck in a straight line. The transposition of music from one key to another is impractical for many musical pieces because the note combinations change. Keyboards were built at various times in the past for just intonation, but they saw little use.

It is possible to approximate to the just scale using other temperaments that are based on other numbers besides 12. The best fits are given by 19, 24, 31 and 53. The equal temperament of 12 is the most practical.

Gurdjieff's Hydrogens: The Ray of Creation

When we consider the octave from the Work's perspective, we consider it in terms of the just scale, with its simple numeric ratios that are presumed to have been chosen by the Absolute. Nevertheless, the equal temperament scale is what rules modern music. Gurdjieff's musical compositions with Thomas De Hartmann, including all the music for the movements, are composed for and played using this scale.

Music can help us to understand some aspects of the octave as long as we bear in mind that the modern musical scale is not exact.

Materiality, Atoms, Forces, and Laws

Figure 11, on the next page, depicts the Ray of Creation adjacent to representations of atoms that correspond to each of its levels. The prime unit is an atom of the Absolute. On the level of All Worlds, atoms are three times as dense, on the level of All Suns, six times as dense, and so on. At the level of the Moon, atoms are 96 times as dense.

The atoms we refer to here are atoms as defined by objective science.* By this definition, an atom is the smallest quantity of a substance that retains all its chemical, physical, biological and cosmic properties, including its psychological properties. The idea of an atom having psychological properties will seem strange—we will discuss its implications later.

Increasing materiality marches in step with an increase in the laws that act at each level. These laws are not individually described statutes per se, they are termed "orders of laws." We can think of an "order of law" as similar to the biblical commandment "thou shalt not kill." It has many possible contexts and nuances: first-degree murder, second-degree murder, manslaughter, negligent homicide, and so on.

Nevertheless these laws are inescapable at the level where they act. For example, a law that applies to plant life is that a plant cannot move its location. This applies to all plants and trees. And it applies to every context of plant or tree. For a plant the law is inescapable. However if we rise slightly higher than plant life and consider insects, it is immediately clear that insects are not subject to this law of immobility.

We have moved up a level and none of the contexts of the law of immobility apply at this level. This concept of what constitutes a law is different to what constitutes a law in, say, modern physics. A law is an unavoidable force that applies at a specific level and every level below that specific level. Thus the only

---

* In Search of the Miraculous by P D Ouspensky, p176

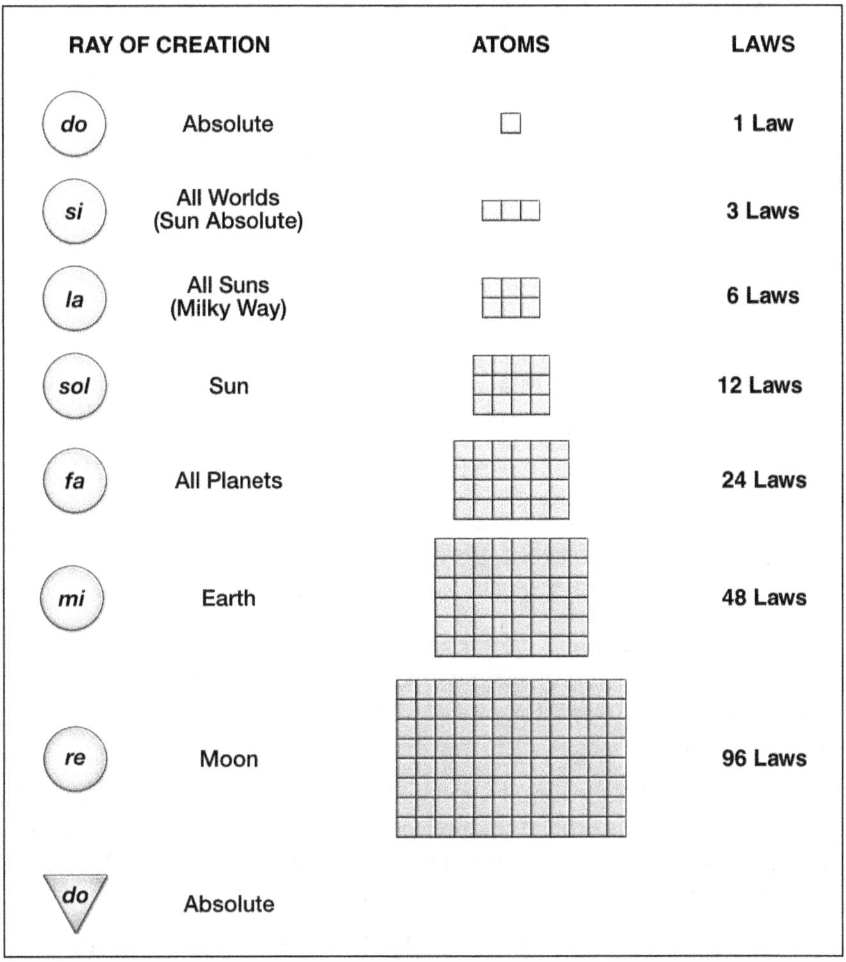

RAY OF CREATION	ATOMS	LAWS
do — Absolute		1 Law
si — All Worlds (Sun Absolute)		3 Laws
la — All Suns (Milky Way)		6 Laws
sol — Sun		12 Laws
fa — All Planets		24 Laws
mi — Earth		48 Laws
re — Moon		96 Laws
do — Absolute		

Figure 11. The Ray, Atoms and Laws

way to escape a law that acts only at the level of the Earth and the Moon is to exist at a higher level than the Earth. In objective science, laws form a hierarchy.

The diagram shows that the number of forces (or laws) is directly proportional to materiality. Accordingly, All Worlds' level is constrained in three ways and has three times the Absolute's materiality. And so it proceeds downwards, doubling with every step until it arrives at the Moon, which is constrained by 96 forces or orders of laws and has a materiality of 96.

The atoms at each level can coexist in the same space. So in our existence on Earth, there are atoms of Moon, Earth, Planets, Sun, All Suns, All Worlds, and

Gurdjieff's Hydrogens: The Ray of Creation

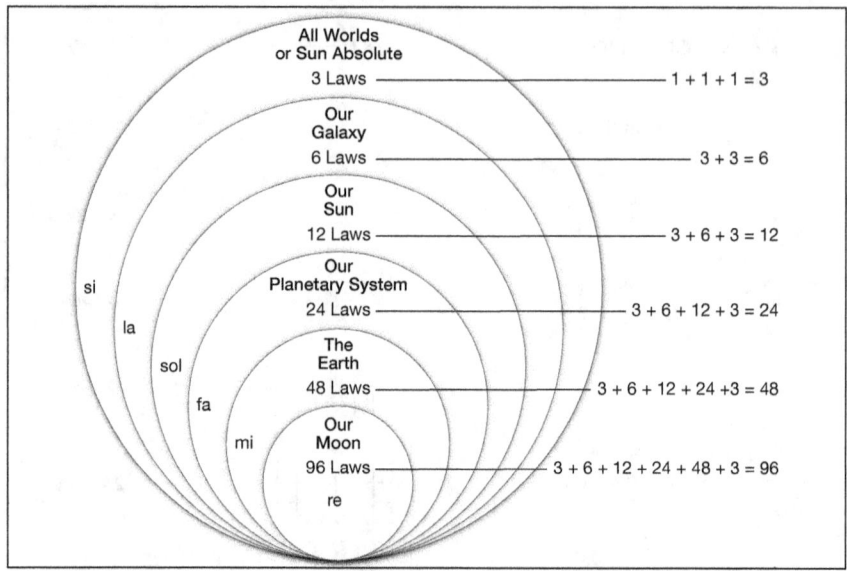

Figure 12. The Ray of Creation From Below

the Absolute, all of which interpenetrate. Atoms of greater materiality are interpenetrated by atoms of less materiality, just as water can and does penetrate sand, and air can penetrate water. These atoms are, as we have previously noted, cosmoses.

Figure 12 represents the Ray of Creation as if it were nested Russian dolls, the higher cosmoses containing the lower ones. Objective science views the Sun Absolute as the universe. It is important to try to grasp the Ray of Creation as it is depicted here, as this is how it manifests.

We inhabit the Earth, which is within the family of Planets, within the solar system, within the Milky Way, which is within the universe, which embodies all galaxies. Spatially, we live within the Sun Absolute and are even subject to its three laws. However, we are unaware of it because we have no direct contact with it.

In *In Search of the Miraculous*, Gurdjieff explains the act of creation as the Absolute manifesting as three separate forces that then unite and by their individual will, and according to their individual decision, create phenomena where they meet.* These phenomena are created worlds. The first world created in this way is the Sun Absolute (All Worlds), the abode of the Absolute.

---

* *In Search of the Miraculous* by P D Ouspensky, p79

At that level (*si*) everything depends upon these three forces that express the Will of the Absolute.

In *The Tales* Gurdjieff writes:*

> "And so, my dear boy, our COMMON FATHER CREATOR ALMIGHTY, having then in the beginning changed the functioning of both these primordial sacred laws, directed the action of their forces from within the Most Holy Sun Absolute into the space of the Universe, whereupon there was obtained the what is called 'Emanation-of-the-Sun-Absolute' and now called, 'Theomertmalogos' or 'Word-God.'"

He asserts that the creation was initiated when the Absolute changed the operation of the Law of Three and the Law of Seven. He also says that the act of creation included actions by these two laws and additionally a specific emanation. We assume that emanation was a substance with a specific vibration, the vibration of the Absolute himself.

The creative act was carried out by a single force, the Will of the Absolute, but gave rise to three forces, the Law of Seven, the Law of Three, and the Theomertmalogos. The Sun Absolute is thus capable of creation (initiating octaves) and transforming substances (Law of Three), and issuing its own emanations, which will be the Theomertmalogos issued with a lower vibration. The Sun Absolute has a lower level of vibration than the Absolute (the note *si*) and that note sounds because the Will of the Absolute fills the interval between *do* and *si*.

The plan for the universe is thus conceived in the Sun Absolute, subject to the Will of the Absolute, but once a lower note was struck, laws were created that were not initiated by the Will of the Absolute, but by a being with a lesser degree of reason, a galaxy. At that level there are six laws, three of which come from the Sun Absolute. Galaxies sound the note *la* in the Ray of Creation in accordance with the Law of Seven that comes from the Absolute.

They participate in any octave created at the level of the Sun Absolute. The three forces from above within a galaxy are now three wills, three consciousnesses, three unities. Each force contains the possibility of any of the three forces, but when the three meet, each will manifest only one principle, either active, passive, or neutralizing. The galaxy also has its own three consciousnesses, its own Law of Three, and its own Law of Seven. Because these forces are local, they are relatively accidental or mechanical (they have a lower vibration level than the level above).

---

* *The Tales*, Chapter XXXIX, The Holy Planet "Purgatory", p756

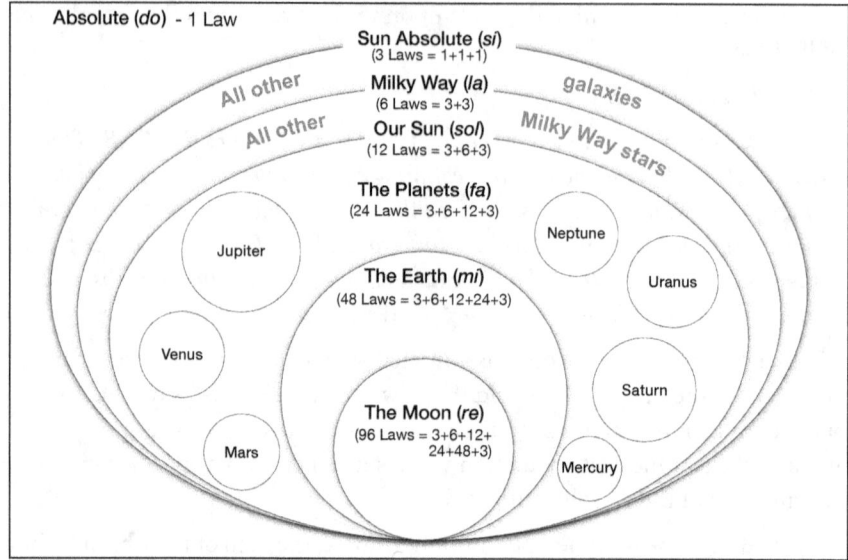

Figure 13. The Ray of Creation From Above

When we observe galaxies (from our human perspective), we note that they vary in form. The Will of the Absolute created these worlds and thus governs them, but it does not govern their creative work; it does not govern the *do* that they sound. We depict the Ray of Creation from above in *Figure 13*. The Absolute is shown surrounding everything, enacting one law, representing three forces or laws united as one. The Sun Absolute has three laws under the control of the Absolute. Each galaxy is under six (3 plus 3) laws, but each of the myriad galaxies is different because three of those laws are individual at that level.

A star within any galaxy is subject to the three laws from the Sun Absolute, six from the galaxy, and three individual laws of its own. And so it proceeds to Planets with 24 forces or laws (3 plus 6 plus 12 plus 3), the Earth (or any other individual planet) with 48 forces or laws (3 plus 6 plus 12 plus 24 plus 3), and finally, the Moon (or any other individual moon) designated by 96 forces or laws (3 plus 6 plus 12 plus 24 plus 48 plus 3).

This is like a hierarchical command structure, such as in an army, where the general's orders are obeyed by the colonel and the lieutenant colonel, and the major and the captain and so on. But the colonel can also give orders to the ranks below, as can the lieutenant colonel and so on down. The colonel commands a regiment, enforcing the general's orders for his regiment, in whatever situation it is in, and giving his own orders. Thus the general's orders govern all the regiments, but the colonel's orders apply to just one. The colonel's orders apply to all the battalions in his regiment, while the lieutenant colonel's commands apply to just his battalion. Thus orders filter down to the individual soldier from different levels of the hierarchy.

It is important to emphasize that this is all theory. The breadth of our galaxy is so great (estimated to be roughly 100,000 light-years) that we do not and cannot see it in its current state. We only know the state of nearby stars—and then only their recent state. The nearest star, Proxima Centauri, is over 4 light-years away. If it were to vanish suddenly, we would not know for 4.24 years. Even when we look at the Sun, we are looking into the past—by about 8 minutes.

We can think of galaxies as large living beings and stars as cells in the body of these beings. We see the cells, but we do not see the body. If we refer back to the atoms shown in *Figure 11*, we can only claim to see atoms of density 12 (as light from the Sun). We have no idea at all of what atoms of density 6 are or how to perceive them.

Estimates of the size of our galaxy currently suggest that it has, perhaps, 400 billion stars. The estimates have increased gradually throughout the last century and continue to increase as astronomers gather more data. The same is true of the estimated number of galaxies. It has continually grown, with the most recent estimate suggesting a figure of 2 trillion.

There is a more precise estimate for the number of cells in the human body. It is in the range 30-100 trillion, with the DNA molecule itself consisting of about 204 billion atoms. And, coincidentally, the number of atoms in a human cell is estimated at 100 trillion. If there is a regular pattern here, maybe the count of suns in a galaxy and galaxies in the universe will end up at the same order of magnitude as cells in the body.

A distinctly different situation occurs at the level of the Sun. The family of Planets in the solar system is not counted in trillions, billions, or even millions. There is a small number of planets (9, if you count Pluto), and it is unlikely that the total will rise significantly, even if a few more are discovered beyond Neptune and Pluto. The same is true of moons. Jupiter has the largest number, but only four have significant size. The others (the estimated number is 63) are more asteroid than moon.

In respect of numerical ratios, the relationship between a sun and its planets is more akin to the relationship between, say, an animal and the organs of its body.

The Lateral Octave From The Sun

We can consider the Ray of Creation from a different perspective by considering an octave that begins at the level of the Sun and descends from there to the lower *do* of the Ray, as illustrated in *Figure 14*.

The note *sol* in an octave is halfway between the higher and lower octave. Thus the Sun vibrates at half the frequency of the Absolute, as Gurdjieff remarks in *The Tales*.*

> ...*the initially given momentum for the fundamental completing process, having lost half the force of its vivifyingness,...*

The theory is that the notes *fa, sol, la* of this lateral octave correspond to organic life on Earth. Specifically, the *mi-fa* interval in the Ray needs to be filled, and the purpose of organic life on Earth is, on the one hand, to fill that interval and on the other to provide sustenance (food of a kind) to the Moon.

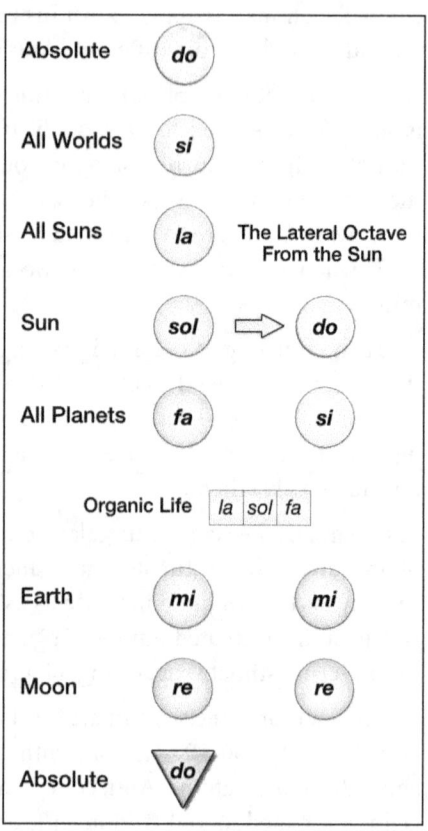

Figure 14. The Lateral Octave

Given the modern view that life evolved almost accidentally from a primordial chemical soup, this theory seems bizarre. But it is not so outlandish from the perspective of a universe that is alive at every level and where everything feeds on everything else.

If one thinks of this from the perspective of the Earth, Nature in all its complexity is simply part of Earth's body. A human being is no more significant to the Earth than a cell of your body is to you. However the Moon is, in theory, an embryonic offspring of the Earth.

It is natural for cells of the Earth's body to feed the growing embryo, even where it results in the death of those cells. It is a curious and perhaps relevant fact that, because of the solar wind, the Earth's outer atmosphere touches the

---

* *The Tales*, Chapter XXXIX, The Holy Planet "Purgatory", p758

Moon's surface for about five days around the time of the full Moon. Thus any substances that accumulate in the higher atmosphere of the Earth might be passed directly to the Moon during that time.

There are two intervals in the Ray of Creation, but the higher interval (*do-si*) is said to be filled by the "Will of the Absolute." The lower (*fa-mi*) interval requires an appropriate mechanism, or it will not be filled.

Our Ray of Creation is one of untold numbers of such rays, made unique by its descent through the Sun to our planet Earth and our Moon. Each planet within the solar system participates in such a Ray, but the only one we are currently able to study is our own. In that particular Ray, the lateral octave that commences with *do* at the Sun's level fills the *mi-fa* interval and organic life on Earth plays a part in that.

From a practical perspective, we can envisage how this might be the case. Organic life (*fa, sol, la* of this lateral octave) forms a relatively thin film that encases the Earth, both on land and in the oceans. It is composed of everything from single-cell life to complex life-forms that permeate the seas, the soil and the lower atmosphere. Influences from the whole universe, including the Sun and the Planets, reach the Earth and are mediated by this thin film of life. Those influences are electromagnetic and magnetic, and they are received by life to the benefit of the Earth.

There may be another role that organic life plays within the solar system. If, as we previously noted, planets are analogous to organs of an animal or human body, the Earth corresponds to this body's stomach. Organic life on Earth might then correspond to the microbiome (the gut bacteria) of that body, absorbing nutrients from beyond the Earth and digesting them for the benefit of the whole body.

Space exploration has provided a wealth of information about other planets, and it now seems that the Earth is the only planet in our solar system with a large and diverse ecosystem of organic life. If organic life exists on other planets, then it most likely consists only of bacterial life.

Three Octaves Within The Ray of Creation

There is a doubling that naturally occurs within an octave in the span of notes from *re* to *do*. If we look back to the numerical depiction of the octave in *Figure 14*, we observe that the numerical gaps between *re-mi*, *mi-sol* and *sol-do* are as follows:

re-mi = 3

Gurdjieff's Hydrogens: The Ray of Creation

mi-sol = 2 + 4 = 6

sol-do = 4 + 5 + 3 = 12

Because of this doubling, and because successive octaves double in vibration from *do* to *do*, we can place three consecutive octaves between *re* and *do*. This is illustrated in *Figure 15*.

These three inner octaves have a particular significance, as we shall see later when discussing the *Hydrogens* in detail. For the moment, we can simply think of these octaves as giving rise to the wide range of substances that exist, from the coarsest to the finest.

If we see ourselves as an image of the Ray of Creation, then we should also detect and identify these three octaves in ourselves. One way we might approach this is to envisage the three intervals as the points where the three different kinds of food enter into us: the lowest *mi-fa* interval for normal food, the middle *mi-fa* interval for air and the highest *mi-fa* interval for impressions. This warrants deeper investigation.

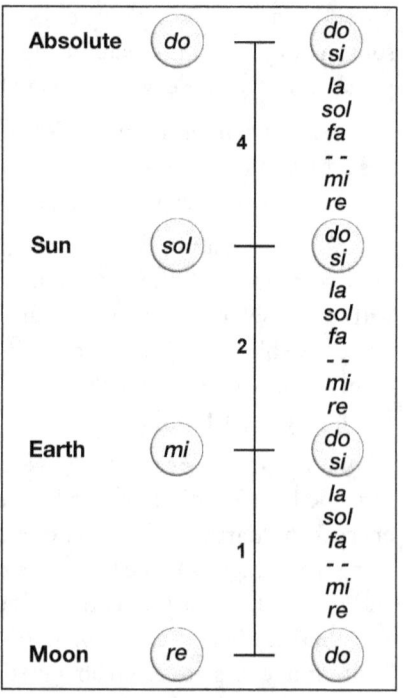

Figure 15. The Three Octaves

Incidentally, it appears that this scheme of three octaves was known to other traditions. The ratio 1:2:4 is found in the design of some cathedrals, including the remarkable cathedral of Chartres. There are 22 notes in the three octaves taken together, the same as the number of chapters in The Old Testament, the number of letters in the Hebrew alphabet and the number of cards in the major arcana of the Tarot. It is possible that these things express this scheme of three octaves in some way.

Knowledge and Information

From the perspective of the Work, man is an organism that, on the one hand, fulfills a cosmic role by creating substances that are useful both to the Earth and the Moon. On the other, he has the possibility of creating substances that may be crystallized within him for his own use.

The Threshold of Objective Science

Man can live a life that has no direct importance to the Work. He is born, grows up, procreates, and ultimately dies to serve the purposes of the Earth and the Moon. It may be that his life offered him no choice in the matter, or that when there was a choice to seek knowledge, he chose otherwise. Such a layer of unconscious humanity is required for the Work to be possible. Gurdjieff explains this in *In Search of the Miraculous*.*

Beginning with the assertion that everything is material (even God is material), he points out that knowledge also is material. In fact, there are many things that we ordinarily do not think of as material that are also material, such as thoughts and feelings. The critical consequence of knowledge being material is that there is a limited supply, just as there is a limited supply of other materials such as gold.

Knowledge must and does possess all the characteristics of materiality. It is thus limited in quantity in any particular place at any specific time. It follows that over a given time, such as a modern man's lifetime, humanity has a specific amount of knowledge available to it, and only that amount. Another important aspect of knowledge is that the consequence of digesting it varies according to quantity.

Thus, if one man or a small group digests a large quantity of it, it can produce good results, but it produces very little or even bad results if consumed in small quantities. It is sometimes said that "a little knowledge is a dangerous thing." In the modern age, with all its technological realities in the multiple diverse fields of engineering, it might seem as though there is a large amount of knowledge available to anyone who wants it. Still, the reality is that even that knowledge—relatively low-quality knowledge, equivalent to the knowledge of the carpenter or the smithy—is possessed only by specialists.

Knowledge has gradations. A child is taught to tie a shoelace, learns to speak, read and write, and acquires various skills. The child accumulates practical knowledge for life, and the process continues to some degree throughout life. The knowledge of the inner world is of a higher gradation. Such knowledge is achieved through self-observation, self-remembering, and intentional suffering. The creation of such knowledge is individual, guided by information received from the Work.

If esoteric knowledge is concentrated in a small number of people, it can produce great results. It is thus better that it is not dispersed among the masses. But even if it were a good idea, most people have no desire for esoteric

---

* *In Search of the Miraculous* by P D Ouspensky, p78

knowledge. This is especially the case in our age where esoteric information is widely published and available at little expense.

An essential aspect of knowledge is that it can only exist within living beings. Among biological life-forms, only man can digest information and transform it into knowledge. Lower life-forms are limited to whatever knowledge they naturally embody as a species but can have no knowledge of personal evolution.

Just as knowledge is limited, so is information.

We can think of the Earth in recent millennia as embodying a more or less constant amount of information. Information is created by the senses of all living beings consuming impressions in the normal course of life. Before reading and writing was a widespread skill, the information men considered important was passed by word of mouth. Now, much of it is simply published on paper or electronically.

Nevertheless, this makes no difference to the quantity of information on Earth, since a man can only ingest so much information in any given second, minute, or hour. And this is the case for all life-forms.

Knowledge is information transformed. Man is exceptional, able to consume information through language in ways unavailable to other creatures—listening to the spoken word, reading, and experiencing art. If such information is not digested, it does not become knowledge, and it is destroyed when a man dies.

Information too has gradations of quality, according to its source:

– A influence

– B influence

The reality is that only man has an interest in B influence information, and even then a relatively small number of men.

Chapter 3

THE LADDER OF HYDROGENS

"And he dreamed that there was a ladder set up on the earth, the top of it reaching to heaven; and the angels of God were ascending and descending on it."

~ Genesis 28:12

Every science needs to propose foundational definitions of its sphere of investigation and application. For objective science, this is everything in the universe. In particular, it needs to define what it means by "substance," what the different categories of "substance" are and how they interact.

The following are fundamental propositions of objective science:

- Everything is material.
- Substances are materials of different densities that vibrate at different frequencies.
- In general, the less dense the substance, the higher the vibration. But the vibration of a substance can vary. For example, a piece of wood or a rock can be warmer or cooler.
- Vibrations proceed through time in octaves, according to the Law of Seven.
- All changes occur according to the Law of Three; three different forces acting through three distinct substances meet and combine.

Contemporary science has a far more complex view of "substances" that relates to context. If a substance is at the atomic level or smaller, it is defined and discussed in terms of quantum mechanics. A little larger than that, and it is described in the context of chemistry; what it reacts with and how. If it is a protein, it is discussed in a microbiological context. If it is a living thing, a cell, or a collection of cells, it is viewed in a biological context. If it is large and inanimate, the context is geological. If it is planet-sized or greater, then we leap to an astronomical context. In objective science, there are no such distinctions in how changes occur; at every level, irrespective of size and scale, there are substances and vibrations, subject to precisely the same two laws: the Law of Seven and the Law of Three.

The Four Elements: EARTH, WATER, AIR, and FIRE

The number of laws that apply at each level of the Ray of Creation denote distinctly different worlds. More laws means more limitations on activity, which makes the reality of life at that level different. The life of a dog and a tree are different because the tree is subject to more laws.

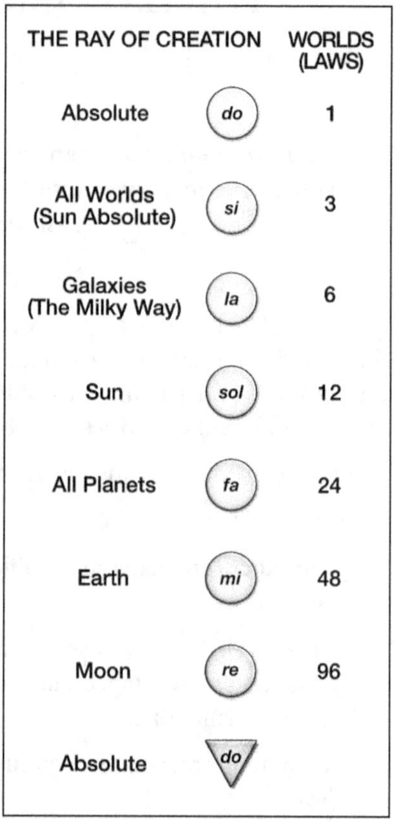

In *In Search of the Miraculous*, Gurdjieff says* that on Earth, our world, forty-eight orders of laws are operating which govern our whole life. If we lived in World 96, at the level of the Moon, twice as many laws would apply. In practice, this would mean that our life and activity would be much more mechanical to the extent that there would be no possibilities of escape.

In worlds 12, 24, 48, and 96, the Will of the Absolute has less and less possibility of manifesting. Just 3 laws come from the Absolute, and all the other laws at any level are mechanical in respect of the laws from the Absolute. There are another nine at the level of the Sun, 21 more at the

THE RAY OF CREATION		WORLDS (LAWS)
Absolute	do	1
All Worlds (Sun Absolute)	si	3
Galaxies (The Milky Way)	la	6
Sun	sol	12
All Planets	fa	24
Earth	mi	48
Moon	re	96
Absolute	do	

Figure 16. Worlds and Laws

level of the Planets, 45 more at the level of Earth, and 93 more at the level of the Moon. The Absolute could only abrogate those laws by destroying the universe because all of those laws are a consequence of His intention. Ouspensky provides the example of a card game, where the law is that the ace beats the

---

* In Search of the Miraculous by P D Ouspensky, p83 - p88

deuce. If the Absolute intervenes in just one game allowing the deuce to beat the ace, then the whole game and its laws are destroyed. Thus such an intervention is impossible. It not uncommon for people to pray for precisely such interventions.

Materiality

Objective science makes no distinction between energy and matter. There is an equivalence between mass and energy in modern physics, enshrined in Einstein's famous equation: $e = mc^2$. Matter can be in various states (solid, liquid, gas, plasma), and energy can manifest in multiple ways (kinetic energy, chemical energy, electrical energy, etc.) Ultimately, physics regards matter and energy as related, but manifesting very differently. Objective science has the single concept of materiality, everything is material and all material vibrates.

There are seven categories of materiality. Gurdjieff states that they are intermixed in the sense that they interpenetrate one another. We tend not to think of matter in this way, although it is the case. We think, for example, of marble as being solid and impenetrable. In fact, water penetrates marble quite easily. Indeed one of the ways that marble and other forms of rock erode is that water penetrates them when it rains, then when it freezes, it turns to ice. The ice then expands and cracks the marble. In general, the principle is that finer matter penetrates coarser matter, and it does so at every level.

If we consider the Moon then, the lowest point of the Ray of Creation, its material will be composed predominantly of atoms of that level, but it will contain some atoms of higher levels. Precisely the same relation between different kinds of matter may be observed in the whole of the universe: the finer matters permeate the coarser ones.

The critical point here is that the atoms of these matters interpenetrate. Atoms of a higher level can occupy the same space of atoms at a lower level. Thus on our Moon, there are "atoms of the Moon," but there can also be atoms of every other kind of matter permeating them. Similarly, there are dense atoms under 96 laws ("atoms of the Moon") on planets and suns. Also, there will be "atoms of Earth" within planets and suns. Such atoms are relatively heavy and dense.

The usual scientific perspective is to think of materiality as solid, liquid, or gaseous. Anything more energetic than a gas tends to be thought of as energy or primarily energy. Before the advent of modern science, stretching back at least to the Ancient Greeks, it was common to classify matter as EARTH, WATER, AIR, and FIRE.

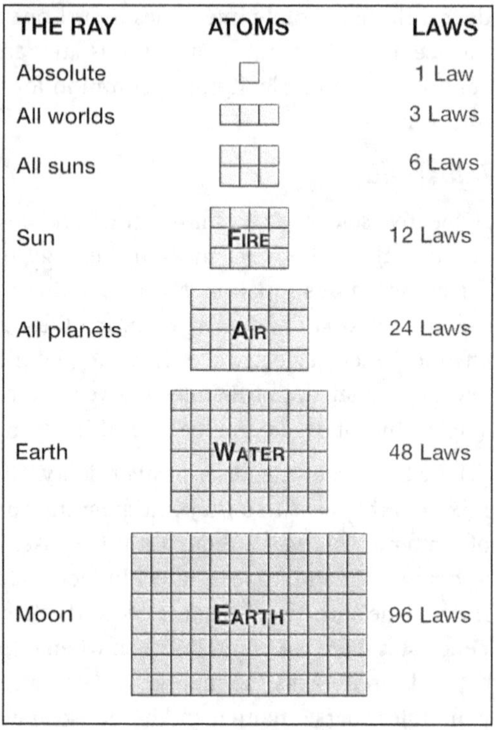

Gurdjieff's Hydrogens: The Ray of Creation

Objective science can adopt the classifications because Earth, Water, Air and Fire,* are gradations of density. It is shown in *Figure 17*, which depicts the seven levels of materiality as atoms of increasing density. So the lowest level, which we think of as solid (like rock, or iron or glass), is what was called Earth by alchemists and is matter at the level of the Moon.

The matter of world 48, our Earth, is of the nature of Water, and indeed our planet is a planet dominated by liquid. We can call the next higher gradation of matter, subject to just 24 orders of laws, Air. It stands at the level of the family of Planets. Planets have atmospheres, and, in part, it is the transparency of our atmosphere that enables the other planets to beam their influence on the life of Earth.

THE RAY	ATOMS	LAWS
Absolute	□	1 Law
All worlds		3 Laws
All suns		6 Laws
Sun	Fire	12 Laws
All planets	Air	24 Laws
Earth	Water	48 Laws
Moon	Earth	96 Laws

Figure 17. Earth, Water, Air and Fire

The matter of world 12, which the alchemists called Fire, is at the level of the Sun and is referred to as plasma, matter in ionic form. It is subject to fewer orders of laws and, although associated with the level of the Sun, it is plentiful on Earth. We know it as fire, as electricity, as light. Plasma was first noticed by Sir William Crookes in 1879—he dubbed it "radiant matter." The term "plasma" was coined by Irving Langmuir in 1928, and since then, science has gradually accumulated more knowledge about plasma and its behavior. So contemporary science thinks of matter as solids, liquids, gases, and plasmas.

If we want now to think about these laws, it is evident that solid matter (Earth) is the most constrained. It is crystalline and remains so unless it is heated or it reacts chemically with some other substance. It can be eroded by the persistent action of water or other liquids.

---

* To avoid verbal confusion, when using the words "earth," "water," "air" and "fire" to denote the four elements we print them in small caps: Earth, Water, Air, Fire.

Liquid matter (WATER) travels more easily and more quickly. It flows, whereas solid matter does not. It is nevertheless contained by its surface tension. Surface tension is a property of liquids that gases do not have. Because of that, gaseous matter (AIR) can disperse by diffusion. Plasma (FIRE) is even less constrained. Plasma does not consist of atoms but of ions. These are charged particles, the smallest of which is an electron (a negatively charged ion) or a proton (a positively charged ion). Atoms and molecules that have an added or missing electron are also ions. Thus plasma has electrical and magnetic properties that denser forms of material do not have.

While we may think that only solid material (EARTH) and perhaps liquids (WATER) have form, objective science insists that there can be form in every level of material. AIR exhibits form with clouds and tornadoes. FIRE (energetic material) can also have form. For example, higher bodies crystallize from energetic matter.

Points of Stability

Objective science views the universe as having points of concentration or stability. In *The Tales*, Gurdjieff writes:

"And sometime later, in the course of these occupations, this Saint Venoma first constated in cosmic laws what later became a famous discovery, and this discovery he first called the 'Law of Falling.'

"This cosmic law which he then discovered, St. Venoma himself formulated thus:

" 'Everything existing in the World falls to the bottom. And the bottom for any part of the Universe is its nearest "stability," and this said "stability" is the place or the point upon which all the lines of force arriving from all directions converge.

" 'The centers of all the suns and of all the planets of our Universe are just such points of "stability." They are the lowest points of those regions of space upon which forces from all directions of the given part of the Universe definitely tend and where they are concentrated. In these points, there is also concentrated the equilibrium which enables suns and planets to maintain their position.'

We know that Planets ("points of stability") include matters in any of the four states: solid, liquid, gas, or plasma. We conceive of planets as being wrapped in an atmosphere and having a magnetic field that encompasses them. In the space between these points of stability (suns and planets), there can only be matters of a higher level.

Logically, the idea of the Absolute, a unity, suggests that the Absolute is omnipresent within this scheme—that atoms of the Absolute may exist anywhere within the universe. It naturally follows that the abode of the Absolute, the Sun Absolute, also pervades the whole universe. It seems that immediately below that level (the galaxies), matter begins to cluster. Thus we can think of space as filled with relatively fine matter, which contemporary science calls plasma or energy (in the form of electromagnetic radiation).

The points of stability are the notes *sol, mi, re* in the Ray of Creation. The Moon exists within the atmosphere of Earth, the Earth (and other planets) within the atmosphere of the Sun, and the Sun itself within the atmosphere of the Milky Way. We can thus regard our sun, individual planets, and moons as points of stability.

These are points where "all the lines of force arriving from all directions converge." We assume that "lines of force" means electromagnetic radiation, whether radio waves, heat, light, X-rays, or cosmic rays. Additionally, denser material (particles, atoms, asteroids, and comets) falls to these points.

He states that "In these points, there is also concentrated the equilibrium which enables suns and planets to maintain their position." The suns, planets, and moons do not fall into other points of stability but orbit around them and hence exhibit an equilibrium (literally meaning "an equal level").

In Gurdjieff's description in *The Tales*, the equilibrium is between the Law of Falling and the Law of Catching up. (Note that Gurdjieff uses the archaic meaning of "catching up,"—the term originally meant to "hold on high." The Law of Catching up is the centrifugal force that comes from orbital motion.)

Orbit and Spin

When we consider the universe from an astronomical perspective, we notice its spherical nature. Suns, planets, and moons are (approximately) spherical, and they spin about their axes. Earth's moon and almost all other moons within our solar system spin exactly once with each orbit and thus always present the same face to the mother planet (as would a space station). The only exception is Hyperion, a moon of Saturn, which rotates chaotically (most likely because of Saturn's twin gravitational influences and its large moon Titan).

The tilt of the axis, about which a planet spins, varies. The Earth, Mars, Saturn, and Neptune are very similar, rotating with an axial tilt (to the plane of the ecliptic) of 23.44º, 25.19º, 26.73º, and 28.32º, respectively. Pluto and

Uranus have large tilts (82.23° and 57.47°), and even the Sun has an axial tilt (of 7.25°).

All the planets have moons except Venus and Mercury. Each one orbits the Sun at specific distances from each other so that no planetary orbits even come close to crossing each other, except for Neptune and Pluto, which do cross. The planets have elliptical orbits, but the Sun is never at the center of the ellipse. The planet with the most eccentric (i.e., least circular) orbit is Pluto.

The Sun itself orbits the Milky Way in one of the galaxy's spiral arms. In fact, all-stars orbit their galaxy (as far as anyone knows). It is suspected (but not proven) that our galaxy orbits other galaxies in a cluster of galaxies. The whole population of galaxies is likely to be in some orbit within the Sun Absolute. It seems as though the whole universe is "falling" and "catching up."

Hydrogen, Carbon, Oxygen and Nitrogen

Gurdjieff chose the word "*Hydrogen*" to denote a category of substances that has a specific range of vibration. He then added the idea that substances need to be distinguished according to which of the three forces of the Law of Three (active, passive, or neutralizing) is acting through the substance. A substance is referred to as *Carbon* if it carries the active force, as *Oxygen* if it carries the passive force, and as *Nitrogen*, if it carries the neutralizing force.

The terms *Hydrogen*, *Carbon*, *Oxygen*, and *Nitrogen* are not Gurdjieff's invention. These words appear in the writings of the Theosophist Madame Blavatsky and may go as far back as Paracelsus or even earlier. Some commentators equate the Theosophical use of these words with the traditional four elements (EARTH, WATER, AIR and FIRE), as does Gurdjieff.

Each of these four words has a specific etymology that may provide a clue to its meaning.

- Carbon comes from the Latin *carbo* meaning "coal" or "glowing coal" or "charcoal." Its Proto-Indo-European (PIE) root word *ker* has the meaning of "heat," " fire," "burn."
- Oxygen is from the Greek root *oxys*, literally meaning "sharp," implying acidic.
- Nitrogen is from Greek *nitron*, meaning carbonate of soda (which contains no nitrogen), and has been used as a cleaning agent for thousands of years (implying perhaps that it is neutralizing). The word *nitre* was also used in the Middle Ages to describe saltpeter (potassium nitrate). If oxygen implies acid, then nitrogen implies alkaline.

Gurdjieff's Hydrogens: The Ray of Creation

— Hydrogen is from the Greek *hydr*, (stem of *hydor*, meaning water).

Thus we can (at a stretch) think of these four as the traditional four elements: *Carbon* (FIRE), *Hydrogen* (WATER), *Oxygen* (AIR), and *Nitrogen* (EARTH). It will not have slipped many people's notice that, as chemical elements, these four also provide the basis of Organic chemistry.

The hydrogen atom has a single electron in its orbital shell, allowing it to readily combine with oxygen to create water (H_2O), with nitrogen to create ammonia (NH_3), and carbon to create methane (CH_4). Of the three elements: carbon is the lightest and most reactive, then comes nitrogen, with oxygen being the heaviest.

The Ray of Creation and The Law of Three

Now that we have defined these terms, we can discuss *Figure 18*. This diagram of Ouspensky's* shows the Ray of Creation interacting based on the Law of Three. So the first three worlds taken together produce a phenomenon that influences the succeeding worlds, and so on. In the first three worlds, the Absolute is *Carbon*, conducting the active force. World 3 is *Oxygen*, conducting the passive force, and World 6 is *Nitrogen*, conducting the neutralizing force.

The sequence of the order of their action is 1, 2, 3, (C, O, N), but in respect of their density, the three must stand in the order 1, 3, 2 (C, N, O) because, in accord with the Law of Three, the higher blends with the lower to produce a result in between the two in respect of density. When substances occupy this order, phenomena are produced. But if the next descending triad is to form, the result of the interaction (N) must now occupy the active force's position (C). So in that triad, the forces resume their natural order 1, 2, 3.

The next triad proceeds in exactly the same manner, resulting in N at the level of the planets, which then become the active force, C of the next triad. Here we arrive at the note *fa* where there is an interval between it and the note *mi*. No interaction can occur here unless there is a substance that can fill the interval. This substance is organic life on Earth, *fa-sol-la* of the lateral octave from the Sun.

The Planets are active force (C), the Earth is passive force (O), and organic life is neutralizing (N). We can envisage a forest, or shoals of fish or man receiving planetary influences, although each will receive different influences. Man receives more complex influences, but even in that case, people differ. The majority of humanity is important only in the mass, in the sense that only the

---

* The Fourth Way by P D Ouspensky, p211

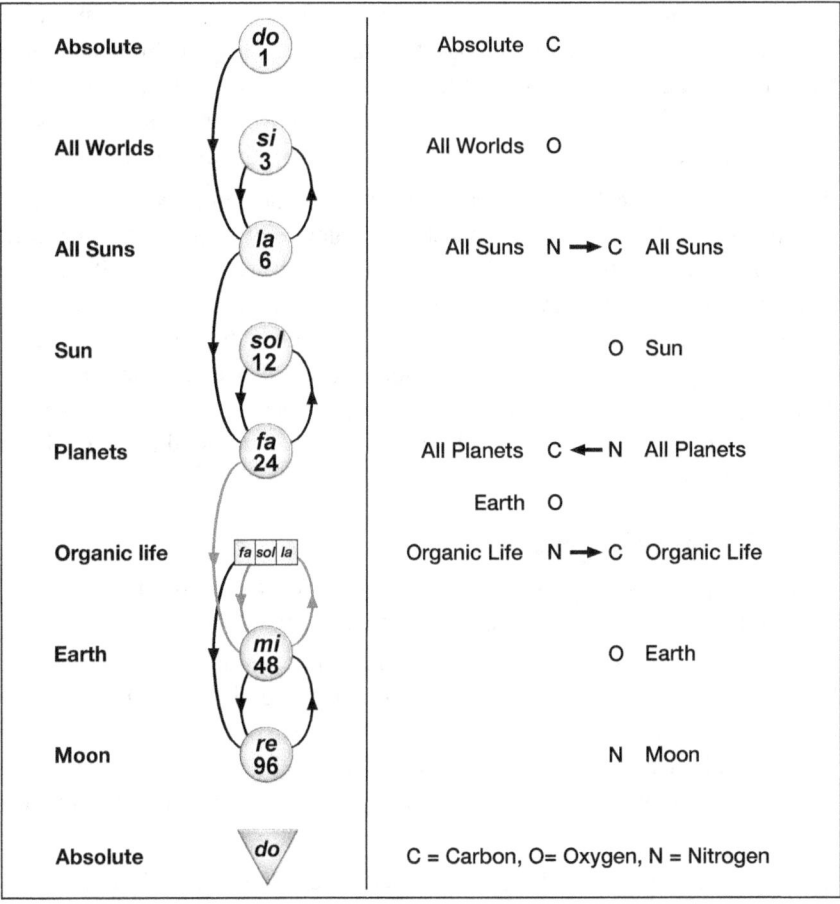

Figure 18. The Ray of Creation and Law of Three

mass receives one or another influence, which is then amplified by imitation. Others may receive finer influences individually.

In general, the phenomenon that occurs here is the Planets and Earth combine to determine the form of Organic life to Earth's benefit. On the one hand, we can think of this as biological evolution and on the other as astrological influence. In either case, it is the Planets influencing the Earth.

The final triad has Organic life as active force (C), the Earth (O) as passive force, and the Moon as neutralizing. This triad is of Organic life (C), combining with the Earth (O) to feed the Moon. By this triad, the influence of Organic life passes to Earth and thence to the Moon. It happens with the death of biological life but may happen in other ways.

Gurdjieff's Hydrogens: The Ray of Creation

It may at first be difficult to see what is happening with these cosmic phenomena. Ouspensky provides this analogy:* A man can, by his will, influence some tissue in a part of his body. But the tissue is composed of cells, so to affect the tissue, man's will must first influence the cells composing the tissue. The tissue is different from cells, but it contains the cells and can be influenced.

It follows that if the Absolute wishes to influence the Sun Absolute, World 3, he must first influence a certain number of galaxies of which it is composed. Let's now consider the nature of the universe in respect of orbits. We notice that galaxies consist of many orbiting stars, and to influence the galaxy, it will be necessary to first influence some stars. The same is the case with the solar system, with planets forming by their motion a spiral layer around the Sun as it orbits the galaxy. No influence from outside the solar system can reach the Sun without first passing through the world of Planets, that spiral layer surrounding the Sun.

The Table of Hydrogens

To deduce classes of substance, we commence with *Figure 19*, showing three octaves that connect the Absolute to the Moon, the first connects the Absolute to the Sun, the second, the Sun to the Earth, and the third, the Earth to the Moon. The octaves double in vibration as they ascend. In the diagram, the Earth-Moon octave goes from *(re)* to *(mi)* spanning one-eighth of the whole octave. The Sun-Earth octave doubles that span, going from *(mi)* to *(sol)*, and finally, the Absolute-Sun octave doubles the previous octave, going from *(sol)* to *(do)*.

As far as the Ray of Creation is concerned, this is the whole of the universe comprising all substances from the densest, vibrating with a low frequency, to the most energetic with very low density and a high frequency of vibration. The definition of a substance is that it has mass of a given density within a range of densities and that it vibrates at a given frequency within a specific range of frequencies.

Thus ice is a substance with a specific density that will vary slightly according to the pressure applied to it. It vibrates at a given frequency as evidenced by it emitting infrared rays and light. In *The Tales*** Gurdjieff writes:

> "You must also know further, that only one cosmic crystallization, existing under the name 'Omnipresent-Okidanokh,' obtains its prime arising—although it also is crystallized from Etherokrilno—from the three

---

* The Fourth Way by P D Ouspensky, p211
** The Tales, Chapter XVII, The Arch-Absurd, p138

The Ladder of Hydrogens

Holy sources of the sacred Theomertmalogos, that is, from the emanation of the Most Holy Sun Absolute.

This statement suggests that the first cosmic crystallization, which gives rise to the table of *Hydrogens*, is present everywhere. Describing the three octaves in the Ray of Creation,* Gurdjieff notes that, although there are six 'intervals' in these three octaves, only three need to be filled from without. The three intervals between *do* and *si* do not. The first of those three is filled by the Will of the Absolute. The second is filled by the influence of the Sun's mass upon radiations passing through it. This can be thought of as the intention of the Sun. However, the Sun's intention derives from what it is (its mass), influenced by radiation from the galaxy or even higher levels, reaching it and passing through it to its point of stability.

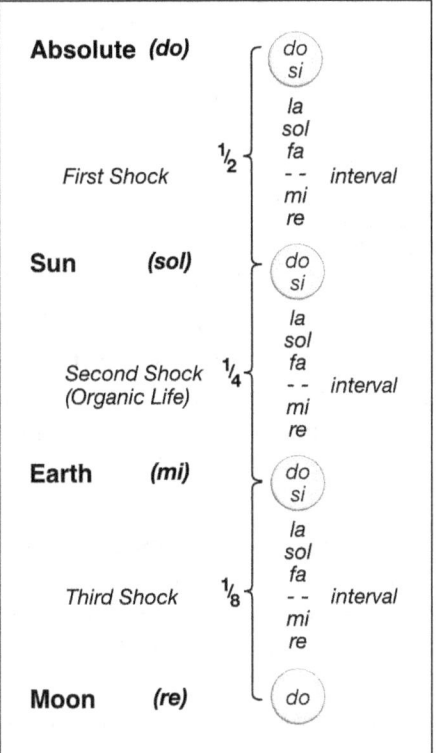

Figure 19. The Three Octaves

Similarly, we can think of the Earth's *do-si* interval being filled by the Earth's intention, which derives from its mass, influenced by radiations from all points of the universe, passing through it to its point of stability. Thus only the *fa-mi* intervals need to be filled. These are filled by 'shocks,' which can be provided by substances from octaves that pass through the same point or parallel octaves, starting from higher points in the Ray of Creation.

We know nothing directly about the shock at the *fa-mi* interval in the first octave, Absolute-Sun. We can envisage it, by analogy, as a high food of "impressions" for the Megalocosmos. However, substances at that level are not what man experiences as impressions. The shock in the second octave, Sun-Earth, is Organic life on Earth, that is, the three notes *la, sol, fa* of the octave which starts in the Sun and ends with the Moon. These three notes constitute

---

* In Search of the Miraculous by P D Ouspensky, p169

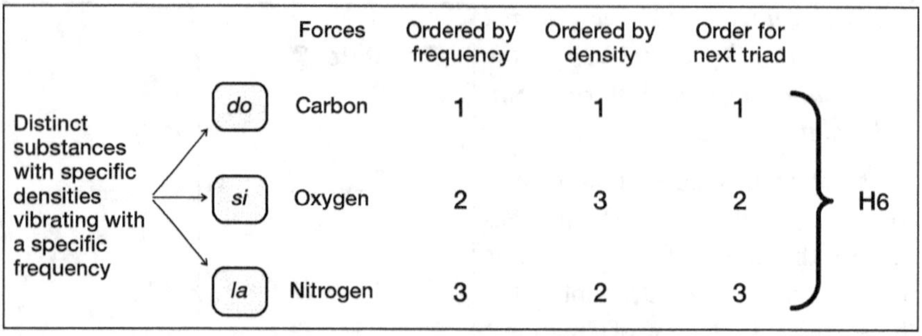

Gurdjieff's Hydrogens: The Ray of Creation

		Forces	Ordered by frequency	Ordered by density	Order for next triad	
Distinct substances with specific densities vibrating with a specific frequency	do	Carbon	1	1	1	
	si	Oxygen	2	3	2	H6
	la	Nitrogen	3	2	3	

Figure 20. The First Class of Hydrogens

breath for the megalocosmos. However, they are not necessarily filled by organic life-forms in other Rays of Creation, only in our Ray. The third interval in the octave Earth-Moon is unknown to us but corresponds to food for the megalocosmos. We do not know how this interval is filled, only that it is filled by activity within stars, planets, and moons.

In *Figure 20*, we depict the first three notes of the Absolute-Sun octave. The note *do* is *Carbon*, denoted by the number 1, *si* is *Oxygen*, denoted by the number 2 and *la* is *Nitrogen*, denoted by the number 3. So *do*, with the highest level of vibration acts on *si* with a lower level of density and vibration to produce *la*, which has an even lower level of vibration. The order in time is 1, 2, 3. However, the Law of Three interaction must result in a substance with a density between *Carbon* and *Oxygen*. Thus, as shown in *Figure 20*, the three substances ordered by density must be in the order 1, 3, 2.

However, when it comes to the subsequent (the next lower) triad, the active force will act through the note *la*. The Law of Three can act in different ways. When it acts in a creative (descending) octave, each interaction must begin with a higher substance interacting as the active force, with one at a lower level of vibration manifesting as passive. Taken together, these three substances in all their possible manifestations constitute a class of substances with a given range of vibrations and densities, which can be designated *Hydrogen* 6 or simply H6.

We might want to think of this first triad as representing the moment of creation, but in reality this triad is always occurring. The Ray of Creation is always in motion, and the triads that transform substances down towards the Moon do so all the time. They are balanced by other triads that raise substances to higher levels.

Following that first triad, subsequent triads proceed according to the combined actions of the Law of Three and Law of Seven. This is illustrated in

Absolute	do	C	1	1	1	⎫					
	si	O	2	3	2	⎬ H6					
	la	N	3	2	3	⎭	C	2	2	2	⎫
	sol						O	4	6	4	⎬ H12
	fa	C	4	4	4	⎫	N	6	4	6	⎭
	--	O	8	12	8	⎬ H24					
	mi	N	12	8	12	⎭	C	8	8	8	⎫
	re						O	16	24	16	⎬ H48
Sun	do	C	16	16	16	⎫	N	24	16	24	⎭
	si	O	32	48	32	⎬ H96					
	la	N	48	32	48	⎭	C	32	32	32	⎫
	sol						O	48	96	48	⎬ H192
	fa	C	64	64	64	⎫	N	96	48	96	⎭
	--	O	128	192	128	⎬ H384					
	mi	N	192	128	192	⎭	C	128	128	128	⎫
	re						O	192	256	192	⎬ H768
Earth	do	C	256	256	256	⎫	N	256	192	256	⎭
	si	O	512	768	512	⎬ H1536					
	la	N	768	512	768	⎭	C	512	512	512	⎫
	sol						O	768	1024	768	⎬ H3072
	fa	C	1024	1024	1024	⎫	N	1024	768	1024	⎭
	--	O	2048	3072	2048	⎬ H6144					
	mi	N	3072	2048	3072	⎭	C	2048	2048	2048	⎫
	re						O	3072	6144	3072	⎬ H12288
Moon	do						N	6144	3072	6144	⎭

Figure 21. Three Octaves of Hydrogens: From the Absolute to the Moon

Figure 21. In the second triad, **la** as the active force is designated as 2, **sol** as 4, and **fa** as 6. This convention of doubling the numbers proceeds through all the other triads as they descend through the three octaves, providing us with twelve variously numbered *Hydrogens*. It may seem strange in the three *fa*-interval-*mi* triads to think of an interval as a substance. However, there must be a substance from a different octave there for the triad to function, although there is no way to know which note from which octave it will be.

Following this diagram, we can imagine the creation unfolding from H6 downwards in cascading triads. Note, that these interactions of substances could be happening in multiple different contexts. This seems to be what is described in the following excerpt from *The Tales:**

> "In order that you may be able to understand, at least approximately, concerning this Omnipresent-Okidanokh also, I must tell you, first of all, that the second fundamental cosmic law—the Sacred Triamazikamno—consists of three independent forces, that is to say, this sacred law manifests in everything, without exception, and everywhere in the Universe, in three separate independent aspects.

---

* The Tales, Chapter XVII, The Arch-Absurd, p138

Gurdjieff's Hydrogens: The Ray of Creation

> "And these three aspects exist in the Universe under the following denominations:
>
> "The first, under the denomination, the 'Holy-Affirming';
>
> "The second, the 'Holy-Denying'; and
>
> "The third, the 'Holy-Reconciling.'
>
> "And this is also why, concerning this sacred law and its three independent forces, the said objective science has, among its formulations, specially concerning this sacred law, the following: 'A law which always flows into a consequence and becomes the cause of subsequent consequences, and always functions by three independent and quite opposite characteristic manifestations, latent within it, in properties neither seen nor sensed.'
>
> "Our sacred Theomertmalogos also, that is, the prime emanation of our Most Holy Sun Absolute, acquires just this same lawfulness at its prime arising; and, during its further actualizations, gives results in accordance with it."

The final paragraph above describes the nature of the separation of the Will of the Absolute into three. In the paragraph that follows,* the descent through the octaves seems to be described:

> "And so, my boy, the Omnipresent-Okidanokh obtains its prime arising in space outside of the Most Holy Sun Absolute itself, from the blending of these three independent forces into one, and during its further involutions it is correspondingly changed, in respect of what is called the 'Vivifyingness of Vibrations' according to its passage through what are called the 'Stopinders' or 'gravity centers' of the fundamental 'common-cosmic sacred Heptaparaparshinokh.'"

In *The Tales*, Gurdjieff's term "Omnipresent-Okidanokh" does not have the same meaning as *Hydrogen*. It refers to specific active elements of world three. In other words, a substance, atoms of which consist of the blending of these three independent forces into one. The density of the Omnipresent-Okidanokh is thus always within the range of the possible densities of atoms of world 3. However, the frequency of vibrations of Omnipresent-Okidanokh can and does vary considerably. Gurdjieff writes:*

> "I repeat, among the number of other already definite cosmic crystallizations, the Omnipresent-Okidanokh unfailingly always participates in both large and small cosmic formations, wherever and

---

* The Tales, Chapter XVII, The Arch-Absurd, p139-140

The Ladder of Hydrogens

under whatever external surrounding conditions they may arise in the Universe.

"This 'common-cosmic Unique-Crystallization' or 'Active-Element' has several peculiarities proper to this element alone, and it is chiefly owing to these peculiarities proper to it that the majority of cosmic phenomena proceed, including, among other things, the said phenomena that take place in the atmosphere of certain planets.

"The first peculiarity is that when a new cosmic unit is being concentrated, then the 'Omnipresent-Active-Element' does not blend, as a whole, with such a new arising, nor is it transformed as a whole in any definite corresponding place—as happens with every other cosmic crystallization in all the said cosmic formations—but immediately on entering as a whole into any cosmic unit, there immediately occurs in it what is called 'Djartklom,' that is to say, it is dispersed into the three fundamental sources from which it obtained its prime arising, and only then do these sources, each separately, give the beginning for an independent concentration of three separate corresponding formations within the given cosmic unit. And in this way, this 'Omnipresent-Active-Element' actualizes at the outset, in every such new arising, the sources for the possible manifestation of its own sacred law of Triamazikamno.

"It must without fail be noticed also, that in every cosmic formation, the said separated sources, both for the perception and for the further utilization of this property of the 'Omnipresent-Active-Element' for the purpose of the corresponding actualizing, exist and continue to have the possibility of functioning as long as the given cosmic unit exists.

"And only after the said cosmic unit has been completely destroyed do these holy sources of the sacred Triamazikamno, localized in the 'Omnipresent-Active-Element-Okidanokh,' reblend and they are again transformed into 'Okidanokh,' but having now another quality of Vivifyingness of Vibrations.

This passage discusses the context of *Hydrogen* interactions when a new being (cosmic unit) is created. We believe it has the following meaning. A new being is created when the note *do* (of that being) is sounded in the act of conception. A whole octave of substances (*Hydrogens*) of varying vibrations participate. Think, for example, of the fertilization of a human ovum. All the substances required to form the physical body of the new being are present. The sperm completes part of that process by donating foreign DNA, and an essence of some origin acts as neutralizing force.

Gurdjieff suggests that this process is accompanied by the Omnipresent-Okidanokh with a corresponding level of vibration entering into the cosmic unit at its formation and undergoing Djartklom. It is *"dispersed into the three fundamental sources from which it obtained its prime arising."* Then *"these sources, each separately, give the beginning for an independent concentration of three separate corresponding formations within the given cosmic unit."*

Those three separate formations most likely relate to the three foods that every cosmic unit consumes. It may be that each of the three parts of the Omnipresent-Okidanokh initiate three specific food octaves in some way. However, Gurdjieff provides no further details, so it is difficult to know for sure.

In man (or other three-brained beings), when impressions are consistently digested consciously, he establishes his "own sacred law of Triamazikamno" symbolized by the inner triangle in the enneagram. In that case, when the man dies, the "holy sources of the sacred Triamazikamno, localized in the Omnipresent-Active-Element-Okidanokh," reblend, but with a higher frequency of vibration than before. If not, then the frequency of vibration may be unaltered or lower.

A similar process to this occurs with the birth of a new star, planet, or moon. It happens at all scales.

The Trogoautoegocratic Perspective

The ladder of *Hydrogens* defines a classification of substances from the highest frequency of vibration to the lowest. This set of substances defines the limits of the Trogoautoegocratic process, about which Gurdjieff writes:*

> *"This Most Great common-cosmic Trogoautoegocratic-process was actualized by our ENDLESS UNIBEING, when our Most Great and Most Holy Sun Absolute had already existed, on which our ALL-GRACIOUS ENDLESS CREATOR had and still has the chief place of His existence.*
>
> *"This system, which maintains everything arisen and existing, was actualized by our ENDLESS CREATOR in order that what is called the 'exchange of substances' or the 'Reciprocal-feeding' of everything that exists, might proceed in the Universe and thereby that the merciless 'Heropass' might not have its maleficent effect on the Sun Absolute.*

---

* The Tales, Chapter XVII, The Arch-Absurd, p136

The Ladder of Hydrogens

Gurdjieff has presented us with a ladder of substances, from the highest to the lowest, which can combine in various ways in accordance with the Law of Three. He continues:

> "*This same Most Great common-cosmic Trogoautoegocratic-process is actualized always and in everything on the basis of the two fundamental cosmic laws, the first of which is called the 'Fundamental-First-degree-Sacred-Heptaparaparshinokh,' and the second the 'Fundamental-First-degree-Sacred-Triamazikamno.'*"

While the ladder of *Hydrogens* depicts the creation of beings at various levels, it does not show all of the possibilities or indicate the subsequent movements of *Hydrogens* within the creation. The diagram is thus a fragmentary snapshot of the universe in motion, which in its totality involves the birth, maintenance, and eventual death of beings at every level.

Gurdjieff described* the whole universe being in motion at every level with substances from the finest matter to the coarsest being transformed at every level and scale. This does not happen via a simple organization. The continual ascent and descent of *Hydrogens* happens through living beings (cosmic units) of different scales.

They can be thought of as transmitting stations in the sense that they transform *Hydrogens* and transmit them into the environment they occupy. This is true of all biological life and true at the scales of galaxies and stars. The transformation is purely mechanical in the sense that the being performs these transformations simply by being alive.

At a point in time, a substance occupies a specific place in a specific octave, which is either ascending or descending, and if we know its context, we will also know if it is on the way to becoming finer or denser. The transmitting stations have a specific lifetime, and thus when that life is complete, it disintegrates. All the component substances become available as food to other beings in the location where it died. Thus every cosmos is involved in some kind of reciprocal exchange: it eats and also is food. This reciprocal exchange takes place in everything, in both organic and inorganic matter.

Everything is in motion, but it is never straight-line movement. It has a twofold direction, falling toward the nearest center of stability (the Law of Falling) and circling around itself (the Law of Catching Up), which is the true law of motion. If you throw a stone, you may think of it as moving in a straight line, but it is not. It has the component of motion you gave it, plus the motion of the

---

* *Views From The Real World* by G I Gurdjieff, Essentuki Lecture 1918

Earth's spin, plus a component of motion due to falling towards the Earth's center.

Additionally, the Earth is falling towards the Sun and also orbiting the Sun. And the Sun is falling towards the center of the Milky Way and also orbiting it. And the Milky Way is no doubt moving in some way too.

Thus the universe is complex both in its motion (spinning, orbiting, and falling) at every level and in the transformation of substances through cosmoses/transforming apparatuses/life-forms working at seven different scales, from the minuscule (at the atomic level) to the massive (at the level of a galaxy). Each cosmos appears to be a well-ordered aggregation of other cosmoses. So cells are made of atoms, life-forms are composed of cells, Great Nature is made of many life-forms, etc. In some way, each of these life-forms consumes substances produced by other life-forms, sometimes made available by their death. This is the Trogoautoegocrat.

Harmony and Balance

Everything in the universe is measurable. Everything has mass and vibrates and is in constant motion, with substances being transformed up to higher levels of vibration or downwards. For this to be practical, there must be harmony, a balance. The amount of substance at every level must remain roughly constant.

If that were not the case, then in time, everything would gather at a particular level, and the universe would run down and stop. Within this system, perpetual motion is a necessity. Something may speed up or slow down, but nothing ever stops.

If we consider a specific cosmos, just like the universe itself, it comprises a ladder of *Hydrogens*. For it to work as a system, *Hydrogens* must ascend and descend as part of an octave that defines the cosmos. The *Hydrogens* entering the cosmos must be balanced with the *Hydrogens* leaving the cosmos. And within the cosmos, there must be efficient circulation mechanisms to bring one *Hydrogen* into contact with another so that a third be generated. We can also think of these circulations as orbits of a kind.

The cosmos will remain perpetually in motion throughout its lifetime. If you removed just one substance from the cosmos, it would be quite likely to destroy it. The cosmos is not static in any way, but it is extremely robust in its dynamic structure. It supplements its inner substances by feeding.

What applies to the individual cosmos applies to the megalocosmos, with the added complication that the megalocosmos needs to supplement substances

The Ladder of Hydrogens

required at every level by feeding on itself. To serve the vast scale of the universe, the Trogoautoegocrat needs to be extraordinarily well balanced. The correct quantity and quality of cosmoses of various kinds need to be present at every level in every location. And if something disturbs the balance, then the system must naturally correct for it. If, for example, an asteroid smashes into the Earth and destroys most of Nature, it must repopulate quickly to restore the balance.

The Reduced *Hydrogens* Scale

In *In Search of the Miraculous*, Gurdjieff explains the formation of the ladder of *Hydrogens* and states that H6 is irresolvable for normal man,* as we have no direct contact with that substance. Consequently, we can, without concern, view it as H3 and adjust all the other numbers accordingly. However, even if we do so, we arrive at an H6 that is still irresolvable for man, so we can make a second adjustment, giving us the rightmost column. These two adjustments are simply a convention that provides an easy-to-use *Hydrogen* scale for discussing the cosmos of man.

		1st Step Down	2nd Step Down
do / si / la	H6	H3	H1
sol / fa	H12	H6	H3
-- / mi / re	H24	H12	H6
do / si / la	H48	H24	H12
sol / fa	H96	H48	H24
-- / mi / re	H192	H96	H48
do / si / la	H384	H192	H96
sol / fa	H768	H384	H192
-- / mi / re	H1536	H768	H384
do / si / la	H3072	H1536	H768
sol / fa	H6144	H3072	H1536
-- / mi / re / do	H12288	H6144	H3072

Figure 22. The Reduced Scale

[Note: the diagram we are using here is from p212 of *The Fourth Way* rather than p174 of *In Search of the Miraculous*. The two diagrams do not agree. Figure 22 is the correct one.]

Thus we have a list of 10 *Hydrogens* that are appropriate for the study of man. Every *Hydrogen* in the rightmost column from H6 down plays or can play some part in the life of man. This deduced scale is the scale that Gurdjieff subsequently uses in discussing substances. It is important to understand that when the Law of Three acts it always blends three distinct *Hydrogens*, for

---

* *In Search of the Miraculous* by P D Ouspensky, p174

example it might involve H12, H24 and H48. In which case the middle *Hydrogen*, H24, will carry the neutralizing force.

Hydrogens as Substances

It is far easier to comprehend what these categories (the *Hydrogens*) represent if we have examples of familiar substances for each *Hydrogen*. In the text of *In Search of the Miraculous*,* Gurdjieff provides such examples. We list them in *Table 2*, on the next page, also showing the halogens and their equivalent *Hydrogens*.

From H3072 up to H192, the examples seem reasonably coherent. Iron is heavy and solid. Wood is less so and has organic origins. Food (for man) tends to be solid, soft rather than hard, water is a liquid, air is gaseous. H96 is less clear. We think of vitamins and hormones as organic chemicals. Vitamin A is $C_{20}H_{20}O$, vitamin C is C_6H_8O, and the hormone dopamine is $C_8H_{11}NO_2$. We could describe all such organic chemicals with such formulae.

However, neither "animal magnetism," the term invented by Mesmer to describe a "healing energy," nor N-rays "discovered" and named by French Physicist Blondlot, is recognized as real phenomena by science. If they exist as described, we would most likely think of them as energy rather than as a chemical substance. With H96, it seems that we are dealing with a substance that is sometimes invisible and difficult to detect but sometimes very much in evidence. However, it is possible that while we may witness vitamins as substances in the form of pills, we are witnessing a *Hydrogen* of a lower level, at the level of H192. It may be that they are transformed when circulating in the bloodstream to an ionic state, which allows them to function to benefit our health. This seems likely. They serve the immune system, which loses all its power immediately when the body dies.

Above this level, we have the *Hydrogens* that Gurdjieff described as matters unknown to physics and chemistry. It is probably better to think of them simply as *Hydrogens*, which modern science has little knowledge of and which, whenever they are detected, are classified as energy.

Gurdjieff says that the halogens* represent successively denser *Hydrogens* from H12, which is hydrogen itself, down to iodine, H192. The sixth halogen, astatine, was not discovered until 1940, long after Gurdjieff gave the lectures recorded by Ouspensky in *In Search of the Miraculous*. The seventh, tennesseeum, was discovered very recently.

---

* *In Search of the Miraculous* by P D Ouspensky, p175

The Ladder of Hydrogens

Hydrogen	Example Substances	Halogens
H1		
H3		
H6		
H12	Hydrogens 48, 24, 12, and 6 are matters unknown to physics and chemistry, matters of our psychic and spiritual life	Hydrogen
H24		Fluorine
H48		Chlorine
H96	Hormones, Vitamins, Animal Magnetism, N-rays	Bromine
H192	Air	Iodine
H384	Water	Astatine
H768	Food	Tennesseeum
H1536	Wood	-
H3072	Iron	-

Table 2. *The Hydrogens with Examples*

Many elements in the final two rows of the periodic table (see *Figure 23* on the next page) have not been discovered naturally but synthesized under experimental conditions. They have very short half-lives and quickly decay into some other element. This is true of astatine, the sixth halogen. It was created by bombarding an isotope of bismuth (20983Bi) with helium nuclei. Once created, it has a half-life of 8.1 hours, after which it decays either into another element (bismuth, polonium, or radon) or a different isotope of astatine. The name "astatine" comes from the Greek word *astatos*, which means "unstable"—a name the element acquired, both because of its short half-life and its lack of stable isotopes.

Gurdjieff notes* that the halogens' atomic weights stand roughly in an octave ratio one to another. Fluorine (F), atomic weight 19 is H24; chlorine (Cl), atomic weight 35.5 is H24; bromine (Br), atomic weight 80 is H96, iodine (I), atomic weight 127 is H192. The fact that the octave ratios are inexact stems from the fact that the atoms of the periodic table are distinctly different from the atoms of objective science.

The atoms of objective science consider all the properties of a substance, its chemical, physical, psychic and cosmic properties. Atoms are defined as the smallest amount of substance that retains all such properties. Chemical and physical properties are relatively easy to understand. Consider an atom of water. Physically it is a liquid, and hence it flows and exhibits surface tension. So an atom of water can be thought of as a small drop of water. But that small

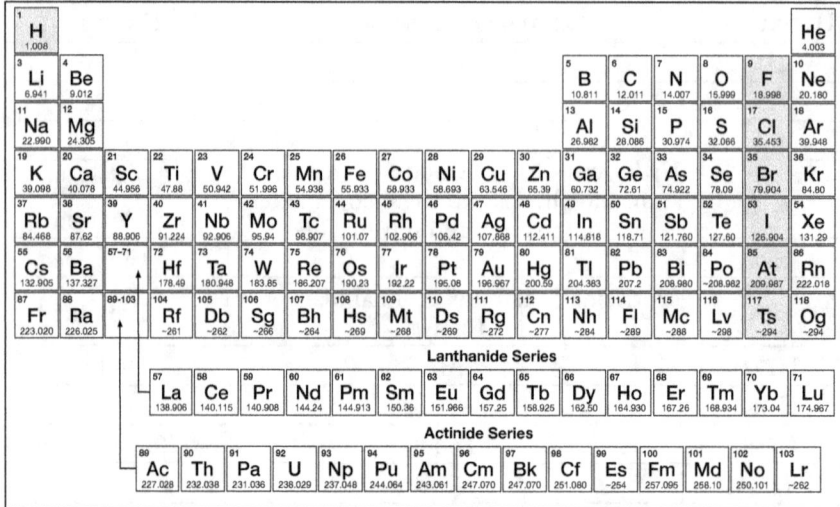

Figure 23. The Periodic Table and The Halogens

drop must also have all the solvent capabilities of water and be able to participate in hydrolysis and hydrogenation.

The cosmic properties of water can be considered to be the role water plays in specific cosmoses. Thus in respect of biological life, water plays a series of roles in each living cosmos. If a drop of water of a particular size cannot fit those roles, it is too small. However, even a small amount of water would be able to play those roles. Finally, according to Gurdjieff,* when we consider water's psychic properties, we are talking about its degree of intelligence. In other words, what beings can it be food for?

The objective science definition of atom could, in theory, refer to a small or large group of identical atoms defined in the periodic table. In reality the individual atoms of the periodic table rarely occur alone, and thus for some elements an independent atom may never exist. Gases typically occur as molecules (N_2, O_2, etc.). In their native state, most metals occur as ores (oxides usually). Fluorine is so reactive (it even reacts with the noble gases krypton, xenon, and radon) that, as far as we know, it is never found in its elemental form in Nature. Thus it is possible that an objective fluorine atom cannot manifest in this universe, only objective atoms that contain fluorine.

The atoms of objective science can refer to all compound matters possessing definite functions in the universe or in the life of man. As well as an atom of water, there can be an atom of air (that is, atmospheric air suitable for man's

---

* *In Search of the Miraculous* by P D Ouspensky, p175

breathing), an atom of bread, an atom of meat, and so on. An atom of water could be defined in the following terms: one-hundredth of a cubic millimeter of water taken at a specific temperature by a thermometer. Such an atom might be visible to the naked eye.

The atoms of objective science relate directly to the Ray of Creation and thus to life itself. Each would have at least one location somewhere in the Ray, perhaps as a fractional part of a fractional part of a fractional part of a fractional part.

The Atoms of Contemporary Science

When we consider contemporary science's idea of an atom, we quickly realize it is just a theoretical model. The simple definition of an atom (via Dictionary.com) is as follows:

> *A unit of matter; the smallest unit of a chemical element. Each atom consists of a nucleus with a positive charge, and a set of electrons that move around the nucleus.*

The "atomic model" described by that definition is very useful for chemistry. It allows for the modeling of a world composed entirely of molecules, which are combinations of various atoms. The various elements are arranged in the periodic table according to their atomic structure. Their position in the table is determined by the number of electrons that an atom of the element is expected to have. The table itself is organized in seven rows according to the electron shells surrounding the atomic nucleus. The electrons in each shell number up to 2, 8, 8, 18, 18, 32, and 32, respectively, corresponding to the seven rows. Thus the first electron shell can accommodate 2 electrons, the second 8 electrons, the third 8, the fourth 18, the fifth 18, the sixth 32 and the seventh 32.

When we mix various elements together under specific circumstances, molecules form. In most instances, the molecules are constituted in a way that can be explained in terms of the number of electrons in the electron shells of each elemental atom.

An elemental atom's capacity to combine with atoms of other elements is called its "valence." The word "valence" was coined from the Latin *valentia*, which means strength or capacity. Chemistry experiments demonstrate that the metals in the leftmost column of the periodic table, which have a single electron in their outer orbit (lithium, sodium, potassium, etc.), are the most reactive. They easily bind with many other elements because they can easily "lose" their "spare" electron.

The most reactive non-metals are the halogens. It is easy to think of the halogens as complements to the very reactive metals since they have just one electron missing from their outer shell and will readily capture the "spare" electrons that the non-metals can provide. The hydrogen atom is unique as it can be thought of as a metal (with a single electron in its outer shell) and a halogen (with a single electron missing from its outer shell).

Figure 24 illustrates the scientific concept graphically. It shows a water molecule as an arrangement of two hydrogen atoms connected to a single oxygen atom. An electron from each hydrogen atom fills the two missing slots in the outer shell of the oxygen atom, which has only six electrons in an outer shell. The shell can accommodate eight.

As if to confirm this idea, in the rightmost column of the periodic table, we find the noble gases which have complete electron shells. Experiments long ago demonstrated that these elements do not easily react with any other substances.

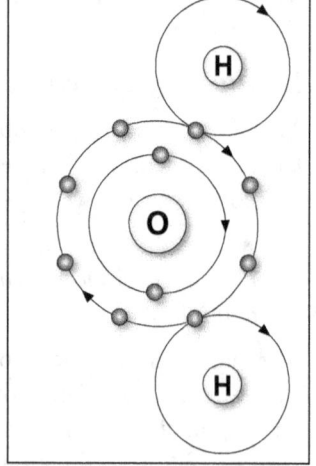

Figure 24. Water Molecule

We need to remember that this is only a model, even if it is an elegant and useful one. *Figure* 24 is two-dimensional and hence geometrically misleading, as the real world has three dimensions. Note also that the "completion of the outer shell" is not the only factor involved in how atoms choose to combine. For example, there are seven elements that are known to form diatomic molecules—molecules of two atoms: hydrogen, nitrogen, oxygen, fluorine, chlorine, bromine, and iodine. While H_2 involves a completed outer shell, none of these other combinations do (N_2, O_2, F_2, Cl_2, Br_2, and I_2).

We also need to note that we think of substances subjectively. We naturally think of substances in the mode that we normally encounter them. So, to us, nitrogen is a gas, water is a liquid, salt is a crystal. All of these substances are capable of being in other distinct states, and in those states, they have properties at the atomic level with which we are less familiar.

Many of the chemical elements and compounds that chemistry has synthesized do not occur naturally in nature—that is, we do not encounter them in that context. They are man-made and only play a part in nature as a feature of the life of man. Man synthesized them and uses them for his own purposes, but nature has no use for them.

Although Gurdjieff is distinctly uncomplimentary about the periodic table, it serves well as a list of elements, their atomic structure, and their density. It is one of the foundations of chemistry, and chemistry is important.

The Inner Ansapalnian Octave

Let's look again at the *Hydrogens* in *Table 2*. It seems that we encounter an apparent "contradiction" or possible source of confusion at the start of the list of halogens as *Hydrogens*. Gurdjieff claims that H12, H24, H48 are undetectable by science, and yet hydrogen (H12), fluorine (H24), and chlorine (H48) were clearly well known to the science of Gurdjieff's era. Although we might not be able to fully detect objective atoms of hydrogen, fluorine, and chlorine, we can detect their physical existence as gaseous matter.

While we can chemically synthesize and collect fluorine gas, it may be that "objective" atoms of fluorine gas play no obvious role anywhere in the Ray of Creation, except in man's world of industrial and experimental chemistry. Perhaps, an "objective" fluorine atom has little relevance or interest for us.

However this seems unlikely, since Gurdjieff directly calls attention to the halogens, both in *In Search of the Miraculous** and in *The Tales*.** From the point of view of atomic structure, this is reasonable, as the atoms of hydrogen, fluorine, chlorine, bromine, iodine, astatine, and tennesseeum each have a single electron missing from their outer shell. Although modern chemistry chooses not to classify hydrogen as a halogen, it is by that definition. Thus there appears to be a unique octave here that commences with hydrogen.

The halogens are notable as the vertical column of elements in the periodic table that includes elements in all three familiar states of matter (solid, liquid, gas) at standard temperature and pressure. The first five halogens can bond directly with hydrogen to form acids, although hydrogen astatide quickly decomposes. Very little is known about the chemical nature of tennesseeum as it is only recently discovered and very unstable.

The elemental halogens are toxic to most life-forms, including most bacteria. The same is true of some of the compounds they form—for example, the hydrochloric acid of our stomachs (HCl) is toxic to most bacteria. Man uses compounds involving the middle halogens (chlorine, bromine, and iodine) as disinfectants. Fluorine is one of the most reactive elements. Indeed, it can re-

---

* In Search of the Miraculous by P D Ouspensky, p175
** The Tales, Chapter XL, Law of Heptaparaparshninokh, p830

Gurdjieff's Hydrogens: The Ray of Creation

Hydrogen	Substances	Halogens	Ansapalnian	Alchemists
H6	-	-	-	-
H12	Higher Emotions	Hydrogen	Planekurab	-
H24	Motion	Fluorine	Alillnofarab	-
H48	Impressions	Chlorine	Krilnomolni-farab	-
H96	Animal Magnetism	Bromine	Talkoprafarab	-
H192	Air	Iodine	Khritofalmono-farab	-
H384	Water	Astatine	Sirioonorifarab	Hydro-oomiak
H768	Food	Tennesseeum	Klananoizu-farab	Petrkarmak

Table 3. *The Inner Ansapalnian Octave*

act with many otherwise-inert materials such as glass and the heavier noble gases.

In *The Tales*, Gurdjieff emphasizes the octave of halogens, which he calls the inner Ansapalnian-octave. He writes:*

> "By the way, you should know that this same hydrogen of theirs is just one of those seven cosmic substances which in their general totality actualize specially for the given solar system what is called the 'inner Ansapalnian-octave' of cosmic substances, which independent octave, in its turn, is a one-seventh independent part of the fundamental 'common-cosmic Ansapalnian-octave.'
>
> "Such an inner independent Ansapalnian-octave is likewise present in that solar system to which our dear Karatas belongs, and we call these seven heterogeneous cosmic substances of different properties:
>
> (1) Planekurab—which is just their hydrogen
>
> (2) Alillonofarab
>
> (3) Krilnomolnifarab
>
> (4) Talkoprafarab
>
> (5) Khritofalmonofarab
>
> (6) Sirioonorifarab
>
> (7) Klananoizufarab

---

* The Tales, Chapter XL, Law of Heptaparaparshninokh, p830

The Ladder of Hydrogens

"And on your planet the genuine learned beings at different periods called by various names these same seven relatively independent crystallizations of different properties or, according to their expression, active elements, which compose the inner Ansapalnian-octave of their own solar system; the contemporary, as they are called, learned chemists there, however, who are already 'learned-of-new-formation-of-the-first-water,' call them:

(1) Hydrogen

(2) Fluorine

(3) Chlorine

(4) Bromine

(5) Iodine

As with all passages from *The Tales*, it pays to read every word. Note that here Gurdjieff calls the substances in the inner Ansapalnian-octave "heterogeneous," clearly stating that each is a mix of some kind. And he also states that this octave is a "one-seventh independent part" of a larger octave: the fundamental "common cosmic Ansapalnian-octave."

He continues...

"For the last two definite crystallizations, they have no names at all because their names did not reach them from their ancestors, and at the present time, they even do not suspect the existence on their planet of these two cosmic substances, although these two cosmic substances are the principal necessary factors for their own existence.

"These two latter cosmic substances, which might be quite tangible and quite accessible in all spheres of their planet, were still known only about two centuries ago among the 'scientific beings' there who were then called 'alchemists'—but whom the contemporary 'comic-scientists' simply call 'occult-charlatans,' considering them to be only 'exploiters of human naïveté'—and were called by them 'Hydro-oomiak' and 'Petrkarmak.'

Gurdjieff describes the final two substances in this octave, "Hydro-oomiak" and "Petrkarmak," as the "principal necessary factors" for our existence. (A factor is, by definition, something that "acts as an agent.") It is worth noting that nowhere does Gurdjieff state that these substances are halogens in the way that we would understand them, either chemically or biochemically.

Figure 25 shows the periodic table from a different perspective. It highlights the atoms that participate in the biochemistry of man. As indicated in the key, the darkest square indicates the principal elements of life (H, C, N, O). Those

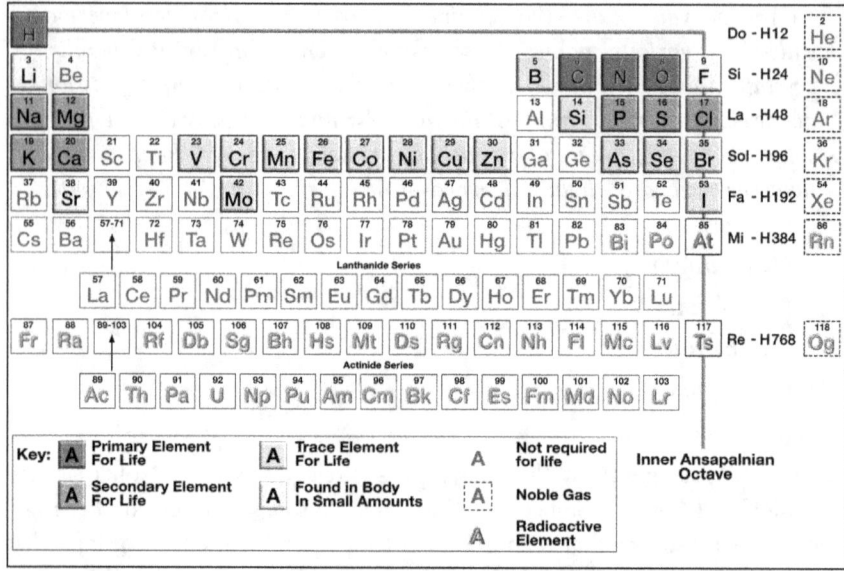

Figure 25. The Periodic Table and Life

elements categorized as secondary (P, S, Cl, Na, Mg, K, Ca) in the illustration are found in significant quantity in the body of man. The trace elements (B, Si, V, Cr, Mn, Fe, Co, Ni, Cu, Zn, As, Se, Br) are found in relatively small quantities. Three other trace elements (Li, F, and Sr) play some role because the body suffers if deprived of them, but no biochemical interaction has been specifically identified involving them.

Figure 25 also identifies the noble gases, none of which play a role in life. Neither do the radioactive elements, which comprise all elements with a higher atomic weight than lead (Pb). On the right-hand side of the table, we show an octave from do (H12) to re (H768), which corresponds to the Inner Ansapalnian Octave. We note that only the atoms found from the note *fa* up to the note *do* play any part in the substances of life.

However, if we take a distinctly different perspective and consider the *Hydrogens* relevant to man, we see in *Table 3* that the octave runs from H6 to H768. If tennesseeum is H768, it has the same vibrational level as food for man, and if astatine is H384, it has the same vibrational level as water. Maybe in some way, they are the "principal necessary factors" for our existence. They are the notes *re* and *mi* of the Inner Ansapalnian octave.

The Ladder of Hydrogens

Objective Atoms

The halogens illustrated in the periodic table do not qualify as atoms of objective science, although they are clearly important as chemical building blocks. Gurdjieff points out that they almost correspond to the Inner Ansapalnian Octave in the periodic table in respect of atomic weight.

The atoms of objective science are defined as the final division of matter in a similar way to the atoms of modern science. Still, there is a difference in the two definitions. Atoms of modern science are not really indivisible; they can be broken up into subatomic particles. They are indivisible except for that, and of course, if they are split in that way, they are no longer atoms. Atoms of objective science could easily be split in a similar way, and if they were, they would no longer be atoms.

There is another distinction. The atoms of objective science apply to specific levels and are under specific orders of laws. Thus only atoms of the Absolute are truly indivisible. Atoms of the next level are a combination of three atoms of the Absolute. They are three times bigger, three times heavier, and move correspondingly slower. The atoms of world 6, are still larger, heavier nd slower. And so it proceeds with atoms of world 12, 24, 48, and 96 being are proportionally bigger, heavier, and slower.

Gurdjieff said* that ordinary science, unaware of the Law of Three, misses many things. For example, it does not distinguish between a substance according to whether no force acts through it (H) or one of the three forces acts through it (C, O, N). He claimed there was a different kind of chemistry, distinct from ordinary chemistry, which studies substances according to their cosmic properties—first according to its place (within a cosmos and an octave) and by the force acting through it. So every substance has four different aspects or states. This special chemistry looks at every substance as having a separate function, even the most complex biochemical substance, for example, as an element.

This can be pondered through the prism of the Trogoautoegocrat. If it is the case that the whole universe functions by feeding at every level and scale, then a substance, whatever it might be, must occupy some role at some level and scale. And thus, it must have a purpose and significance.

The final detail that needs to be pondered relates to the fact that many substances are invisible to our senses. If we are to understand them, their purpose, and significance, we need to know about their nature and how to detect them.

---

* *In Search of the Miraculous* by P D Ouspensky, p89

Gurdjieff's Hydrogens: The Ray of Creation

These substances were referred to collectively in ancient times as FIRE. Nowadays, we call them plasma.

Chapter 4

THE ESSENCE OF PLASMA

—∞—

"Reading furnishes the mind only with materials of knowledge; it is thinking that makes what we read ours."

~ *John Locke*

There are four states of matter: solid, liquid, gas and plasma. We probably know very little about the last one. No doubt this is partly because we were never taught much about it at school, but also because the science in this area has never attracted much attention in the scientific media. Nevertheless, we need to examine it in some depth so that we have some foundation for the cosmology discussions we will enter into later in this book. Indeed it will soon become fairly clear that any detailed study of objective science requires us to have knowledge in this area.

Most of the observable universe is plasma. The current estimate from contemporary science is that the solar system is 99% plasma and that the interplanetary medium and interstellar medium are nearly 100% plasma. Indeed, most of our manifestations, our thoughts, our feelings and our deepest aspirations are manifestations of plasma. So...

What is Plasma?

The Wikipedia entry reads:

> *Plasma (from Ancient Greek πλάσμα, meaning 'moldable substance') is one of the four fundamental states of matter, and was first described by chemist Irving Langmuir in the 1920s. It consists of a gas of ions—atoms which have some of their orbital electrons removed—and free electrons. Plasma can be artificially generated by heating a neutral gas or subjecting it to a strong electromagnetic field to the point where an ionized gaseous substance becomes increasingly electrically conductive. The resulting*

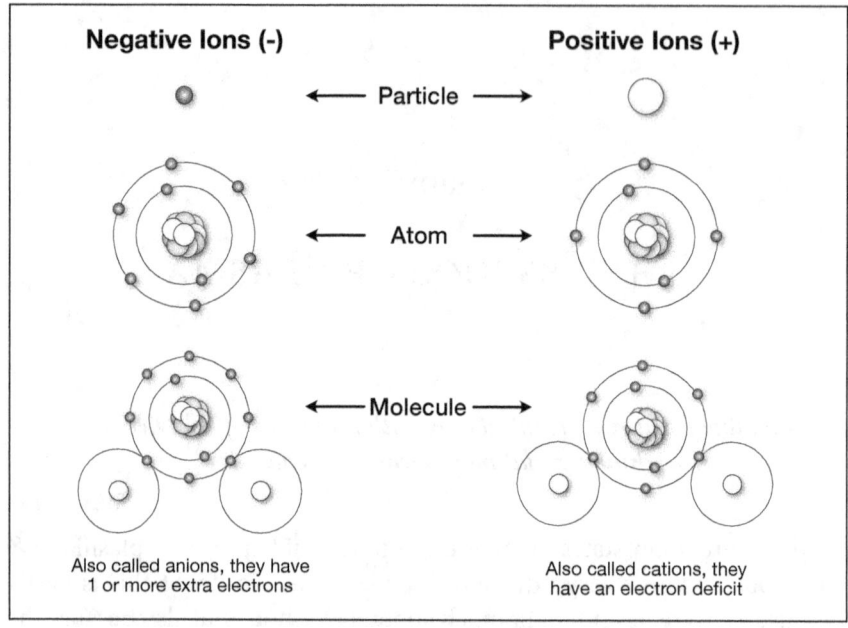

Figure 26. Ions as Particles, Atoms or Molecules.

charged ions and electrons become influenced by long-range electromagnetic fields, making the plasma dynamics more sensitive to these fields than a neutral gas.

That's a reasonable definition to start out with. Plasma is defined as gaseous because you are mostly likely to encounter plasma in a gaseous state. However, it is worth noting that if you pass an electric current through a sodium chloride solution, sodium ions (Na+) are attracted to the cathode and chlorine ions (Cl-) to the anode. Those ions are plasma permeating water.

Figure 26 above should clarify the concept of ions. In respect of subatomic particles, we commonly encounter positively charged protons and negatively charged electrons, which carry electrical charges of exactly equal magnitude. Electricity, which is a plasma flow, is typically modeled as a stream of electrons.

Atoms and molecules can also be ions by virtue of carrying an extra electron, and hence being negatively charged, or missing an electron, and being positively charged. Thinking simply, no matter whether we are referring to an atom or a molecule, the net electric charge is the difference between the number of protons and the number of electrons. The negatively charged

The Essence of Plasma

hydroxide ion, OH⁻ provides an example of a molecular ion, as does the positively charged ammonium ion NH_4^+.

The terms anion and cation are also used to describe ions, an anion being negatively charged—an ion with more electrons than protons, and a cation being the opposite. Anions are attracted to what is termed the "anode" of an electric device—the anode is positively charged. Similarly cations (positively charged particles) are attracted to the cathode which is negatively charged.

The etymology of the words anode and cathode are curious. "Cathode" is from the Greek *kathodos* meaning "the way down," whereas "anode" is from the Greek *anodos* meaning "the way up." A negatively charged ion, is thus literally "a thing ascending," and a positively charged ion is literally "a thing descending."

The following excerpt from *The Tales* (p156, Chapter XVIII, The Arch-preposterous) seems to conceptually relate two parts of the the Omnipresent-Okidanokh to anions and cations.

> "'Owing to the pulling of this lever, that process has begun in this vacuum whereby in the separate parts of the Omnipresent-Okidanokh there proceeds what is called the "striving-to-reblend-into-a-whole."
>
> "'But since, intentionally by an "able-Reason"—in the present case myself—the participation of that third part of Okidanokh existing under the name of "Parijrahatnatioose" is artificially excluded from the said process, then this process proceeds there just now between only two of its parts, namely, between those two independent parts which science names "Anodnatious" and "Cathodnatious."

Etymologically, the word "Anodnatious" can be taken to mean "born of the anode," and Cathodnatious, "born of the cathode."

Waves: Propagation & Radiation

To move the discussion of plasma forward, we need to refresh our minds about some basic aspects of physics. We will begin with a discussion of the nature of waves. Our ability to conceptualize waves may be limited by the fact that many of the waves we encounter are not visible. The waves we most commonly notice are waves on the surface of water. We may have observed the wave motion of a spring when playing with a Slinky, but we do not directly observe the wave motion of sound or the wave motion of light.

By definition, a wave is a "propagating dynamic event" with the propagation occurring through a medium. In the case of sound, waves propagate through

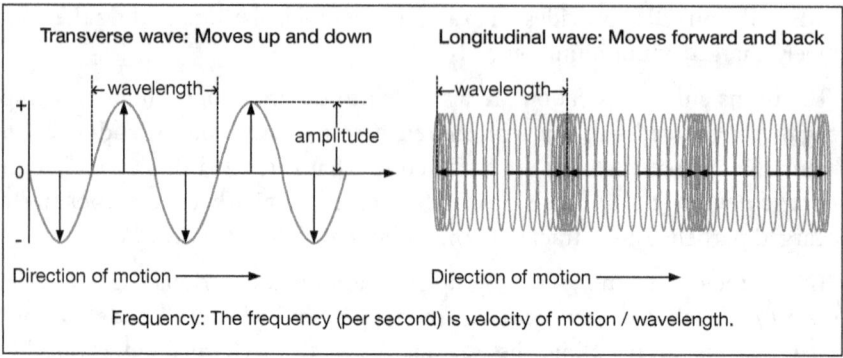

Figure 27. Types of Wave

the air. As regards light, the view of objective science is that it propagates through the aether. A wave carries energy of some form from one place to another.

There are two types of waves as illustrated in *Figure 27* above:

- **Transverse wave:** This is where the wave motion is up and down. Consider the simple example of putting a wave into a rope with a flick of the wrist. You move the end of the rope up and down and a wave moves along the rope, transferring some energy from your hand to the end of the rope.

- **Longitudinal wave:** This is where the motion of the wave is backward and forward as you can demonstrate to yourself using a Slinky. Again, you can see the wave.

While we do not see the transverse waves of electromagnetic radiation (EMR), we can detect them experimentally and use their nature to explain the refraction of light.

As for invisible longitudinal waves, consider the Newton's Cradle shown in *Figure 28*. When the swinging ball A hits the stationary ball B, its momentum is transferred through that ball and also the next two, causing ball E to move out along an arc.

This is a clear example of a wave traveling only longitudinally. The force passes from ball A to ball E by means

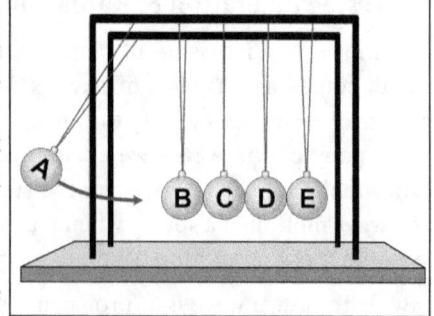

Figure 28. Newton's Cradle

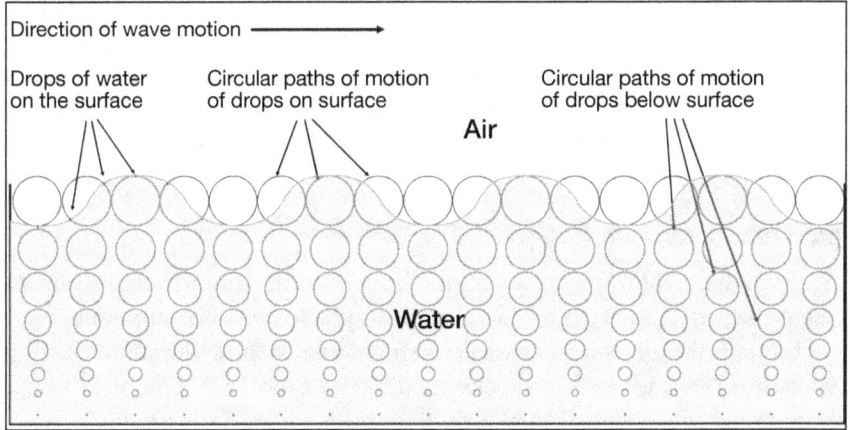

Figure 29. The Movement of Waves on Water

of a longitudinal wave. Although it is too fast to notice, the body of balls B, C, and D experience a wave of compression as the longitudinal waves passes through each of them.

Now consider waves on water. This appears to be another example of transverse waves, but it is a little more complicated than that. Waves on water can be created simply by throwing a stone into the water, in which case they radiate out from a point, or they can be created by the action of the wind on the surface of the water, in which case they move in one direction. If you have experience of sailing at sea, you know the waves can be quite high in a place when the wind is not blowing. The force imparted by the wind somewhere else is propagating energy to your location.

In fact, waves on water provide an example of a combination of transverse and longitudinal waves. Note that it is power that travels along a wave. If you observe a cork floating on the surface of the sea, you notice that despite the waves, the cork does not move much. In fact neither does the water. If we imagine the water being composed of millions of individual drops of water, experimental observation clearly demonstrates that the drops of water exhibit a kind of circular motion, as illustrated in *Figure 29*.

That circular motion is visible on the surface and a similar kind of circular motion is happening below the surface, as illustrated. The movement is less below the surface and, as the depth increases, the circular motion tails off. The point to note here is that the waves are formed by both forward and backward motion and upward and downward motion.

An earthquake may occur in Indonesia sending a tsunami across the Indian Ocean to Africa. When the tsunami waves arrive on the beaches of Africa, the inundation they cause may be devastating, but the water that is driven inland by the tsunami is local sea water. It did not travel there from Indonesia, but some of the energy of the earthquake did and it was propagated by all the water in between.

Electromagnetic Radiation

We now have sufficient information about waves to begin a discussion of electromagnetic radiation (EMR). Sound is a phenomenon of vibrations passing through the air, waves on water are a phenomenon of vibrations passing through water, and seismic shocks are a phenomenon of vibrations passing through the solid crust of the Earth. So what is the medium through which electromagnetic waves pass?

Since electromagnetic waves happily pass through a vacuum (near-vacuum) as easily as they pass through the air, these waves are not being propagated by atoms. Neither are they propagated by plasma, since they happily pass through spaces that are devoid of plasma. The only answer that anyone has ever provided to this question is aether.

James Clerk Maxwell, who laid the foundation of our modern understanding of electricity, assumed this to be the case and never thought otherwise. It was later physicists, who chose to deny the existence of aether, preferring to believe that EMR waves, unlike all other waves in physics, don't require a medium through which to propagate.

In respect of our understanding of electricity it is difficult to overstate Maxwell's importance. He invented the mathematic equations that model most of the activities of electrical engineering, including power generation, electric motors, wireless communication and radar. Maxwell's equations describe how electric and magnetic fields are generated by charges, currents, and changes to the fields. One far-reaching consequence of Maxwell's equations was that they demonstrated how fluctuating electric and magnetic fields propagate at a constant speed (c, the speed of light) in a vacuum.

The full EMR spectrum is shown in *Figure 30*. It spans wavelengths of thousands of miles down to wavelengths in the femtometer range that are no wider than the width of a proton. In general, the cause of EMR is the acceleration of positive or negative charge, in one way or another. All EMR is caused by plasma.

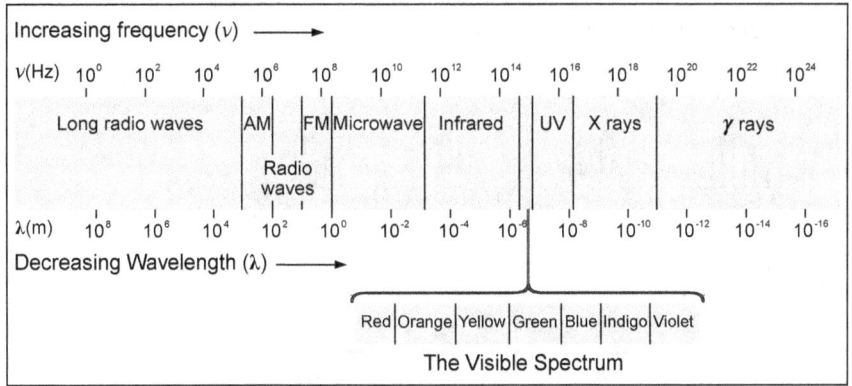

Figure 30. The Spectrum of Electromagnetic Radiation

When astronomers view the universe they are detecting EMR, using Earth-bound telescopes and devices mounted on satellites and space probes. They detect EMR of all wavelengths, arriving from every part of the cosmos. Almost all the data mankind has collected from the universe constitute records of EMR and patterns of EMR over time and space. From our perspective, and according to our data, the universe is plasma in motion.

Almost all the EMR spectrum, up to and including the microwave range, is used in radio communications of one sort or another. The radio waves used in broadcasting and communications are created by an electronic transmitter producing a controlled alternating current. The oscillation frequency of the current determines the frequency of the radio wave produced when the current passes through an antenna. An antenna is typically a dipole comprising two connected metal rods through which the current oscillates to create the radio wave. Sound or any other kind of data is added to radio waves simply by varying the amplitude of the wave.

Exactly the same mechanism is used to create radio waves across the whole range of frequencies, until we approach the infrared band. Infrared radiation is created through the action of thermal energy—the energy of heat. Thermal energy is believed to be and defined to be the kinetic energy of atoms or molecules moving within matter.

All matter with a temperature above absolute zero is comprised of atoms or molecules that have some kinetic energy. Their movements lead to exchanges of kinetic energy, atom to atom. We call this activity "conduction" in solid matter and "convection" in liquids and gases. If you heat matter up then the average kinetic energy of its atoms and/or molecules increases.

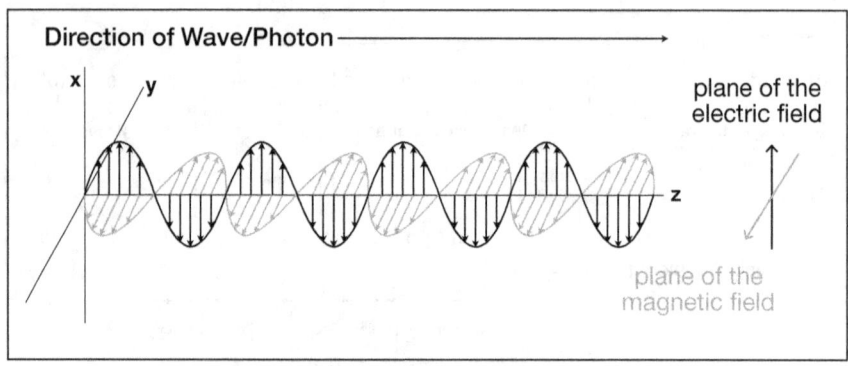

Figure 31. An Electromagnetic Wave

Infrared radiation is formed in the following way. The kinetic interactions provoke charge acceleration of ions and dipole oscillations (interactions between positive and negative ions). Coupled in this way, oscillating electric and magnetic fields generate photons which are radiated away as infrared radiation. Think of this as behaving like a radio transmitter except that the aerial is very very small.

With visible light, the theory is that EMR is caused by the collision of photons or electrons with atoms. The electrons jump from one orbital shell to another. This is a very small movement of a negative charge from one place to another and it causes the emission of a photon. A further photon will also be emitted when it falls back to its previous orbit. This oscillation creates light of a specific wavelength. It is for this reason that atomic elements which have characteristic patterns of electrons in their shells emit a characteristic pattern of EMR, by which they can be identified through spectroscopic analysis.

Once we move beyond the visible wavelengths, to UV, X-rays and Gamma rays, EMR is created by particle collision. For example, X-rays are created in an X ray tube by bombarding a tungsten target with electrons. Gamma rays are created within suns by the energetic interactions of very fast moving particles. Thus they are also generated by powerful particle accelerators on Earth, the most prominent of which is CERN's Large Hadron Collider (LHC).

Whatever the wavelength, EMR always moves at the same speed through a 'vacuum' and exhibits the same wave form as shown in *Figure 31*. This is the wave model originally described in Maxwell's equations. It comprises two orthogonal fields that move in step, the magnetic field and the electric field. EMR naturally travels in a straight line, but will bend by the process of wave diffraction when it encounters the edge of an object. The extent of the bending

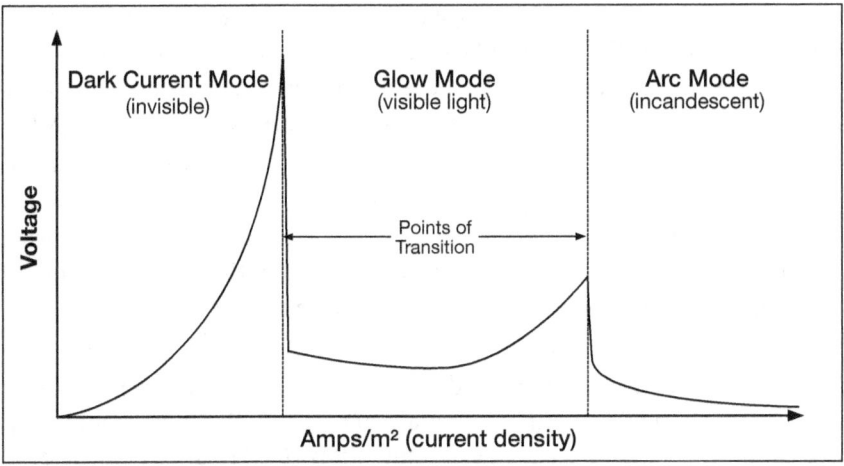

Figure 32. The Three States of Plasma

depends on the wavelength. Thus long-wave EMR can bend around hills, but microwaves cannot.

As far as we know, EMR is the only means by which anything moves across the universe quickly, and it's important to note that what is being transferred is energy in the form of photons which can and do carry information. EMR is also how most sentient creatures perceive their environment.

The Three States of Plasma

All substances composed of charged particles are plasma. Unlike EARTH, WATER or AIR, each of which has just one state, plasma has three possible states, Dark, Glowing and Arc:

- Dark plasma is plasma you cannot see, such as the flow of electricity along a wire.
- Glowing plasma is visible, such as the flame of a match or the glow of a neon tube.
- Arc plasma is incandescent like lightning in a thunder storm or the electric arc in arc welding.

These three types of plasma are distinct and the quantitative difference between them is illustrated graphically in *Figure 32* above. Note the two axes of the graph: voltage and amperage. The first, dark state, is the relatively low power situation, where plasma simply flows from one place to another. The glow mode occurs when the power increases to a point where photon energy

(EMR) starts to be generated. While it takes a high voltage to excite the plasma into glow mode, as the graph indicates, as long as the amperage (the current density) is strong enough, a high voltage is not required to maintain it. In Arc Mode, the extremely energetic plasma generates extremely short wave EMR.

What is immediately clear from the graph is that the change between two adjacent plasma modes is abrupt. This is reminiscent of the abrupt change that occurs between a solid and a liquid, or a liquid and a gas. To appreciate what the graph is trying to communicate, let us here explain the concepts of voltage and amperage, which readers who never studied electricity may not be familiar with.

Consider a stream that flows sedately through fields and then suddenly arrives at the top of a cliff. It becomes a waterfall and descends rapidly to hit the ground beneath. Let's imagine that the cliff is 100 ft high. Any single drop of water that enters the waterfall will fall no more nor less than 100 ft. In that situation, each drop of water has a potential energy corresponding to a height of 100 feet. Voltage is the same concept as this, but applied to electricity.

It expresses a potential difference in the force of electrical charge between one point and another, just as a drop of 100 feet expresses a potential difference in the force of gravity, between two heights.

Pursuing this analogy, amperage is the force of the stream. If it hasn't rained in recent days, then the 100ft waterfall may just be a trickle. However a downpour can swell the stream causing a far more impressive cascade. The potential difference has not changed, but the flow has changed and the volume of water passing over the waterfall is far greater.

So amperage indicates the volume of the electrons flowing per second. To be precise, on this particular graph, the amperage axis indicates the volume of electrons flowing through an area of a square meter—i.e. through a surface.

The relationship between plasma and EMR is clearly important. It may help if we summarize what we have discussed about EMR here:

- Waves are a means by which energy is propagated from one place to another. Activity within plasma gives rise to EMR (electromagnetic waves).

- EMR is a transverse wave, which combines a magnetic wave with an electric wave.

- EMR is created by the oscillation of electric charge at some scale. The wavelength of the EMR is proportional to the oscillation. Extremely high

energy activity (in stars for example) gives rise to very short wavelength EMR, such as gamma rays.
- All waves propagate through a medium. EMR propagates through an invisible aether with which, as far as we know, all of endless space is filled. In *The Tales*, Gurdjieff refers to aether as "etherokrilno."
- The energy which EMR propagates is conceptualized as a photon. A photon is the energy of a single wave of EMR.

An Excursion into Fundamental Physics

To discuss plasma in greater detail we need to step back and provide a broad overview of other areas of physics. In particular we need to discuss the fundamental meaning of the concepts: force, power, energy and work. This will provide us with a better context for discussing the nature of plasma. Please read this section of the chapter carefully and, when necessary, refer back to it:

Fields and Space

When we use the word *field* we refer to a mathematically defined finite region of space. It is a way to mathematically represent a region of space and what it contains. For example, a magnetic field, represented mathematically, has at every point within the region under study some magnetic force that may produce an effect. Diagrammatically we may represent such a magnetic field by lines emerging from one or other pole of a magnet. These lines represent the field and they may even align with a two-dimensional pattern of iron filings that can be observed in reality.

When we use the word *space* we refer to a mathematical concept expressed in terms of three orthogonal dimensions, length, breadth and height, often represented diagrammatically by x, y and z axes. These dimensions proceed outwards to infinity in every direction.

They can be used to measure aspects of a field (a region of space), by arbitrarily taking a point of origin for the three dimensions. For example, to measure the dimensions of a box take one corner of the box as origin and employ a unit (centimeters for example) to express the three dimensions.

Note that, in various scientific speculations we may encounter the idea that "space is curved." By definition, space can never be curved as it is mathematically defined not to be. However, something that manifests somewhere within space may involve curvature.

In scientific speculations we may also encounter the idea of multiple dimensions. It is certainly possible to mathematically invent systems that have more than three dimensions of length. However, to our knowledge, nobody has yet demonstrated such dimensions manifesting in reality. Spaces involving more than three dimensions of length are simply a mathematical idea, and they may even have practical application in mathematic manipulations, but they do not describe reality. In general, man observes the world as three-dimensional space. Thus the mathematics of space and man's perception of the reality of space are closely aligned.

SI Units

In what follows, you will need to be familiar with the basic units of science. The international system of units (Système international d'unités or SI units) is based on the metric system, and provides the sciences with a common a set of units that is adhered to almost everywhere. It establishes a foundation that can be used both mathematically and in discourse. There are seven base units and 22 derived units. The seven base units are:

- **Time.** The unit is the *second*. Its symbol is *s*.
- **Length.** The unit is the *meter*. Its symbol is *m*.
- **Mass.** The unit is the *kilogram*. Its symbol is *kg*.
- **Electric current.** The unit is the *ampere*. Its symbol is *A*.
- **Temperature.** The unit is the *kelvin*. Its symbol is *K*.
- **Amount of a substance.** The unit is the *mole*. Its symbol is *mol*.
- **Luminous intensity.** The unit is the *candela*. Its symbol is *cd*.

We will be using these and other units for the unit analysis of equations. Please become familiar with them. Incidentally, by convention with SI units, if they are named for a particular scientist, the unit's symbol is capitalized.

Armed with standard units, we can proceed to discuss basic concepts of physics, specifying the units by which they are measured.

Concepts of Mechanics

First we will consider several fundamental concepts that apply to mechanics, the area of physics concerned with the motion of objects. We can think of mechanics as the domain of solids, liquids and gases. The important concepts here are:

- **Matter:** Matter is physical substance that occupies some area of space. All objects are, by definition, matter. Matter can be very dense or very rarefied. Note also that one kind of matter (or substance) can interpenetrate another, as for example, water can interpenetrate sand. The SI unit for matter is the *mole*. One mole of a substance is defined to be exactly $6.02214076 \times 10^{23}$ particles of the substance (that happens to be the exact number of particles found in 12.000 grams of carbon-12). The particles in a *mole* may be atoms, molecules, ions, or even electrons.
- **Mass:** Mass is a property of a physical body, an object. It is defined to be the resistance of the object to acceleration (that is, to a change in its state of motion). The term "inertia" has the same meaning as mass. The SI unit of mass is the *kilogram*. It is important to understand that mass is not the same as weight. The mass of an object is the same everywhere in the universe, whereas weight varies according to location. The two concepts are sometimes confused.
- **Weight:** The weight of an object is the force acting on the object due to gravity. Thus, for example, the weight of an object on Earth will be different than its weight on the Moon. Weight is an example of a force. Consequently, the SI unit for weight is the same as the SI unit for force, the *newton*. Confusion between weight and mass can arise for those who understand the word kilogram to mean weight as when buying groceries; it is also a measure of mass.
- **Force:** In general, a force is something that can bring about a change of state. In the context of mechanics, a force is something that, when applied, changes the motion of an object. The SI unit for force is the *newton*. It is defined to be the force required to accelerate a mass of one *kilogram* at the rate of one meter per second per second.
- **Friction:** Friction is a resisting force, which opposes the relative motion of solid surfaces, fluid layers or material elements. Again, the SI unit of force is the *newton*. As there are different contexts where friction arises, so physics defines distinct types of friction:
    a. **Dry friction:** The friction that opposes the lateral motion of two solid surfaces that are in contact.
    b. **Lubricated friction:** This is the same context as that above except that a lubricant fluid separates the two solid surfaces.
    c. **Fluid friction:** The friction between the layers of two fluids that are moving relative to one another.

d. **Skin friction:** The force resisting the motion of a fluid across the surface of a body of some kind.
   e. **Internal friction:** The force resisting the motion of parts of a solid material that is undergoing deformation of some kind.
- **Work:** In mechanics, work is the outcome of applying a force to an object. The force causes motion (or change of motion) and thus work is done. For example, an object is held above the ground and then dropped. The work done by the gravitational force on the object is to move it from its location to the ground. The SI unit of **work** is the *joule*. In any context where work is done there will always be some kind of resisting force, or friction, involved, even if very small.
- **Power:** Power is the amount of work done in a particular time, the rate of doing work or the rate of transferring energy. The basic unit of power is the *watt*. A watt is precisely one *joule* per second.
- **Energy:** Finally we get to energy, about which we will have much more to say. For the moment we can assert that is a quantitative property that can be transferred from one object to another, or from one context to another. In the way that we measure it, it is identical to **work**, which so far we have defined purely in the context of mechanics. Energy provides the capacity to do work and hence if we measure the work it does we have measured the energy expended in doing the work. The SI unit of energy is, of course, the *joule*.

Some Basic Equations of Mechanics

It may help the reader if we also define these basic concepts of physics by several simple well-known equations. For those who detest mathematics and physics, please bear with us and try to grasp what is being stated. These equations are not complicated and they will provide a foundation for understanding plasma. We will not only provide equations but also carry out what is called "unit analysis" on these equations. The point of unit analysis is to demonstrate how units of measurement equate to each other on each side of an equation.

We can begin with the simple equation for work, which is its definition. Work is the application of a force to move an object a specific distance:

$$\text{Work} = \text{Force} \times \text{Distance, or}$$

$$W = F.d$$

Applying unit analysis we see that *joules* (W) equate to *newtons* (F) multiplied by *meters* (d):

$$[j] = [n].[m]$$

The unit analysis convention we apply is to enclose the units for a particular term in the equation in parentheses.

A simple equation for force, taken from mechanics is:

Force = Mass x Acceleration, or

F=m.a

Applying unit analysis, we see this as *newtons* (n) equating to *kilograms* (kg) multiplied by *meters* per *time* per *time* (m/s/s):

$$[n]=[kg].[m/s/s]$$

For normal physics this equation is fine, but from the perspective of objective science, it does not tell the full story. The scenario is that an event occurred where a force was applied to an object, which brought about an increase in its velocity. Since an event occurred, three forces were involved. In the equation only two forces are apparent, an active force which overcomes the inertia of the object. Nevertheless, if we ponder the situation we realize that three forces are indeed involved. What is missing from the equation is the second force.

The active force is the pushing force that is applied to the object. The passive force is friction, the resistance of the medium within which the object resides. If the object is in almost empty space the resistance to motion will be almost zero. If moving through the air, friction (air resistance) will be greater. If moving through a liquid it will be greater still. The third force, the neutralizing force is simply the inertia of the object: its mass. The active force is expended, the motion of the object changes and the object's mass, the third (neutralizing) force is unchanged by the event.

The equation is incomplete because it includes no term to denote the resistance of the medium, but will be almost accurate in situations where friction is minimal.

The Varieties of Energy

When it comes to energy we soon realize that we need to do more than focus on a simple equation. Energy is complex, as there are many forms of energy, and in some situations one kind of energy can be converted to another. So let's consider energy first by classifying different types of energy under one of two headings: potential energy and kinetic energy.

Gurdjieff's Hydrogens: The Ray of Creation

Potential energy is energy that has no component of motion. It does work when a force acts through the substances (the *Hydrogen*) that contains it. The following are types of potential energy:

- **Gravitational energy:** This is the energy stored in an object by virtue of its distance from another object, as in the example of Newton's apple on a tree, which has gravitational energy because it is several feet above the ground. The greater the distance and the greater the mass of the object, the more gravitational energy is stored. When Newton's apple fell, some of its potential energy was converted into the energy of motion, and it acquired momentum. One way that we harness gravitational energy is by using hydroelectric turbines to generate electricity.

- **Mechanical energy (internal):** This is the energy stored in objects by virtue of tension. A compressed spring or stretched elastic are examples of mechanical energy. We harness mechanical energy in mechanical watches, which gradually release the energy of a spring to move the mechanism of the watch.

- **Chemical energy:** This is the energy stored by the atomic bonds within molecules. Natural gas, petroleum, coal and wood store chemical energy. We burn these substances to produce heat.

- **Magnetic energy:** This is the energy stored in a magnet, which is responsible for its magnetic field. If you place one magnet close to another, a force of attraction or repulsion occurs, depending on the orientation of the two magnetic fields.

- **Electrical energy (static):** This is the energy stored by virtue of the difference of electrical potential between an object and its surroundings. We model it as static ions (plasma) on the surface of an object. The energy in a charged battery is the same. We grade batteries by **voltage**, a measure of the electric current that can flow between the cathode and the anode. If we wished to grade them by the amount of electrical energy they contained, we would grade them by *joules*.

- **Nuclear energy:** This is the energy that binds the nucleus of an atom together. Energy is released when the nucleus splits apart (nuclear fission) or when nuclei combine (nuclear fusion).

Kinetic energy is energy due to motion. It always involves something moving from one place to another. The list of different types is as follows:

- **Mechanical energy (moving):** It takes energy to set an object in motion and it expends energy as it overcomes the resistance to its motion. Wind is an example of a substance containing mechanical energy. It can transfer that energy to the blades of a windmill.

- **Mechanical energy (spinning):** Note that mechanical energy can also be the energy of an object spinning. All celestial bodies appear to spin except those, like our Moon, which are restrained from spinning by gravitational forces. We use the spinning of a turbine to generate electricity.
- **Thermal energy:** This is heat, the energy that comes from the vibrational movement of atoms and molecules in a substance. Heat increases when these particles move faster. The circulation of thermal energy is important in many aspects of the life of man. At a large scale, in some areas of the planet we use geothermal energy to generate electricity.
- **Sound energy:** This is the movement of energy through substances in longitudinal waves. Sound is generated by the vibration of objects. Just as our eyes only see a range of electromagnetic energy, our ears only experience a range of vibrational energy. The phenomenon of resonance alerts us to the fact that objects resonate to specific rates of vibration.
- **Electrical energy (direct):** This is the flow of charged ions (usually electrons) through a wire or a conducting medium. Direct current flows and lightning are both examples of this.
- **Electrical energy (alternating):** This is the flow of charged ions (usually electrons) in a transverse wave. The stream of ions flows first in one direction (up the wire) and then in the opposite direction down the wire. The direction of the current simply alternates. Generators convert mechanical energy into this type of energy.
- **Radiant energy:** This is EMR, electromagnetic energy that travels in transverse waves. Radiant energy includes visible light, x-rays, gamma rays, and radio waves.

We note here that kinetic energy comes in three forms: flowing, moving in waves or spinning.

A useful way to think about potential energy and kinetic energy is to consider the behavior of a pendulum as illustrated in *Figure 33* on the next page. When the pendulum bob is pulled out to the side from the vertical, the force of gravity acting on it has been increased. It is now no longer at height h above the ground, but at height h + i.

When released it has zero velocity, but the force of gravity acts and its velocity gradually increases until it acquires a value, v, at its lowest point. The additional force of gravity that it was given has now been used up. However it now has kinetic energy, the momentum of the mass of the pendulum bob moving with velocity v. As it moves in an upward arc, this kinetic energy diminishes, but the force of gravity increases. Eventually it reaches the furthest point of the arc where it has velocity zero. If there were no resistance from the air and no friction to overcome at the fixed point from which the pendulum hangs, this exchange between potential and kinetic energy would persist forever.

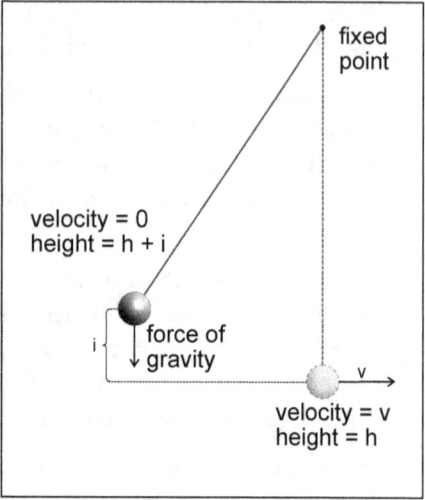

Figure 33. A Pendulum

The action of a pendulum is a useful example to keep in mind when considering energy transformation. It is a simple oscillation and, if it were frictionless, when set in motion it would do no work whatever. Nothing would be transformed.

Electrical Concepts and Equations

We now need to define and discuss the basic terms of electricity, so that we can relate them to forces, power and energy.

- **Volt and Ampere:** These two units are best taken together as they are defined in terms of each other. One volt is the **potential difference** between two points of a conducting wire when an electric current of one **ampere** consumes one **watt** of power. The **ampere** is thus the unit of electrical flow. We have already referred to the analogical idea of thinking of voltage (or potential difference) as water from a stream falling from the top of a cliff to the ground below under the force of gravity. The voltage is the height of the cliff. The amperage is the volume of flow of the stream. The SI units here are **V** and **A**.
- **Watt:** The watt is the basic unit of power. It can be defined in respect of mechanics as the rate at which work is done when an object's velocity is held constant at one meter per second against a constant resistance of one **newton**. In electrical terms, it is the rate at which electrical work is

done when a current of one **ampere** flows across a potential difference of one **volt**. A watt, SI unit **W**, is one **joule** per second.

- **Coulomb:** The coulomb is the unit of electrical charge. It can be applied to a charge of static electricity or of flowing electricity. When expressed as flowing electricity, a coulomb is a flow of one **amp** per **second**. Its SI unit is **C**.
- **Ohm:** The ohm is the electrical resistance between two points of a conductor when a constant potential difference of one volt, applied to these points, produces in the conductor a current of one ampere. Its SI unit is Ω (the Greek O).
- **Farad:** The farad is the unit of electrical capacitance, the ability of a body to store an amount of electrical charge. A farad is defined as the capacitance across which, when charged with one **coulomb**, there is a potential difference of one **volt**. Its SI unit is **F**.

Taken together these are the primary concepts and units required to measure the activity of plasma. In physics, mechanics (including fluid mechanics) studies the interactions of solids, liquids and gases; electronics studies the interactions of plasma. Note that there is a common unit (the **watt**) which measures power in both domains.

Beginning with the **volt**, we can say that when there is a difference in charge between two points we can measure that in **voltage**. If a current flows between those two points then that flow is measured in **amperes**. We measure the speed of that flow in **coulombs**, i.e in terms of **amps** per **second**. In most circumstances there is some resistance (a kind of friction) to the flow of charge. This helps define the **ampere** since the **ohm** is what the **ampere** overcomes to do work. The **watt** measures that work. The **farad** is required because we need to measure the charge stored in a capacitor.

If we look at these definitions as simple equations of relationship, we have:

$$\text{Watts} = \text{Volts} \times \text{Amps or } [W]=[V].[A]$$
$$\text{Ohms} = \text{Volts} / \text{Amps or } [\Omega] = [V]/[A]$$
$$\text{Coulombs} = \text{Amps} \times \text{time or } [C] = [A].[S]$$
$$\text{Farads} = \text{Coulombs} / \text{Volts or } [F] = [V]/[C]$$

Einstein's Energy Equation

Let us now turn our attention to Einstein's famous equation: $e=mc^2$. This important equation has been validated in the context of radioactive decay. A radioactive substance loses weight as some of its atoms decay to become lighter elements. The truth of the equation can thus be tested and verified by measuring the energy output during a period of radioactive decay and comparing that to the loss of mass.

$$\text{Energy} = \text{mass} \times \text{speed-of-light} \times \text{speed-of-light}$$

$$e = m.c^2$$

This equation declares that energy and mass are equivalent in some way. They are equated by the square of a constant that applies to the movement of (plasma) energy by means of EMR: the speed of light. If we understand what the implications of this are, we will better understand our universe.

Applying unit analysis, we see this as joules (j) equating to kilograms (kg) multiplied by meters per time (m/s) (i.e a velocity) and multiplied again by meters per time (m/s):

$$[j]=[kg].[m/s].[m/s]$$

If mass is a property of matter (and it is) then, according to this view, energy is also a property of matter. This implies, for example, that a ray of light is matter, even if very rarified matter. It also means that if we add some energy to some matter it will increase its mass. This proposition is validated by particle accelerators, which accelerate particles to very high speeds (i.e they increase the energy of the particle). When this is done the mass of the particle increases.

Note that we have learned to think of matter in terms only of three states, solid, liquid, and gas. Everything else is usually thought of as energy, Gurdjieff offered a different view which we discussed at length in Chapters 2 and 3. We'll repeat it here in summary.

Energy and Changes in Materiality

Gurdjieff explained the Ray of Creation as seven worlds, each residing within the one above. When descending from the Absolute, each level manifests as a denser materiality—heavier, slower, and subject to more laws. Naturally, there are ranges of density within each level, but the levels are distinct. Solid is not liquid, liquid is not gas, and so on. If we consider ice and water, while it is true that ice and water are identical in respect of atomic composition, their materiality and the laws they are under are clearly different. The different degrees of materiality depend directly on the qualities and properties of the energy manifested at each point.

The Essence of Plasma

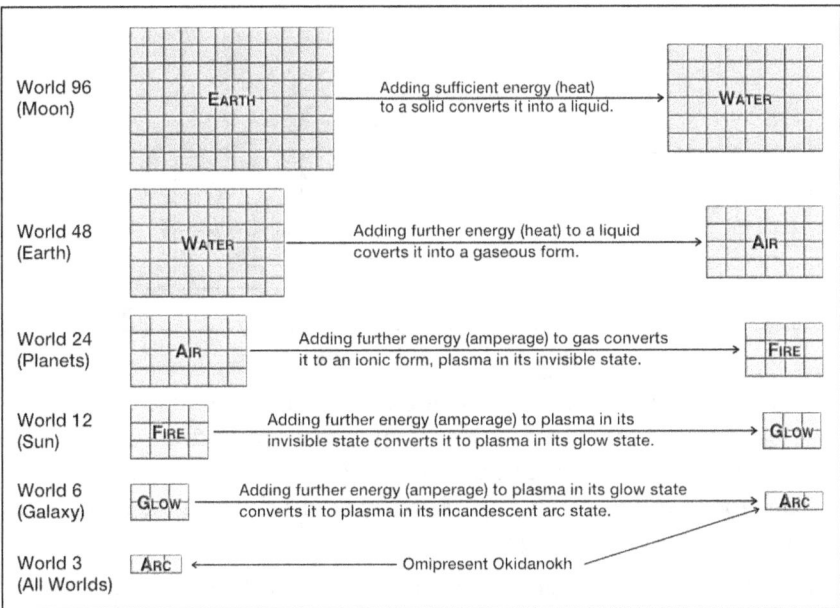

Figure 34. Atoms and Materiality

We illustrate the situation in *Figure 34*. Atoms are arranged in order of greatest materiality (or density) from the top down. Consider an atom of ice. We add energy to it in the form of heat, and the materiality changes from EARTH to WATER. So the ice becomes liquid water. The critical point to note is that to raise a substance from a lower level of materiality to a higher one, energy is required just to change the materiality. If we add energy in the form of heat to ice to reach its melting point, it does not melt. We need to add further energy to cause it to melt.

It is not difficult to determine how much energy is required experimentally. To melt a one gram block of ice requires 167 joules of energy. If you provide precisely that amount of heat, the substance changes' characteristics, but the temperature does not. An input of energy is always required when the materiality of a substance changes. As *Figure 34* illustrates, the same occurs when we change the materiality of liquid to become a gas or gas to become plasma in its lowest invisible state. The two higher modes of plasma, glow mode and arc mode, are also changes of materiality that require an input of energy in order to occur.

Consider the conversion of water to plasma. Simple electrolysis (the addition of electricity) separates the water molecule into hydrogen and oxygen ions in solution. Suppose we now put the hydrogen or any other gas in a vacuum tube

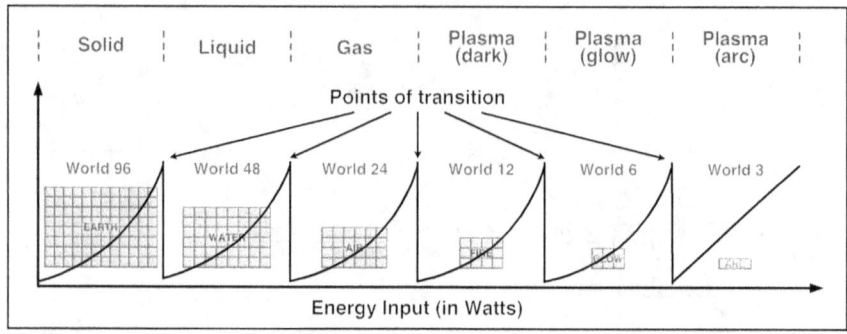

Figure 35. Six Levels of Materiality

and pass a current through it. In that case, we can witness the change from invisible plasma to visible plasma—the movement of electrons through the tube exciting the gas to emit light. Raise the current further, and an arc light will be produced. The energy required to change levels of plasma is illustrated by the graph in *Figure 32, page 121*.

In *Figure 35* above, we represent the same thing but show it for the six levels of materiality below the Absolute. The pattern repeats in that more and more energy is added to the substance until a transformation occurs, when an amount of energy is consumed by the transition. The energy builds up again within the substance until another point of transition arrives, and again energy is consumed by the transition.

The Atoms of Objective Science

We can summarize here the distinction between the atoms of objective science and the atoms of modern science. Conceptually they are distinct. This is abundantly clear when you juxtapose descriptions of their characteristics as in *Table 4*.

The most important points highlighted by this comparison are:

- Objective science is "theistic," because it embraces and defines an Absolute intelligence. Modern science is atheistic or agnostic and excludes any higher intelligence than man from consideration.
- Objective science is a unified theory. Modern science does not currently have a unified theory.
- Objective science is materialistic. Modern science is materialistic within specific realms. It has no explanations for the materiality of some phenomena, particularly psychological phenomena.

The Essence of Plasma

Atom (Objective Science)	Atom (Modern science)
Defined by size, chemical, physical, cosmic and psychological qualities.	Defined by standard atomic model (electrons, protons neutrons).
Different types defined by place and function with octaves.	Different types defined by the Periodic Table of Elements.
Atoms can contain multiple elements and even multiple different molecules.	Atoms are always elements. There are isotopes.
Applies to all substances in the universe. All substances have atoms.	Defines only the chemical building blocks of other substances.
All atoms can manifest.	Do not necessarily manifest physically as a single atom. (Sometimes manifests only as a molecule.)
Seven levels of materiality from the world 1 to world 96. Physically corresponding to the Absolute, arc plasma, glow plasma, dark plasma, gas, liquid, water.	Materiality can vary. Solid, liquid, gas and ionic forms are acknowledged. However the substance is not regarded as distinct when it moves from one form to another.
It is subject to specific orders of laws according to its level of materiality.	Is not specifically regarded as being under different laws according to its materiality. Still its states of materiality and their constraints are understood.
Matter is substance, energy is force acting through a substance.	Energy and matter are regarded as distinct but related and matter can be converted to energy under specific conditions.
Interactions only occur by the Law of Three. This is the only way that any event can occur at any level of scale. Some such interactions might be described as chemical or biological or physiological or astronomical.	Interactions are physical (particle/quantum physics or Einsteinian physics) and chemical. However there is no unified theory that covers all interactions from macrocosm to microcosm.
It is either alive or part of something living.	Is not alive per se.
It is characterized by its specific place or point in the universe and the octaves it consequentially participates in.	There is no concept of an octave in modern science. Chains of interactions are recognized but not octaves.
Substance and energy are indestructible within the creation.	Matter and energy can neither be created nor destroyed.
In principle everything is material, even absolute intelligence is material.	In principle everything is built from atoms.

Table 4. A Comparison of Atoms

Planck's Constant

Before we proceed to discuss electricity and magnetism, we need to complete the EMR picture by describing Planck's constant, which is fundamental to EMR. An aspect of physics, like the speed of light, Planck's constant is taken to be one of the primary constraints of the observed universe. Planck encountered it while investigating "black body" radiation. Conceptually, a black body is an idealized physical body that absorbs all EMR regardless of frequency or angle of incidence. That is why it is black—i.e. has no color. It reflects no visible color of any frequency. So it emits "black-body radiation," which is another way of saying that it emits radiation of every wavelength.

Experimental observation demonstrated that when an (almost ideal) black body was heated up, the spectral distribution of radiation followed a particular pattern. Planck attempted to explain those observations. In doing so, he hypothesized that the EMR of the black body object was caused by a set of harmonic oscillators, one for every possible frequency. This eventually led him to the idea that, at the basic level, light was emitted in small packets (quanta).

This laid the foundation for quantum mechanics. He had concluded that black body wavelengths could only be modeled as having discrete values. Conceptually this was important. He demonstrated that a mathematical model involving spectral continuity simply did not work.

He formulated the idea of a photon, the basic energy packet of light. Planck determined the value for that packet, which is now referred to as Planck's constant, h. It is equal to 6.63×10^{-34} Joule seconds.

Planck's constant appears in many equations, most notably in what is called the Planck-Einstein relation, which is:

$$E = h.\upsilon$$

Here, E is the energy of each packet (or 'quanta') of light, measured in Joules; υ is the frequency (in cycles per second, or hertz); and h is of course Planck's constant. It is important, for the sake of precision to know that this formulation is a simplification, presumably for the sake of elegance.

The equation indicates the amount of energy in a time of exactly one second, so the constant h relates specifically to one second of time. Planck's more exact formulation of this energy equation was:

$$\delta E = h_\delta . \delta t . \upsilon$$

Here, the δ indicates a finite change (a delta). So δE is the change in energy over a given time, again h_δ is Planck's constant for that time period, δt is the measured amount of time and v is the wave frequency. We can transpose the equation into one for power.

Power is the work done (energy) in a specific time, which we can write as:

$$P = \delta E/\delta t = h_\delta . \delta t . v/\delta t$$

Thus, cancelling the δt we get:

$$P = h_\delta . v$$

This says that the power delivered by a single wave of EMR is a multiple of a constant h_δ which applies specifically to a single interval of time multiplied by the wavelength. Applying unit analysis to the equation, we get:

$$[J/s] = [J/\delta] . [\delta /s]$$

So Power (P), measured in Joules per second, is equal to a constant h_δ which relates to a time interval, multiplied by the number of waves that occurred in that time interval. The value of Planck's constant for the time interval of one second is 6.626×10^{-34} Joule seconds.

This equation provides us with a measurement of the energy of a single photon, and since photons are deemed to have no mass, it gives us the ability to calculate the energy of a photon with a wavelength of a specific frequency. The highest frequencies so far recorded are for gamma rays, with wavelengths of the order of 10^{-13} meters (i.e. frequencies of 10^{13} Hertz). These are caused by by highly energetic events such as supernovae in distant galaxies.

Theoretically, there can never be a wavelength less than the Planck length. A wavelength that small would mean frequencies of a much much higher order (10^{34} Hertz) than ever measured, and it may be that it is impossible to create them. Indeed the limit may be much lower.

If we knew the highest possible frequency of a particle of the Absolute we would know that limit, because that limit was set by Him in the moment of creation. If we knew it we would be able to measure the energy and power of the Absolute.

Plasma and Static Electricity

EMR is a phenomenon of plasma. It is a means by which plasma passes Energy from one location to another. Electricity is also, clearly, a phenomenon of plasma.

We can think of electricity as having two basic forms. The first is static electricity, where a collection of ions are motionless in some location, at a point or on an object. The second is electricity in motion, when a current (a collection of ions) is flowing along a given path. This flow can be a direct current or an alternating current.

Static electricity is potential energy; an electric current is kinetic energy.

The question is "how does this electrical energy manifest?" With static electricity we can explain this by considering two adjacent points that are charged either positively or negatively and reside in a medium that is not conductive. The various possibilities are illustrated in *Figure 36*, where the amount of charge on each point is given the values q_1 and q_2, and the distance between the two points is r.

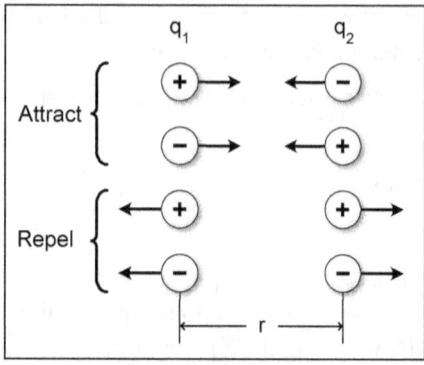

Figure 36. Coulomb's Law

$$F = k_e \cdot \frac{q_1 \cdot q_2}{r^2}$$

Coulombs Law applies to precisely this situation. The reality is that when q_1 is of opposite charge to q_2, there is a force of attraction between the two points. When the charges are both positive or both negative, there is a force of repulsion. Coulomb's law provides the formula for calculating this charge:

Here F is force, q_1 and q_2 are charges, r is the distance separating the charges and k_e is Coulomb's constant, which has a value of about 8.99×10^9. The force F thus has a negative value for attraction, and a positive one for repulsion.

$$F = G \cdot \frac{m_1 \cdot m_2}{r^2}$$

If we apply unit analysis we get this:

$$[F] = ([N].[m^2/C^2]).[C.C/m^2]$$

It is hard not to notice the similarity between Coulomb's formula and Newton's formula for the force of gravity:

Where F is force, m_1 and m_2 are the masses of two objects, r is the distance separating the masses, and G is the gravitational constant, which has the value 6.67×10^{-11}. Unit analysis gives us:

$$[F] = ([N].[m^2/kg^2]).[kg.kg/m^2]$$

This similarity has caused quite a few plasma physicists to wonder whether gravity is an electromagnetic effect of some kind that is not yet understood.

The Essence of Plasma

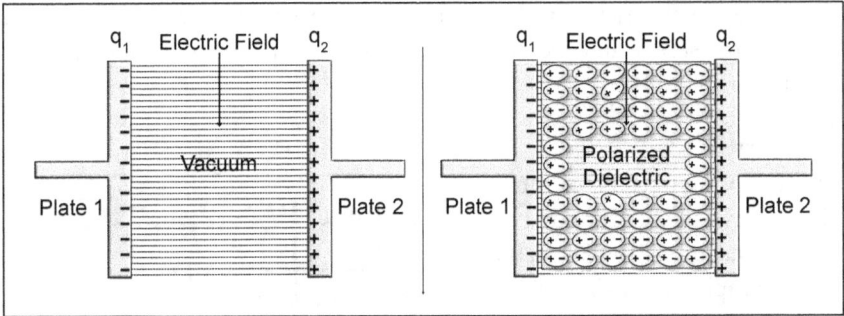

Figure 37. Capacitors with a Vacuum or Dielectric Material

Some theorists argue that this is so, but a coherent explanation of how it could be has yet to emerge.

Capacitors

In *Figure 37*, we depict a capacitor in two different ways. A capacitor is a device that employs an electric field to store an amount of electric charge. The device on the left hand side is simply two conductive plates with a negative charge q_1 on Plate 1, opposite to a positive charge q_2 on Plate 2. The attractive electric field created by the two separated charges permeates the vacuum between the two plates, holding the charges in place. A vacuum is a very effective insulator and thus there will be no flow of charge between the plates.

On the right hand side we show the same arrangement of two plates with charges q_1 and q_2, but this time the plates are separated by a dielectric (i.e. an insulating substance) rather than a vacuum. The attractive electric field still acts between the two plates, but now it passes through the dielectric. Dielectrics that are effective insulators can be thought of as being composed of dipole particles (very small magnet-like particles, as small as molecules or groups of molecules). Under normal circumstances they tend to be chaotically organized—in the sense that each individual dipole may be aligned in any direction.

When the dielectric is placed between charged plates, those dipoles tend to align themselves in harmony with the prevailing electric field. The extent to which they do that depends on the nature of the substance itself, and also other factors, such as the strength of the electric field, the air pressure and the ambient temperature. The dielectric thus responds to the electric field that passes through it, to some degree.

Conceptually, batteries can be thought of as large capacitors—large in the sense that they are designed to store a large amount of electricity. However they store the electricity by chemical means. For practical reasons they need to discharge slowly and the best way to achieve that is through a controlled chemical reaction.

Conductors

It is tempting to think of substances in a binary way as being conductive, like metals, or insulators, like plastic, but reality is more complex than that. First of all the electrical conductivity of a substance varies with temperature and pressure, so discussions of conductivity need to take account of that. At normal temperatures (20°C) silver is the most electrically conductive metal and mercury is the least.

The group of elements generally referred to as metalloids, Boron, Silicon, Germanium, Arsenic, Antimony, Tellurium and Polonium, is found in the Periodic Table in what is sometimes called the "metalloid staircase."

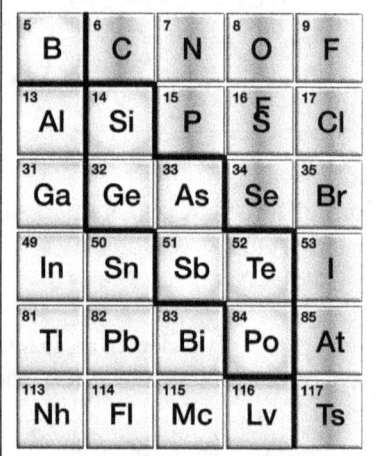

Figure 38. Metalloid Staircase

Referred to as semiconductors, they are extremely poor conductors (by orders of magnitude compared to metals), but they can be slightly modified, by a process called doping, in a way that enables miniature circuits to be etched on their surface. The vast industry of microelectronics evolved from this capability.

Only a few of the metals; iron, nickel and cobalt are magnetic. Some alloys that are based on these metals are also magnetic as are some alloys based on rare earth metals. Magnetic materials are classified as "hard" meaning that once magnetized they tend to retain their magnetism, or "soft" meaning that they can quickly lose their magnetism.

Electricity and Magnetism

A magnetic field and an electric field are entirely different phenomena. It is important to know the difference, because in many circumstances an electric field and a magnetic field will exist within the same space. However, they can be considered apart as illustrated in *Figure 39*.

The Essence of Plasma

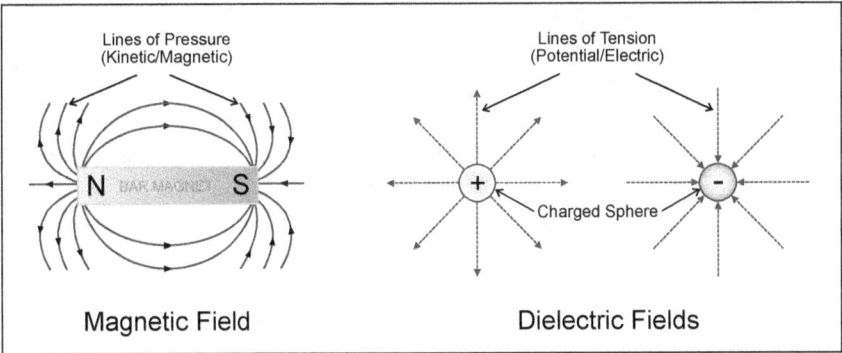

Figure 39. Magnetic Fields and Electric Fields

On the left side of the diagram we show a common bar magnet. The important point to note is that the magnetic field (the lines of force) acts between the two poles of the magnet as shown. They do not connect to the surrounding environment in any way.

On the right hand side we show the situation where there is a static electric charge on a sphere, producing an electric field but no magnetic field. Two possibilities are shown. When the sphere is positively charged in relation to its surrounding environment the force of the electric (or dielectric) field will act to repel the environment; when the sphere is negatively charged, it will act to attract the environment.

Magnetic Materials

There is no discernible difference between a magnetic field generated by a magnet and one generated by an electromagnet. As such, magnets are substances that can crystallize a magnetic field in place. At a microscopic level, magnetic materials are composed of magnetic domains, each of which is a dipole (a small magnet with both a north and south pole). When you magnetize the material, by subjecting it to a magnetic field, all the domains (dipoles) line up with each other. You can observe this happening with a microscope.

It is similar to what happens in a dielectric (as illustrated in *Figure 37*), the difference being that in the case of magnetized material, you get a magnet, whereas with a dielectric, you do not. Magnets always have a north and south pole. If you break a magnet in two, you get two magnets with a north and south pole. And if you keep on dividing them up, you create smaller and smaller magnets until you get down to the magnetic domains themselves.

If you'd like to know what causes the magnetic field, you are not alone. There are theories but no decisively good explanation. It is likely that it has to do with the motion of electrons since when electrons are in motion they create magnetic fields and all such fields have a north and south pole. At an atomic level some molecules are strong dipoles so it's easy to imagine that the molecules create circuits that the electrons travel round and this creates the magnetic field. If that's the case, nobody has yet demonstrated it.

An Electric Current

We can now turn our attention to electric currents. By definition an electric current is a plasma flow of some kind, usually along a conductor such as a copper wire, as we illustrate in *Figure 40*. This shows a cross-section of a wire passing through a dielectric medium (such as a naked copper wire passing through the air). A direct current flows along the wire and it creates the lines of force shown in the diagram.

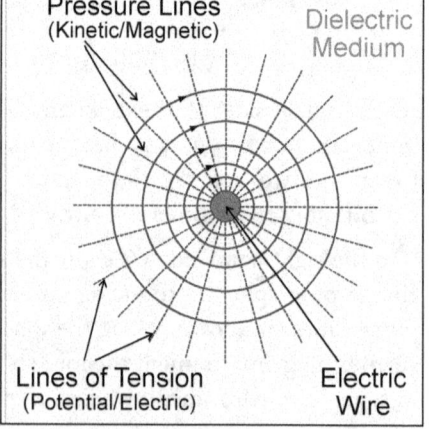

Figure 40. A Wire and Fields

You will notice that each of the circular lines has an arrow-head, indicating its direction. Those in the diagram indicate a clockwise motion and thus we can deduce that the electric current is flowing into the page.

By convention—it is simply an arbitrary designation—flows of electrical current are deemed to be movements of positive charge (i.e. positive ions). Thus, when a flow of electrons (negatively charged particles) moves from B to A, the current is said to be flowing from A to B. There is a practical technique for remembering the direction of the current in a diagram like *Figure 40* called "the right hand rule." To use this rule, clench the fist of the right hand and point the thumb out to the side. When you do so, the curl of your fingers indicates the direction of the circular magnetic lines round the wire and the direction out along your thumb indicates the direction of the current flow.

Figure 41 represents the idealized context of a current flowing along a wire, unrelated to anything else (as if there were nothing else in the universe). The circular lines around the wire denote the magnetic field caused by the flow of current. This magnetic field should be thought of as in motion (kinetic energy)

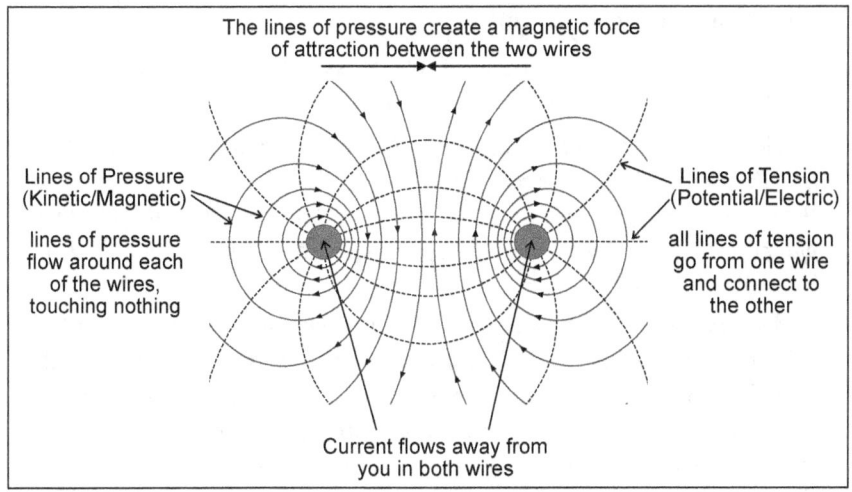

Figure 41. Current Through Two Adjacent Wires

and it exerts a magnetic force. The field strength (at any point) is measured in **Tesla** or in **Amps per meter** (A/m).

The dotted radial lines denote the electric field created by the flow of current. The electric field is orthogonal to the magnetic field, as it is in EMR. Electric fields are always orthogonal to magnetic fields. That's how the universe is. The electric field's radial lines, which appear to extend to infinity, will eventually link to something in their neighborhood. They are measured in **volts per meter**.

Electric field lines denote potential energy, just as water in a lake at the top of a cliff has potential energy. The magnetic field lines denote kinetic energy, just as water falling down from the top of the cliff has kinetic energy.

Now consider *Figure 41* above showing a cross-section of two parallel wires with the current flowing in the same direction, into the page. (You can deduce the direction of the current from the arrow heads by applying the right hand rule.) In these circumstances the two wires are magnetically attracted to each other. The force of attraction is referred to as the Lorentz force. The lines of pressure are compressed between the wires and extended on the opposite sides of the wire.

If the two currents in the wires were traveling in opposite directions they would repel rather than attract each other.

GURDJIEFF'S HYDROGENS: THE RAY OF CREATION

Emanations and Radiations

We now have sufficient information to discuss the difference between emanations and radiations. In *The Tales*, Gurdjieff writes:

> "Here you might as well, I think, be told, by the way, about an interesting fact I noticed, which occurred in the history of their existence concerning the strangeness of the psyche of the ordinary three-brained beings of that planet which has taken your fancy, in respect of what they call their 'scientific-speculations.'
>
> "And that is, that during the period of my many-centuried observation and study of their psyche I had occasion to constate several times that though 'science' appeared among them almost from the very beginning of their arising, and, it may be said, periodically, like everything else there, rose to a more or less high degree of perfection, and that though during these and other periods, many millions of three-brained beings called there 'scientists' must have arisen and been again destroyed, yet with the single exception of a certain Chinese man named Choon-Kil-Tez, about whom I shall tell you later in detail, not once has the thought entered the head of a single one of them there that between these two cosmic phenomena which they call 'emanation' and 'radiation' there is any difference whatever.
>
> "Not a single one of those 'sorry-scientists' has ever thought that the difference between these two cosmic processes is just about the same as that which the highly esteemed Mullah Nassr Eddin once expressed in the following words:
>
> "'They are as much alike as the beard of the famous English Shakespeare and the no less famous French Armagnac.'

The etymology of the word "emanation," is as follows:

> "act of flowing or issuing from an origin; emission; what issues, flows, or is given out from a substance or body;" 1560s, from Late Latin *emanationem*, noun from past-participle stem of Latin *emanare* "flow out, spring out of," figuratively "arise, proceed from," from assimilated form of *ex* "out" + *manare* "to flow."

The meaning is clear. It means flowing out.

The etymology of "radiation."

> mid-15c., "act or process of radiating," from Middle French *radiation* and from Latin *radiationem* "a shining, radiation," noun

---
* The Tales, Chapter XVII, The Arch-Absurd, p141

from past participle stem of *radiare* "to beam, shine, gleam; make beaming," from *radius* "beam of light; spoke of a wheel." Meaning "rays or beams emitted" is from 1560s.

So, in origin, radiation relates to shining, and from that perspective it refers directly to EMR. Emanations, outward flows, refer to anything emitted by a substance or creature, except for radiations, which do not flow. There is an obvious difference between the two, in that radiations, once emitted, travel vast distances, but emanations are local. Even in the case of the Sun, which emanates solar winds; they cease to flow at the heliopause, which lies beyond the orbit of Neptune, and what was emanated loops back to the Sun.

Magnetic fields and electric fields are both emanations, fields of force that have limited extent. EMR is radiation.

Electromagnetic Fields

In *Figure 42* we show a simple circuit design that can create an electromagnetic field. A battery provides a direct current to the circuit. The circuit includes a coil of wire of twelve loops. Collectively these loops create an electromagnetic field as illustrated.

Note that the movement of the current will generate a small electromagnetic field around the wire for the whole length of the circuit, but it is a very weak field. When you coil the wires, each loop of the coil increases the strength of the electromagnetic field. A coiled wire of this kind is called a solenoid. You can increase the strength of the electromagnetic field by wrapping the coils of wire (suitably insulated) around a metal bar made from a soft magnetic substance such as cast iron.

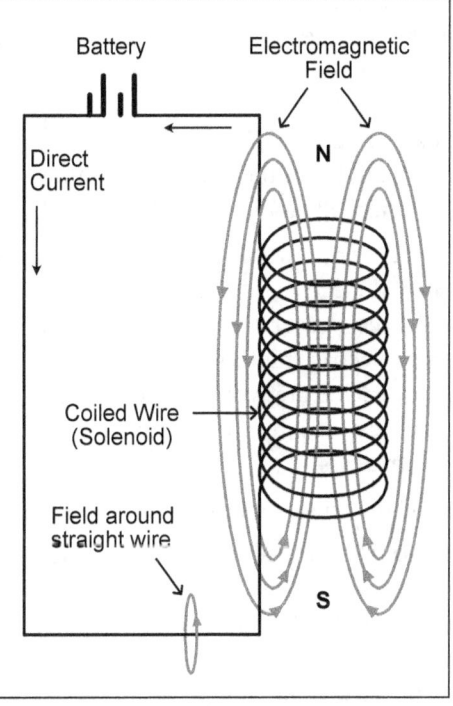

Figure 42. Electromagnetic Field

Gurdjieff's Hydrogens: The Ray of Creation

When we stated that a magnetic field stores a kind of kinetic energy (energy of motion), if you didn't already know, you might have wondered how we could make use of that energy. The answer is simple: you convert it into electricity.

Induction

If you have an electric wire that is connected to an ammeter (a device for measuring electric current) and you just wave a bar magnet close to the wire, you will see the needle on the ammeter move. It registers that a current is flowing.

What happened when you waved the magnet is that its movement of induced the flow of a current through the wire. This phenomenon, referred to as induction, was discovered by Michael Faraday in 1831 and it is the most important electrical phenomenon by far.

Because of it whole range of fundamental electrical devices: dynamos, electrical generators, transformers and electric motors are possible. They enable pretty much everything electrical from food mixers to electric cars.

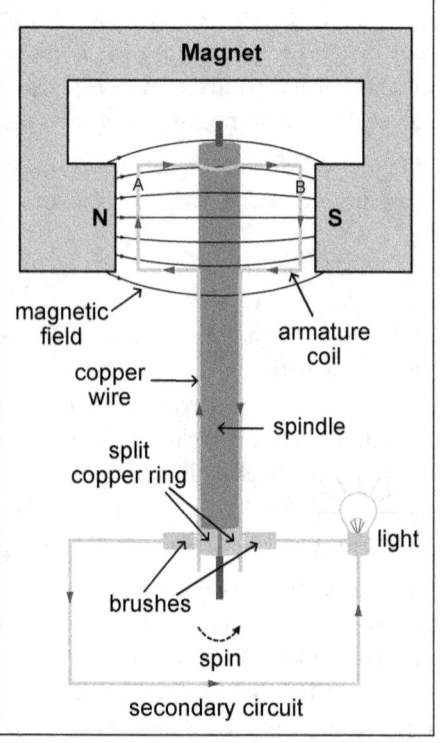

Figure 43. Simple Dynamo

So how do we make use of induction?

This is illustrated in *Figure 43*, which shows a simple electrical generator (or dynamo). The mechanism consists of a magnet with opposite poles (N and S) facing each other. Between those poles, attached to a spindle, an armature coil spins. The armature coil is a simple rectangle of copper wire, which is connected along the spindle to a split copper ring.

As the armature coil spins within the magnetic field, a current is induced in it, which travels down to the split copper ring where it is transferred via brush contact to the circuit with the light bulb.

The reason for the split copper ring is that with every half-spin, the side of the armature labelled 'A' moves away from the influence of the north side (N) of the magnetic field into the influence of the south side (S), and the side labelled 'B' does the opposite. As this happens the direction of current changes and moves in the opposite direction.

Induction takes place because, although the magnet itself is fixed, the rectangular armature coil is moving. Because of that the strength of the magnetic field around the wire itself is changing. It is at its strongest when sides A and B are exactly are directly facing a magnetic pole (as shown in the diagram where side A faces N and side B faces S), and at its weakest when sides A and B are exactly in the middle of the two magnetic poles.

The split copper ring ensures that a direct current passes into the secondary circuit. If that mechanism were removed, this dynamo would still produce a current, but it would be an alternating current.

The final point to note is that, in order to generate the electric current it is necessary to spin the spindle on which the armature is mounted. Thus, in order to get electricity out of the system, you have to put kinetic energy into the system. The principle explained by *Figure 43* is exactly the same principle that is employed in hydroelectric, geothermal, wind, nuclear and coal or gas fired power generation. It is only the source of the power that is different and the fact that it is done at large scale..

Form in Plasma

In our view you now have sufficient information about plasma and electricity for you to be able to consider its place in objective science. Mankind makes use of electricity in a wide variety of clever ways. Ever since Benjamin Franklin flew his famous kite in a thunderstorm, our knowledge of electricity has grown, and nowadays electrical and electronic engineering have a huge influence on the life of man.

Mankind's technology aside, plasma occupies its own place in the universe. We may think of plasma as less constrained than AIR. Proceeding logically, we might think of solid substances as the least free, liquids as freer, gaseous substances as having greater freedom still and plasma as having the most freedom of all — running wild and permeating the whole of empty space.

In fact plasma is also subject to laws and is probably more constrained than you imagine. Put simply, there is form in plasma.

Gurdjieff's Hydrogens: The Ray of Creation

Figure 44. Plasma Crystallized in Zero Gravity

This is amply demonstrated by the image in *Figure 44*. It shows a photographic still from a Youtube video* of an experiment carried out by Russian cosmonauts on the International Space Station.

What you see is plasma that has formed a crystal lattice around micron-sized particles of dust. The Russians had two identical chambers that had video cameras inside them so that they could observe. One was down on Earth, and hence subject to gravity, and the other was on the space station in a zero gravity environment.

First they filled the chamber with plasma and then introduced the dust. They were able to watch the plasma crystallize around the dust. At one point an empty void opened up in the center of the chamber. The experiment on Earth showed nothing, because gravity simply squashed the plasma crystals flat.

Cosmonaut Pavel Vinogradov commented, "It appears that plasma isn't really a fluid in the physical sense. Instead it has a crystalline lattice-like structure. Seeing this changed our views on the fourth state of matter." *

Cosmonaut Sergei Krikalev added, "Dust particles that are suspended within plasma acquire an electric charge and repel each other. Because of that repulsion their space within the chamber is limited and a crystal-lattice-like structure is formed. In space the plasma opens up and forms a 3D structure. Vortices in the plasma resemble a galaxy. When the plasma is cooled cryogenically a DNA-like shape appears."

---

* https://www.youtube.com/watch?v=R4Z_-WbDs4U

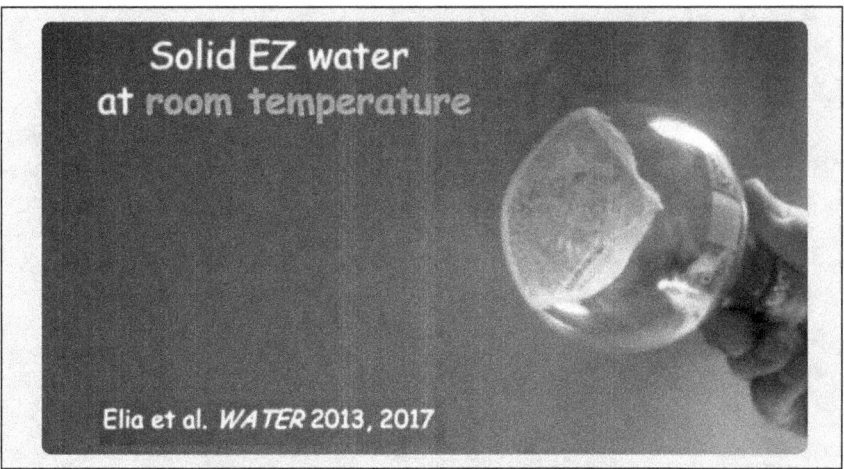

Figure 45. The Fourth Phase of Water

So, plasma is clearly able to manifest in a solid form.

You are probably unaware that there is plasma state of water—to be clear, a form of water that is ionic. Research carried out by Jerry Pollack Ph D into this has been proceeding at The University of Washington, Seattle. There is a Youtube video* of a lecture he gave at a conference in 2019 on *The Physics, Chemistry and Biology of Water*, in Bad Soden, Germany.

It is surprising in many ways, including the fact that there is a form of water (a plasma state, which Pollack calls EZ Water) that it is solid at room temperature. We show a photograph of this in *Figure 45*. The substance at the bottom of the flask shown in the still from the video is water. Pollack thinks it's likely that clouds are also water in its "fourth phase."

One aspect of this lecture that's worth noting here is that the molecular structure of the fourth phase of water would be expressed chemically as H_3O_2 not H_2O. It seems to be the case that water quite readily organizes itself to include plasma layers, usually when in contact with a surface.

Double Layers of Plasma

What is referred to as a plasma double layer is a structure of two parallel layers of plasma which have opposite electrical charge. This is illustrated in *Figure 46*. Typically, double layers form in situations where one region of plasma is surrounded by another with significantly different average charge. They act as a kind of skin that keeps the two regions apart.

---

* https://www.youtube.com/watch?v=nSbg3cuZNRQ

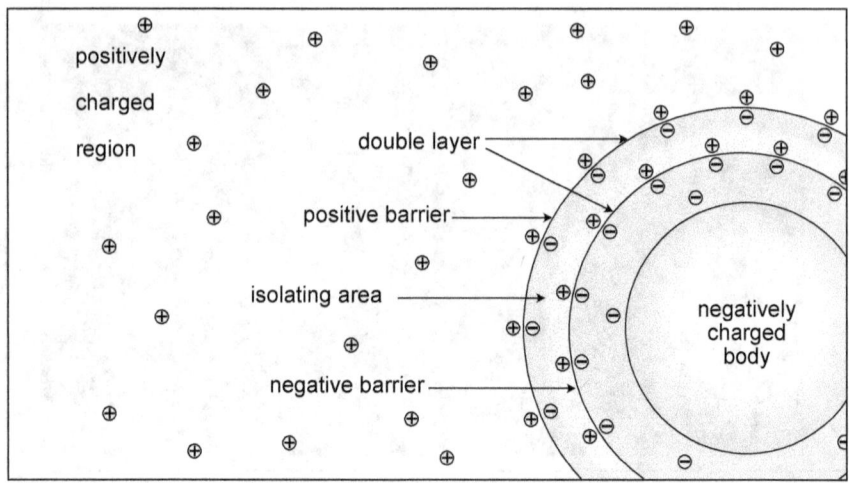

Figure 46. Plasma Double Layer

The sheets of charge give rise to a strong electric field. Ions which come in contact with the sheet (on either side) will be accelerated or slowed by it, and tend to bounce off or align with the field. Double layers can stack up, although they are always very thin compared to the area they surround. They are commonly associated with plasma flows.

They are encountered in various laboratory contexts and it may be that double layers are responsible for the phenomenon of ball-lightning. Ball lightning is a rare event. Occasionally, after a lightning strike, a ball of light, usually no more than a foot in diameter, is observed. It appears to float parallel to the ground, immune to gravity. It usually dissipates quickly.

In our view, the importance of double layers is the role they play in plasma behavior at the largest scale. There are double layers that surround the Earth's magnetosphere and, in theory, there are double layers that surround the heliosphere.

Chapter 5

The Diagram of Everything Living

"What is possible for individual man is impossible for the masses."
~ *Gurdjieff*

Before we discuss the Diagram of Everything Living, or as it is also called, the Step Diagram, we need to discuss the question: What is life?

Modern science confines life to the biological domain, regarding all other things else as dead or inert. The only exception to this theoretical boundary is to be found in the Gaia theory proposed by James Lovelock. It suggests that living organisms interact with Earth's inorganic matter to create a complex, synergistic and self-regulating system that maintains life.

Objective science has a fundamentally different view that goes much further than that.

Cosmic Units, Life-Forms

When objective science uses the term "cosmos," it means an ordered world, a system of a kind, but specifically one that is a living being. A cosmos consumes three different substances:

- **Food:** What we think of as nourishment, assimilated through the digestive tract of man. Comparable structures for assimilating food exist in other multicellular life-forms.
- **Air:** This is assimilated via man's respiration system and via a comparable structure in other multicellular life-forms.
- **Impressions:** These enter through the senses and travel via the nervous system to one or more brains.

Gurdjieff's Hydrogens: The Ray of Creation

A cosmos is mortal. It is created (born) as a relatively independent unit, experiences a lifetime, and eventually dies. On death, it disintegrates into separate substances, which become food for other cosmoses. Gurdjieff explains cosmoses from this perspective in *In Search of the Miraculous*.*

He emphasizes the need to adopt a cosmic perspective—to classify life-forms in respect of their existence within the cosmos of which they are part. Normal science classifies life according to external traits such as bones, teeth, limbs, or groupings: mammals, rodents, snakes, etc. In objective science, classifications are made based on the cosmic traits of a being.

These are precise traits of being that apply to all living things, which, Gurdjieff claimed, allow us to establish the class and the species of a given creature with precision. These are the three traits:

1. What the creature eats.
2. What the creature breathes.
3. The medium in which the creature lives.

The concept of "cosmic traits of being" deserves consideration. When we discuss being, we normally focus on the being of man. This proves to be a complex topic because man spans a range of activities that are simply impossible for other life-forms on Earth. He constructs huge buildings, flies, drives vehicles, farms, uses electricity, and is remarkably innovative. All of this is clear from man's impact on his environment compared to any other species. Nevertheless, despite his remarkable versatility, man cannot change what substances he can digest, or what he breathes, or the medium in which he lives.

- He feeds on $H768$. Anything lower than $H768$ he could not digest. Anything higher would lack the right nutrients.

- He breathes $H192$. In the main, he breathes oxygen extracted from the air. He is unable to breathe like a fish, which breathes oxygen extracted from water.

- He lives in $H192$, in the air. But in truth, he lives mainly in shelters to protect him from the elements and other life-forms (insects particularly). He lives in the anthroposphere, a world he shapes himself.

Suppose we set man to one side for a moment and consider other life-forms. In that case, it is clear that they have very little environmental impact beyond their inherent existence—that is, what they eat, what they breathe, and their

---

* *In Search of the Miraculous* by P D Ouspensky, p320

disintegration at death. Their contribution to the Trogoautoegocrat is predictably fixed, whereas man's can vary wildly.

Before we pursue this idea further, we need to remind ourselves that a *Hydrogen* is a broad grouping of substances. Any given *Hydrogen*, such as H768, comprises many different specific substances. Not all H768 substances can be food for man. Some will be poisons, and some will simply be excreted if eaten because the body has no use for them. Also, the meals man serves himself will not all be H768. Meat can have bones; fruits can have seeds that man cannot digest, and so on. Man is obliged to pick out the H768 on his plate or leave it to his digestive system to do so.

Gurdjieff comments* that although H1536 can be food for dogs, flour worms, fish, and trees, it is different types of H1536 in each case. Commercial dog food can sometimes include H1536 in the form of slaughterhouse waste, mill sweepings from flour mills, fermentation waste from distilleries, and peanut shells. Flour worms will eat H1536 in the form of rotten flour. Aside from other fish, fish consume algae and plankton, the cell walls of which are H1536. Trees consume a variety of chemicals (phosphates and nitrates) from the soil, which are H1536.

Flowers produce the nectar bees feed on specifically to attract them. It is a watery solution of fructose and glucose (sometimes also sucrose). Glucose and fructose are H192, monosaccharides that pass through the small intestine into the bloodstream and provide energy to cells. Bees breathe air like man and live in the medium of air when foraging for food. However, they inhabit a hive, a dense ecosystem that would be impossible for man or any animal.

Trees and plants are curious because they span environments. The medium the roots live in is H3072, but the leaves and branches live in H192. Gurdjieff mentions* that they breathe only partly H192 and partly H96. The H192 is easy to understand; plants breathe in carbon dioxide (H192) during the day, exhaling oxygen (H192). If they also breathe H96 in some form, it is probably part of the process of photosynthesis. In outline, by the Law of Three, H96 (sunlight or a derivative of sunlight) combines with water H384 to produce oxygen H192. However, in reality, the process has multiple stages. Gurdjieff is simply commenting* that plants breathe in H96.

Many insects (butterflies, moths, flies, etc.) have a life that involves a dramatic change, feeding on a specific *Hydrogen* (H1536, plant leaf, in the case of the butterfly) as larvae and feeding on nectar (H192) as adults.

---

* *In Search of the Miraculous* by P D Ouspensky, p320

Dragonfly larvae are particularly interesting in this respect, living for three years underwater (i.e., in the medium of water, H384), where they feed on insects, crustaceans, and even fish larvae. Then changing medium completely as adults—living in the air, H192, and feeding on airborne insects such as midges and mosquitoes.

The Trogoautoegocrat

As far as we know, in the intervening years between when Gurdjieff first taught these ideas and now, no-one has pursued the study of biology from this perspective. Thus for many life-forms, we do not have an accurate knowledge of their food, breath, and the medium in which they live. This may not seem important, but as will become clearer in later chapters, it is. For other life-forms, we do not even have a reasonable classification of what constitutes their food.

The idea of a Trogoautoegocratic universe has broader implications than may be immediately apparent. It has to work so that any substance created by any being for whatever purpose and later discarded will become food for something else. When we study areas of the Earth that are still relatively wild, we usually discover great ecological variety. Rather than individual species evolving, it appears that an ecology evolves.

In respect of what feeds on what, the Trogoautoegocrat defines an ecology. This clearly involves everything, from the chemical composition of rocks beneath the soil, the soil itself and its constituents, local bacteria, plants, and all animal life. It also depends on the land formations (mountains, rivers, valleys, plains), weather, and the latitudes that the land covers. And that is just the Earth. The theory is that the Trogoautoegocrat manifests at every level of every note in the Ray of Creation.

Intelligence

Gurdjieff insists that all substances have intelligence.* Thus, both substances and creatures possess intelligence, and that intelligence is defined by what it can serve as food for. Thus Gurdjieff notes** that a baked potato (a substance) is more intelligent than a raw potato (a life-form). The raw potato can be food for animals such as pigs, but not for man, while a baked potato can be food for man.

Gurdjieff's definition of intelligence is startling because it bears no relationship to the meaning we normally assign to this word when we relate it to the

---

* *In Search of the Miraculous* by P D Ouspensky, p176
** *In Search of the Miraculous* by P D Ouspensky, p321

The Diagram of Everything Living

mind, reason, and understanding. In other life-forms, we think of intelligence as a creature's ability to adapt to changing circumstances. We have no concept of a substance having intelligence.

Gurdjieff defines intelligence as a quality of a substance. In doing so, he relates it specifically to the Law of Three (rather than to a being). His example of the potato makes this clear. If something can be food for a being, it means that once ingested, a specific *Hydrogen* within it can combine with a higher *Hydrogen* and hence ascend to become one level higher. Since all interactions are instances of the operation of the Law of Three, any digestion of information (H48) by a higher *Hydrogen* (H12) causes the information itself to rise to H24 (knowledge) and thus become more intelligent.

The Step Diagram

Having examined Gurdjieff's assertions about life-forms from one perspective, we can look at it from another. *Figure 47* on the next page illustrates the Step Diagram. Gurdjieff also referred to this as "The Diagram of Everything Living," and J G Bennett called it "The Hieroglyph of Life." The fundamental implication of the diagram is that everything is alive.

The Step Diagram represents life-forms arranged into specific classes, each of which forms an ascending staircase, from left (lower life-forms) to right (higher life-forms). According to Gurdjieff*, the squares in the diagram express what the class of creature is (its average level of being) and denote a feeding relationship between each creature's class. This feeding relationship does not relate to what a specific creature eats but to how classes of beings feed each other in some other way. Each square shows three *Hydrogens*. The middle *Hydrogen* denotes what the class of beings is; the top one denotes the class of being that feeds on it, and the bottom one denotes the class of being that is food for it.

According to the diagram, *Plants* are H192, they feed on *Metals* (H768), and they are food for *Vertebrates* (H48). *Man* is H24; he feeds on *Invertebrates* (H96) and is food for *Archangels* (H6). Some of the classes of life-forms in the diagram are familiar to us and reasonably easy to understand. The *Vertebrates* class is large, including fish, amphibians, reptiles, mammals, birds, all of whom possess a backbone or spinal column. Gurdjieff also refers to such creatures as two-brained, meaning that they have both a moving/instinctive brain and an emotional brain.

---

* *In Search of the Miraculous* by P D Ouspensky, p322

Gurdjieff's Hydrogens: The Ray of Creation

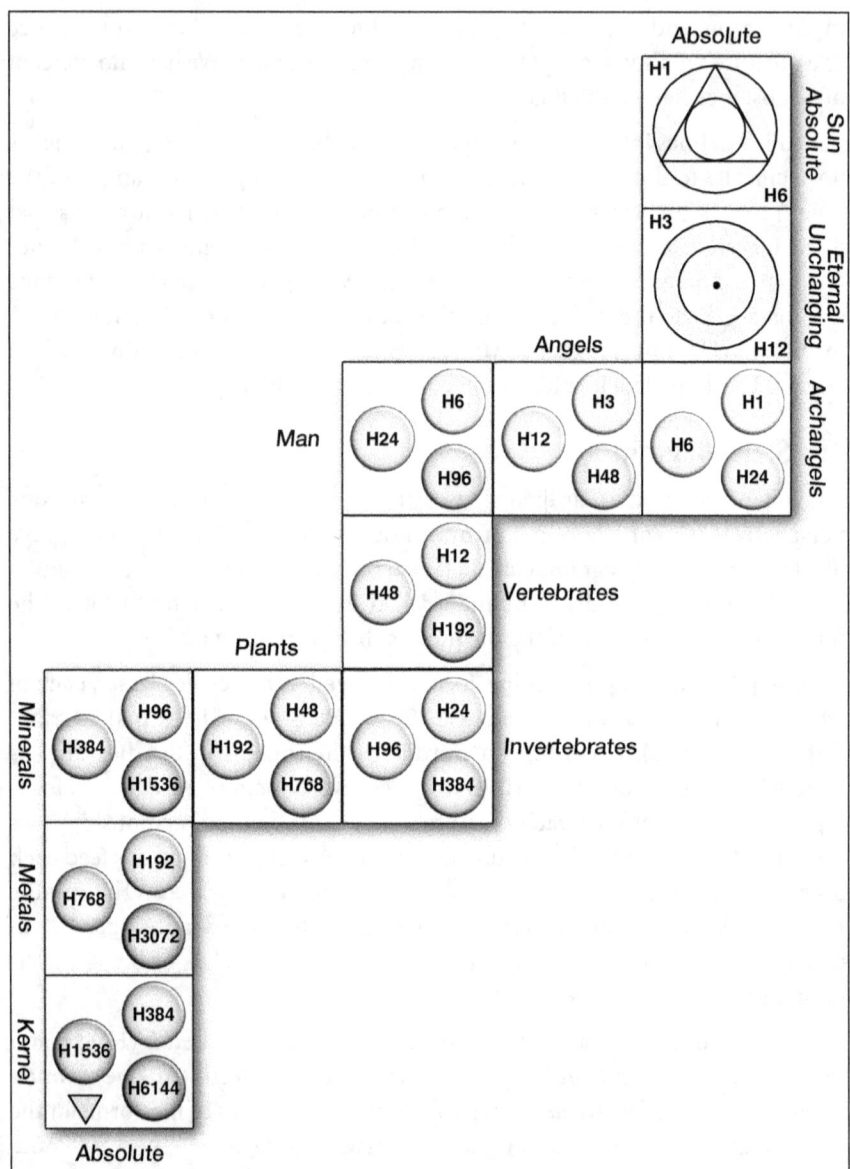

Figure 47. The Step Diagram

Invertebrates usually have exoskeletons rather than backbones and comprise an even broader range than *Vertebrates* accounting for 95 percent of animal species, including arthropods, mollusks, annelids, insects, crustaceans, and many more. Most but not all have exoskeletons. Gurdjieff refers to such crea-

The Diagram of Everything Living

tures as one-brained, fully equipped with a moving instinctive brain but lacking an emotional brain. *Plants* are distinctly different from animals, unable to change location at will, and feeding (that is, synthesizing carbohydrates) by photosynthesis. There is also a wide variety, including trees, shrubs, herbs, grasses, ferns, mosses, and algae.

We are familiar with organic life and thus can easily understand those classifications. However, that is not so with the three life-forms lower down, which we call *Kernel*, *Metals*, and *Minerals*. In *In Search of the Miraculous*, Gurdjieff chooses not to give a name to the lowest square* because we never come in contact with it directly. For the sake of having a label, we chose to name it "*Kernel*," because it is the seed of planets and suns. *Kernel* contains an inverted triangle indicating that this class of life-forms is adjacent to the Absolute in its form as The Holy Firm.

We find *Man* in the fifth square from the top, with the four higher squares referring to higher life-forms than we are probably unfamiliar with. Gurdjieff chose to name the next two squares "*Angels*" and "*Archangels*."

The top two squares have a different visual character. Gurdjieff labelled them "*Eternal Unchanging*" and "*Absolute*." We changed this slightly, because the symbol it contains (the circle within a triangle surrounded by a circle) signifies the *Sun Absolute*, so we choose to label it that way. The reason for this will become apparent when we relate the Step Diagram to the Ray of Creation.

For neither of the classes of these top two squares is there any class of being that feeds on it. The *Hydrogen* shown in the lower right hand corner of each square designates the class of being that it feeds on and the other *Hydrogen*, what the being is. Thus *Sun Absolute* feeds on *Archangels* and *Eternal Unchanging* feeds on *Angels*.*

It is impossible to go much further with this diagram without providing an idea of what Gurdjieff means when he states that a particular class of life-forms is a specific *Hydrogen*. Gurdjieff is assigning a *Hydrogen* to be the center of gravity of that class of being in some way. Our view is that the chosen *Hydrogen* represents the psychic life of that particular class of being.

So with *Man*, as a class of being, he is H24. However, H24 is the average *Hydrogen* of the three *Hydrogens* of man's psyche: H12, H24, H48. We can extend this scheme to other squares in the Step Diagram. However, before we

---
* *In Search of the Miraculous* by P D Ouspensky, p323
** Note that from here onwards, we italicize the names of the Step Diagram squares, as follows: *Kernel*, *Metals*, *Minerals*, *Plants*, *Invertebrates*, *Vertebrates*, *Man*, etc.

proceed any further in this direction, it is important to discuss the form of a cosmos.

The Enneagram of a Cosmos

In *In Search of the Miraculous*, Gurdjieff describes a symbol called the enneagram, a circle divided by nine points, as illustrated in *Figure 48*. He states that* the symbol is a perfect synthesis of the Law of Three and the Law of Seven, containing all the particularities of the laws it represents. He notes that every completed process, specifically every cosmos and every organism, can be described by an enneagram.

However, not every one of these enneagrams has an inner triangle. The inner triangle denotes the presence of higher *Hydrogens* in the organism it represents. There is such a triangle in the enneagrams of plants that impact the psyche of man. Examples include hemp, poppy, tea, coffee, tobacco, and so on—plants which Gurdjieff refers to as "Polormedekhtic" in *The Tales*.**

The enneagram is remarkable for those who are capable of using it. All knowledge can be represented in it and be interpreted using it. What a man understands he can put on the enneagram. What he cannot put on the enneagram, he does not understand. For a man fully able to make use of it, libraries and books are unnecessary. Such a man can learn something new simply by using it.

The Milky Way, man, and microbe are cosmoses of dramatically different scale, but all are created to the same pattern, which an enneagram can represent. Within these cosmoses, substances circulate to enable all the functions and processes of the cosmos. These substances are ingested (in some way) as food, air, and impressions, and unused residues are excreted (in some way).

Figure 48 shows the enneagram of man, with the three octaves of food, breath, and impressions marked around the circumference according to their position in their octaves, with the numbers indicating their *Hydrogen*. Thus $mi48$ indicates a substance at the level of H48 sounding the note *mi* in the octave of breath. A substance ascends in its octave when it reaches a specific location where it can combine with two corresponding *Hydrogens* so that a transformation happens by the Law of Three.

Thus, at the commencement of the food octave, H768 (food) strikes the note *do* and meets with H192 (hydrochloric acid and water) in the stomach, ac-

---

* In Search of the Miraculous by P D Ouspensky, p293
** The Tales, Chapter IX, The Cause of the Genesis of the Moon, p86

The Diagram of Everything Living

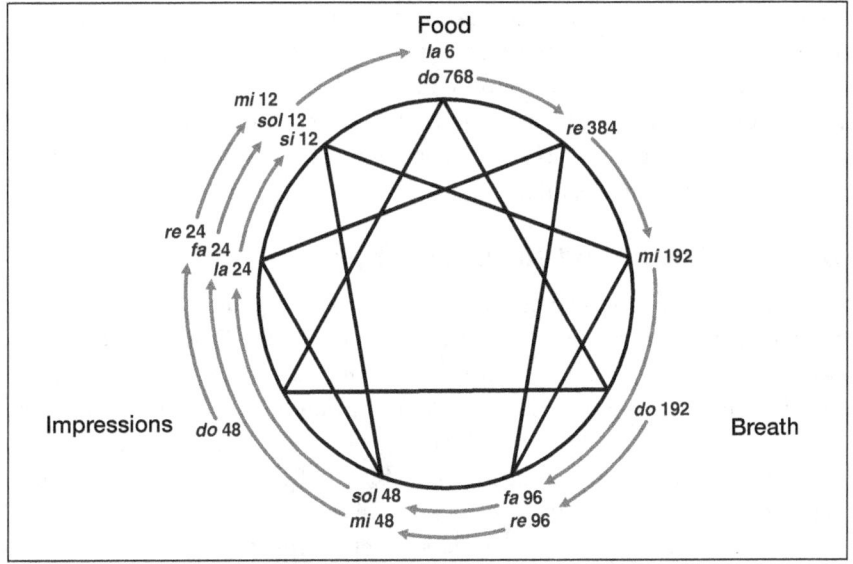

Figure 48. Food Octaves on The Enneagram

companied by digestive enzymes (H384 - Amylase, Lingual Lipase, Gastric Lipase, Kallikrein, and Pepsin). Thus, the H768 is transformed to H384, which we usually refer to as "chyme." This is a simplification. H768 consists of three types of food: carbohydrates, fats, and proteins. Each is transformed in separate applications of the Law of Three.

These substances are broken down as much as possible and liquified by chewing and by the action of the stomach, which also destroys any bacteria that accompany the food. Thus many individual transformations are occurring in the mouth, esophagus, and stomach. The body provides the structures to enable it: lips, mouth, teeth, tongue, esophagus, and stomach, all to exactly the right scale.

If we analyze the subsequent transformations in this octave and the transformations in the other two octaves of breath and impressions, it becomes clear what a remarkable system this is. It includes the means to transport substances from one place to another (the alimentary canal, the lymph system, veins and arteries, nervous system, and so on). It has organs that furnish the right *Hydrogens* in the right quantities at the right places for all these octaves to proceed. Add the need to discard substances that are not required, and you have some perspective about the extraordinary complexity of the systems that work together to enable life.

And incidentally, this system needs to defend itself against pathogens, repair itself when damaged, and replace its vast population of cells on an ongoing basis. And with all that, we are only talking about the cosmos being able to exist. There is also a set of systems to enable movement and the acquisition of physical skills in man's case. The acquisition of emotional and intellectual skills can be added to the mix.

The Step Diagram and The Law of Three

It is important now to distinguish between a cosmos feeding as just described and what Gurdjieff means when, in discussing the Step Diagram, he talks of classes of beings serving as food for or feeding on other classes of beings. Clearly, this is not the same kind of feeding. Thinking simply, when it comes to classes of beings, some *Invertebrates*, such as mosquitoes, feed on *Vertebrates* and *Man*. In fact, feeding on *Man*'s blood, they consume H_{192}, a much higher food than *Man* normally consumes. In general organic life gets its food where it can without caring which class of being provides it.

One way to approach the Step Diagram in respect of classes of cosmoses feeding on other classes is in terms of the Law of Three. In ascending, *Man* as H_{24} can only combine with H_6 (*Archangels*). So "serving as food for" simply means being passive force (O_{24}) in a triad where the active force (C_6) is *Archangels*. Similarly, if "feeds on" means active force, then *Man* (as C_{24}) must be active force in a triad where *Invertebrates* (O_{96}) are passive.

Consider *Figure 49*, which illustrates the Step Diagram differently, simply as a column of squares, showing the *Hydrogens* for each square—what it is, what it feeds on, and what feeds on it. We have shaded alternate squares to highlight the fact that there are two chains of squares. Tracking the white squares, from the bottom square: *Kernel* serves as food for *Minerals*, which serve as food for *Invertebrates*, which serve as food for *Man*, which serve as food for *Archangels*, which serve as food for the *Sun Absolute*. These squares do not interact with the shaded squares, which show the following chain: *Metals* serve as food for *Plants*, which serve as food for *Vertebrates*, which serve as food for *Angels*.

On the right side of *Figure 49*, we illustrate those two distinct chains differently. At the top, labelled "body of reason," we show just the white squares, and for each, we show the three "*Hydrogens* of the psyche." Thus for *Man*, we show H_{12} (the substance of Reason in the higher parts of centers), H_{24} (the fuel of feeling in the middle parts of centers), and H_{48} (mechanical energy in the lower parts of centers). When the centers are working in harmony, these three *Hydrogens* form *Man*'s mental body, the third of the four possible bodies of

The Diagram of Everything Living

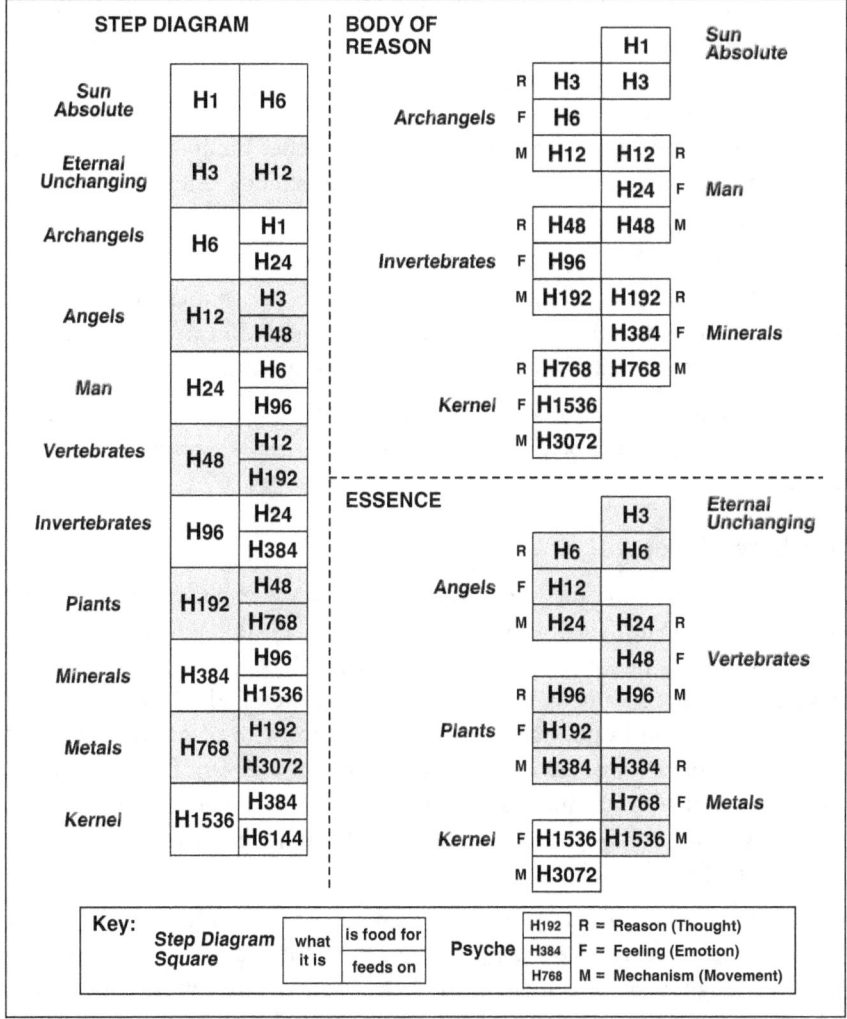

Figure 49. The Step Diagram: Alternate Squares

Man,* and the highest point to which a man can ascend within the solar system.

In the diagram, we group the *Hydrogens* of the psyche for each of the squares in the Step Diagram together. We line them up so that, for example for *Man*, H12 (Reason in *Man*) lines up with the H12 (Mechanism in *Archangels*). By doing this, we can envisage this whole chain linking together all its participants in a single living body. *Archangels* are its psyche, *Man* its nervous

---
* In Search of the Miraculous by P D Ouspensky, p93-94

system, *Invertebrates* its breathing system, *Minerals* are its digestive system, and *Kernel* provides its bones—its foundation. The *Sun Absolute* is the highest part of this being its "I." *Archangels* represent the possibilities of its reason, fed by *Man*'s thoughts and impressions. *Invertebrates* are its breathing system and its immune system. Their highest manifestation is their ability to create images (imagination). *Minerals*, as soil and as rock, digest and excrete. *Kernel* is the foundation.

We group the shaded areas under the title of "essence." Its highest part is the *Eternal Unchanging*, a galaxy. *Angels* are its psyche. *Vertebrates* are its nervous system. *Plants* are its breathing and immune system. *Metals* are its digestive system. *Kernel* is also its foundation but in a slightly different way.

We can think of these as two composite symbiotic beings composed of the classes of beings in the Step Diagram. The first exhibits Reason, the second, Beauty.

The Lateral Octave

We can now examine the Step Diagram from the perspective of the lateral octave from the Sun. The normal view of our planet is that it has a solid rocky crust, two-thirds of which is covered by water. Much of the land and sea is covered by soil or sediment, which is alive with bacteria. Plant life is anchored in the layer of soil in both cases. Seawater is alive with plankton, small and microscopic organisms. There is a large population of *Invertebrates*, including mollusks, earthworms, insects, shellfish, etc. Then come the *Vertebrates*, including fish, amphibians, reptiles, birds, and mammals. Surmounting it all is *Man*, king of the food chain.

The Step Diagram paints a picture where biological life on Earth fills out only four of its squares. All the squares to the left of them are alive despite being usually thought of as inert matter. And to the right are the invisible life-forms, which Ouspensky associated with Planets (*Angels*) and the Sun (*Archangels*).

If we consider the Ray of Creation from the Sun to the Moon, as illustrated in *Figure 50*, it becomes clear that the Step Diagram is related to the lateral octave from the Sun in some way. Gurdjieff stated that organic life on Earth comprises the notes *la, sol, fa* of this lateral octave,[*] and thus they fill the *mi-fa* interval in the Ray of Creation, between the Earth and the family of Planets, allowing the influence of the Planets to reach Earth.

---

[*] In Search of the Miraculous by P D Ouspensky, p139

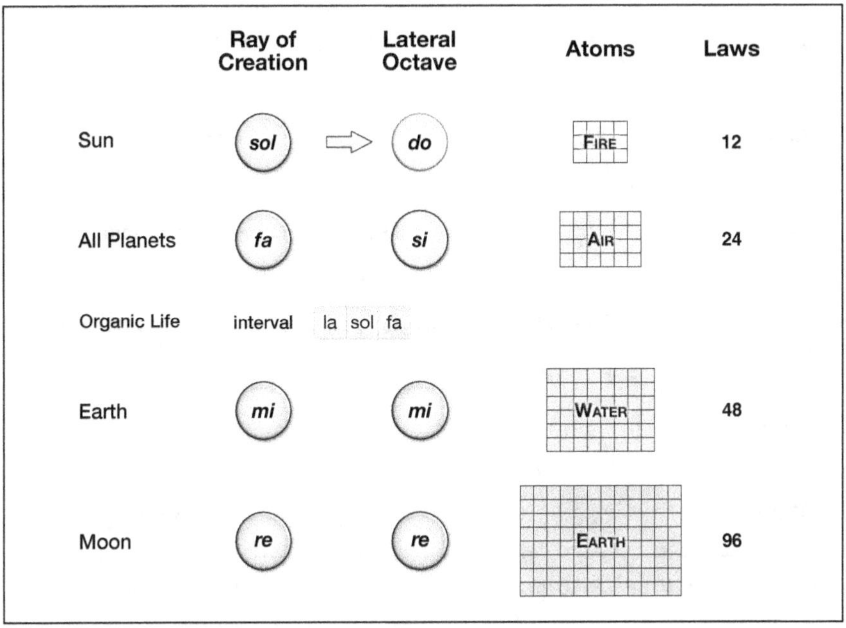

The Diagram of Everything Living

	Ray of Creation	Lateral Octave	Atoms	Laws
Sun	sol ⇒	do	Fire	12
All Planets	fa	si	Air	24
Organic Life	interval	la sol fa		
Earth	mi	mi	Water	48
Moon	re	re	Earth	96

Figure 50. The Lateral Octave From The Sun

To have a concept of what that might mean, we need to consider the Earth and the Planets as living beings. From that perspective, the wide variety of organic life that constitutes Nature is trillions upon trillions of cells in the body of the Earth. And when some species appears out of control—for example, clouds of locusts destroying massive areas of vegetation, or sub-Saharan tribes and their grazing animals extending the Sahara south through the over-grazing of vegetation—perhaps we can think of those things as diseases in the body of the Earth. However, as we do not know the Earth's life processes, those apparently destructive developments could just as easily be a necessary cure for a disease that we do not yet understand.

It is reasonable to conclude that the Earth, as a living being, is composed at least in part, of all the life-forms of the Step Diagram up to the level of *Archangels*.

If we consider ourselves in a similar light, while we may think of ourselves as individuals, each of us is an aggregation of trillions of cells. The majority of those cells, mainly bacteria, are symbiotes and do not even carry our DNA. We have little knowledge of the life of those cells. Mild diseases come and go within our body over the years and, unless a disease is severe, we may not even notice.

Gurdjieff's Hydrogens: The Ray of Creation

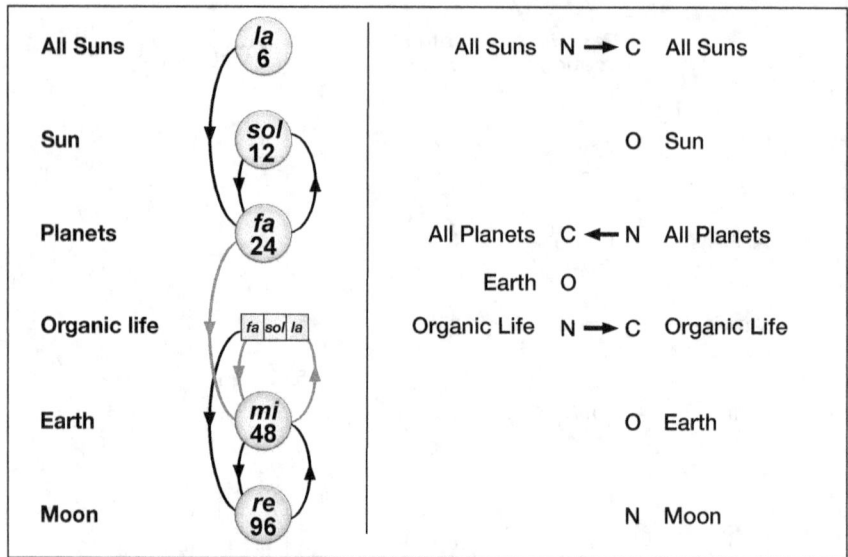

Figure 51. Organic Life on Earth and the mi-fa Interval

Figure 51 shows the filling of the interval from the perspective of the Law of Three. The Planets are the active force, the Earth is passive force, and organic life is neutralizing. It is important to recognize here that it is not all the classes of the Step Diagram that fill the interval; it is just organic life. It is also important to realize that organic life is the result of this triad. Thus before organic life appeared, there must have been something on Earth to carry the neutralizing force for organic life to begin to happen. Most likely, this was bacteria.

We should not conclude that it is only organic life that can fill this interval between the family of Planets and one of its members. Organic life is how it is achieved for Earth. The evidence suggests that it is achieved in different ways for Venus, Jupiter, Mars, and the other planets where organic life is absent. If so, other planets will still be host to *Kernel, Metals, Minerals, Angels,* and *Archangels*. Similarly, it must be the case that the moons of other planets do not, in their current stage of gestation, require to be fed by some form of organic life.

Earth, Water, Air and Fire

Let's turn our attention now to atoms and laws. In *Figure 52*, we show the nine lower squares of the Step Diagram. We can take the diagram's stepped nature to reflect the four elements: Earth, Water, Air, and Fire.

The Diagram of Everything Living

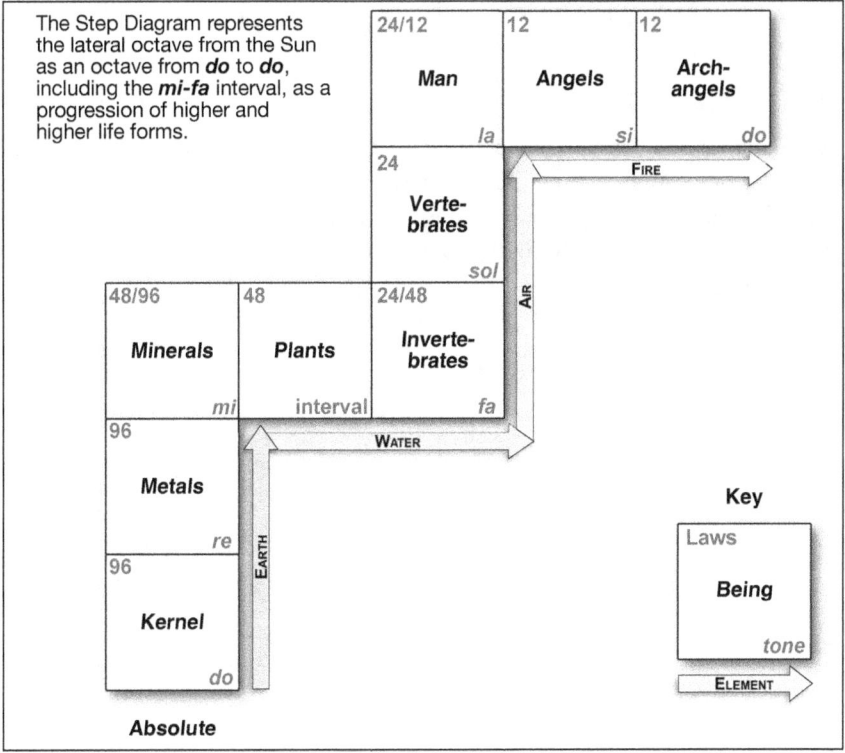

Figure 52. Earth, Water, Air, Fire

The initial vertical rise (*Kernel-Metals-Minerals*) is Earth (solid matter), and thus it comprises life-forms subject to 96 laws, at the level of the Moon. The horizontal triple (*Minerals-Plants-Invertebrates*) is Water and implies life-forms under 48 laws at our planet's level. The second vertical rise (*Invertebrates-Vertebrates-Man*) is Air, under 24 laws at the level of the Planets. And finally, the upper horizontal triple (*Man-Angels-Archangels*) is Fire under 12 laws at the level of the Sun.

Rocks are obviously constrained by more laws than plants—laws about growth, circulation, feeding, breathing, and perceiving. *Plants* being immobile, are constrained in motion and how they can eat, breathe and perceive. *Invertebrates*, *Vertebrates*, and *Man*, similarly have their own constraints. The assumption must be that *Angels* and *Archangels*, whatever they may be, are less physically constrained than *Man* and feed in a different way.

In the diagram, we have assigned the notes of the lateral octave from the Sun to each square. In doing so, we have assumed that *Plants* represents the *mi-fa* interval in that lateral octave. This is the point at which biological life enters

Gurdjieff's Hydrogens: The Ray of Creation

the octave. Initially, biological life in the form of bacteria fills the *mi-fa* interval in this octave.

Further Exploration

This chapter has provided us with a structure that allows us to examine the classes of being that occupy the nine lower squares of the Step Diagram. We have a framework. In the four chapters that follow, we discuss each of these four realms one by one, discussing possibilities and painting the picture with data from many sources.

Chapter 6

The Domain of Earth

"The path to paradise begins in hell."

~ *Dante Alighieri*

We now discuss the Step Diagram as a set of triplets, the first of which–the world of EARTH–comprises the diagram's three lower squares, as shown in *Figure 53*.

We are unused to considering anything in this realm as alive. Nevertheless, we can assume that any kind of activity, from chemical reactions at a microscopic level to eruptions and earthquakes, is a sign of life. Here we are in the domain of geology, and we can thus make hay with data that geologists have gathered.

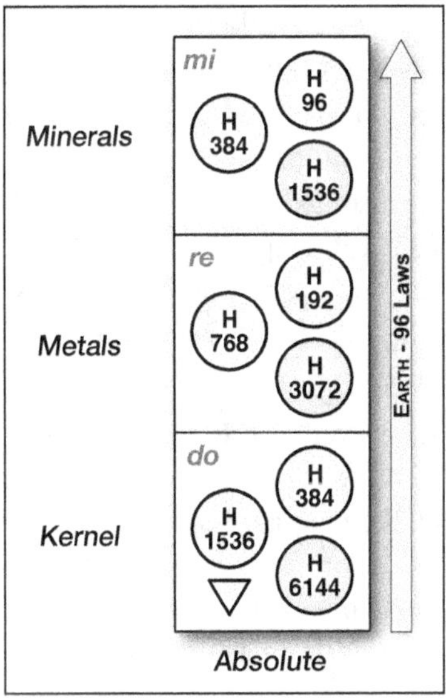

The average density of our planet is estimated to be 5.515 g/cm3, while the average density of the Earth's crust at the surface is far less, at around 3.0 g/cm3. Geologists thus conclude that the Earth's core is densely compressed material, possibly rich in heavier elements.

Figure 53. The EARTH Triplet

It is reasonable to assume it is hot due to the compression if nothing else. Also the decay of radioactive material will generate heat and it is likely that radioactive matter, which is generally denser, gravitates towards the center. However. Although geologists happily calculate and debate underground temperature gradients—there is very little direct evidence of what the Earth's core is comprised. Wikipedia offers a figure of 7000°K for Earth's center which means a temperature gradient above 1000C per 100 km depth.

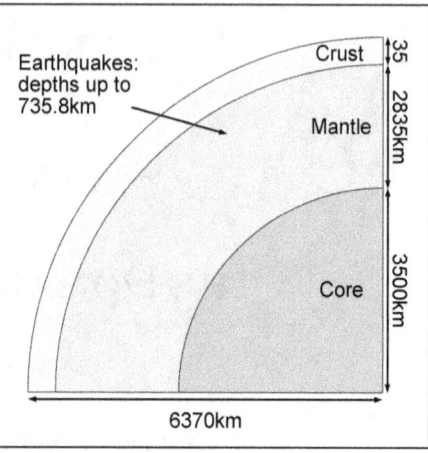

Figure 54. The Earth's Core

The deepest borehole ever drilled reached a depth of 12.3km. It was a Russian scientific endeavor, the Kola Superdeep Borehole project which took place on the Kola Peninsula near the Norwegian border. The borehole was 23cm in diameter. At those depths below 12km, the drillers encountered high temperatures (180+°C), greater than expected porosity and an unexpected decrease in density. The combination of conditions made further drilling beyond 12.3km unfeasible. A German project, the KTB superdeep borehole at Windischeschenbach in northern Bavaria, drilled to a depth of 9.1km, encountering temperatures of 260+°C (500+°F).

Figure 54 illustrates a very simple geological model of the believed internal structure of the Earth. The center is a solid core, believed for no obvious reason to be mostly iron—the core's outer part is supposedly liquid iron. This is surrounded by the magma of the mantle, which is believed to be about 2835km deep. Above that lies the thin solid crust, no deeper than 35km and only 8km deep in some places. The crust is broken up into tectonic plates, which "ride" on top of the mantle and rub up against each other. Earthquakes occur primarily at the points where these plates meet.

However, some earthquakes (called deep-focus earthquakes) can occur at very low depths. For example, in 2013, beneath the Okhotsk Sea, an earthquake of magnitude 8.3 occurred at a depth of 609 km. The deepest earthquake on record occurred beneath Vanuatu at a depth of 735.8 km in 2004. This implies that at those depths, in those places, two very solid masses of rock were pushing against each other.

The Cycle of Rock

Figure 55 illustrates what geologists call the cycle of rock, the movement of rock of various types over long periods of time. The right hand side of the di-

The Domain of Earth

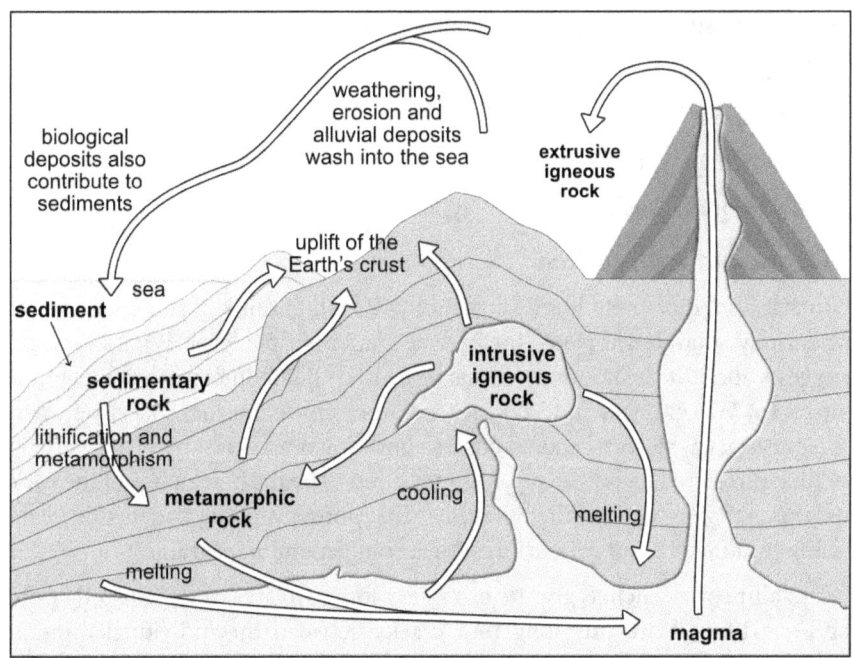

Figure 55. The EARTH's Cycle of Rock

agram shows magma rising to the Earth's surface through volcanic activity. Over time it deposits layers of igneous rock in lava flows. As also illustrated, magma can make its way upwards but fail to reach the surface. When it does so and cools sufficiently, it forms a volume of what is called intrusive igneous rock. This will be surrounded by metamorphic rock or, occasionally, sedimentary rock.

Through weathering and erosion, layers of rock of every kind on the Earth's surface are gradually worn away. Rivers wash a good deal of this into the sea, where it forms sediment. There are also biological deposits, mainly bone and shells, that gather in the sediment. Layers of sediment are compacted over thousands of years and eventually become rock through a process called lithification. As the sedimentary layers sink deeper, they come under higher pressure, and the temperature rises. Eventually, a different process called metamorphism comes into play. This changes the nature of the rock. Because of the uplift of the Earth's crust by tectonic plates, some sedimentary rock and metamorphic rock will rise to the Earth's surface. Other metamorphic rock, the lower layers, is transformed into magma by melting.

In outline, that is the slow but continual interaction between the top of the mantle and the Earth's crust.

We can classify rock into four distinct categories:

- Magma
- Igneous rock
- Sedimentary rock
- Metamorphic rock

Magma and Igneous Rock

Rock is a solid or crystallized substance. We tend to think of stones and rocks as entirely solid, but it is not the case at a microscopic level. What Gurdjieff suggests about heavier atoms (atoms of EARTH) being interspersed by lighter atoms (of WATER, AIR, and FIRE) is true. Rocks have internal spaces between their crystals, which can and do contain both water and air. This characteristic of rock is referred to as porosity. It is measured to be the percentage of the rock or stone that is not solid/crystalline. The porosity of rock depends on a number of factors, but primarily on the arrangement of its grains.

Crystalline rock such as granite has a very low porosity (<1%). The only pore spaces within it are tiny long thin cracks between the individual mineral grains. By contrast, sandstones have very high porosity in the range of 10–35% because the individual sand or mineral grains don't fit together well. It is more of a collection of crystals than a crystalline mass.

Magma is by definition molten rock and hence no longer crystalline when below ground. It exists as lava "momentarily" during volcanic eruptions but soon cools to form a solid crystalline mass. Geologists suggest that about 20% of the Earth's mantle is magma.

Because temperature increases with depth, it is reasonable to expect that substances at greater depths and under great pressure will be magma of some kind. Geologists categorize magma by analyzing the igneous rock it forms when it cools. They classify magmas according to the percentage of silica (SiO_2) they contain, the proportions of Iron and Magnesium, the temperature (it ranges between 900 and 1500°C), and the viscosity.

Different magmas occur in different locations giving rise to different geological formations. It is possible that there are types of magma that never get to the surface.

Igneous rock, magma that has cooled, is rich in silicon and usually includes amounts of oxygen, iron, magnesium, calcium, sodium, and potassium. The dominance of silicon, usually as silica (silicon dioxide), is notable, varying between 44% to 69%. If we think of biological life as carbon-based, then perhaps rock life-forms are silicon-based. Silicon sits one row down in the periodic table and has the same number of electrons missing from its outer orbit as carbon.

Igneous rock can form in different ways. It can form in a granular, crystalline manner, or, without crystallization, it forms natural glasses. On the surface, such rock is found in a wide range of geological settings, not just in lava flows. The majority of surface igneous rock is intrusive igneous rock formed beneath the surface that later became exposed. Because intrusive igneous rock cools very slowly, the rock is coarse-grained, as opposed to the fine grained igneous rock of lava flows. The central cores of whole mountain ranges consist of intrusive igneous rocks—usually granite.

The continental crust (the Earth's landmasses) is composed primarily of sedimentary rocks, resting on a crystalline foundation of metamorphic and intrusive igneous rocks. The metamorphic and igneous rock rises to the surface by a combination of erosion and uplift.

The uplift process has a more profound impact than it may initially seem. For example, peridotite (a common type of igneous rock) will be hotter at greater depths. If it rises during convection within the mantle, it will cool slightly and expand. If it rises sufficiently, the pressure on it reduces, and it can begin to melt. The melt droplets may accumulate into larger volumes and be intruded further upward. The point is that, although such processes are very slow in human terms, convection is a constant process within the Earth's crust and causes change.

Earth's surface on land comprises only about 15% igneous rock, while the surface beneath the ocean is almost all igneous, covered by a thin layer of sediment. Decompression melting gives rise to the ocean crust at mid-ocean ridges, which push the continents apart. Volcanism also makes a contribution where tectonic plates meet beneath the ocean. This may be the process by which the Earth's volume increases.

Other known influences on rock behavior occur from the action of carbon dioxide and water. The presence of carbon dioxide reduces the temperature at which rocks begin to liquify (called the solidus temperature). Water does too and has a much greater impact. For example, peridotite begins to melt near 800°C at a depth of about 100 kilometers in the presence of sufficient water. In the absence of water it will not melt until about 1,500°C. Thus the ocean has a distinctly different impact on the rocks beneath it than the land surface. It also preserves the rock from a great deal of erosion from trees, ice and wind.

Sedimentary Rock.

Figure 55, the Cycle of Rock, is simplistic in respect of sedimentary rock. This is compacted rock composed from deposits of small mineral or organic particles. Once sufficient pressure is applied, the particles fuse to create rock.

Particles may emerge from different sources, dust blown on the wind, erosion by wind and rain, shells reduced to particles of sand by waves on a beach, mineral precipitation in water, and, of course, dust ejected from volcanoes. Such sediment gathers at the bottom of the sea and the bottom of lakes, on river beds, or even on plains.

Some rock is biological in origin. Limestone is formed from the bones of organisms, particularly corals and mollusks. Coal deposits are believed primarily to be the result of the compression of peat, which itself is composed of dead plant material. Peat forms in marshlands, which may eventually become buried beneath the surface.

Oil is also believed to derive from organic life compressed and eventually reduced to hydrocarbons as layers form above it. Because it is liquid or gaseous, it naturally rises and forms oil deposits when it gets trapped beneath an impermeable rocky layer. Deposits of chert (a sedimentary rock) form from the accumulation of the siliceous skeletons of microscopic organisms.

Fossils are most commonly found in sedimentary rock because such rock forms at temperatures and pressures lower than igneous and metamorphic rock, and hence does not destroy fossil remnants. In normal circumstances, dead organisms are consumed completely by scavengers or bacteria. Fossils form in situations where such processes are constrained—when the sedimentation rate is high or where oxygen is in short supply, and hence bacterial activity is less.

As is visible in many photographs of cliffs—pictures of the Grand Canyon provide a good example—sedimentary rocks form in layers called beds or strata, which can be anywhere between centimeters and several meters thick. The rate at which sediment deposits differs according to location. In some tidal flats, top layer sediment can build-up by more than a meter in one day. On the deep ocean floor, it is more likely to be a few millimeters per year. Layers form when the nature of the sediment laid down suddenly changes in composition.

Metamorphic Rock

The process of metamorphism forms metamorphic rock. The process usually acts on sedimentary rock, but it can also act on igneous rock and mixtures of igneous and sedimentary rock. In general, metamorphism takes place at temperatures of 150°C and pressures of 100 megapascals or greater. Metamorphic rock is classified by texture and by chemical and mineral content. It may be formed simply by sinking deep enough beneath the surface to encounter the right temperatures and pressure. It can also form through tectonic processes. About 12% of the Earth's land surface is metamorphic rock.

Metamorphism can fundamentally change rock at the particle level. For example, small calcite crystals in sedimentary limestone and chalk are recrystallized into larger crystals in the metamorphic rock we know as marble. Metamorphosed sandstone recrystallizes the original quartz sand grains into very compact quartzite. The high temperatures cause the atoms and ions in solid crystals to migrate, thus reorganizing the crystals.

During metamorphism, chemical reactions can be provoked by heat, leading to the formation of other minerals. Chemicals from neighboring rocks can change the rock's chemical composition. Alternatively change may be provoked by water transporting chemicals from distant rock formations. Water will also conduct some chemicals away.

The chemical alteration of a rock by hydrothermal or other subterranean fluids is called metasomatism and can occur in different contexts. For example, it can happen as a direct interaction between igneous magma and sedimentary rock, with chemicals exchanged or introduced from one to the other.

By considering these various interactions, the transport of substances from one place to another bringing about fundamental changes in rock of various types, we may get a sense of the Earth's interior being alive in some way. Magma may be liquid at a high enough temperature and pressure, but the blood that brings nutrients to rock is water. And if there is water, it will contain gases—oxygen and carbon dioxide in particular. So this blood perhaps carries necessary gases for use where needed. This blood circulation may be slow, but it is ubiquitous beneath the ground.

We now consider the lower three squares of the Step Diagram as life-forms, one by one.

Kernel Life

When speaking about the *Kernel* square in the Step Diagram, Gurdjieff noted there was no simple way to name it because man never encounters it. However, he noted that the square comes in contact with the Absolute as "Holy the Firm." The small triangle at the bottom of this square indicates contact with the Absolute.

Gurdjieff's Hydrogens: The Ray of Creation

In *In Search of the Miraculous* Ouspensky reports a conversation with Gurdjieff about H6144.* This *Hydrogen* had piqued Ouspensky's curiosity because Gurdjieff never mentioned it when deducing the ladder of *Hydrogens*. Gurdjieff remarked that it was an incomplete *Hydrogen* "without the Holy Ghost."

The lowest *Hydrogen* Gurdjieff had mentioned was H3072, which is composed of C512, O1536, and N2048. To form a denser substance, N1024 must become C1024. However, there are no substances O3072 and N2048 with which to form a triad. Thus new *Hydrogens* cease at this point, and C1024 condenses into H6144.

Even though H6144 cannot descend any further, it is food for *Kernel* beings. It may not be able to descend, but it can ascend, and logically it must do so, or else the being represented by the *Kernel* square will not "feed." This situation repeats in the *Metals* square above this one. The H3072, which it feeds on, does not correspond to a class of beings.

We assign life-forms of this square to the region below the Earth's mantle, a sphere of radius approximately 3600km. That core is believed to be solid because it transmits transverse seismic waves, which anything molten or liquid would be unable to do. That is the total of empirical knowledge about it.

The Law of Falling

In discussing this lowest square, we believe it helpful to quote "The Law of Falling." In *The Tales*, Beelzebub relates:**

> "This cosmic law which he then discovered, St. Venoma himself formulated thus:
>
> "'Everything existing in the World falls to the bottom. And the bottom for any part of the Universe is its nearest "stability," and this said "stability" is the place or the point upon which all the lines of force arriving from all directions converge.
>
> "'The centers of all the suns and of all the planets of our Universe are just such points of "stability." They are the lowest points of those regions of space upon which forces from all directions of the given part of the Universe definitely tend and where they are concentrated. In these points, there is also concentrated the equilibrium which enables suns and planets to maintain their position.'

---

* *In Search of the Miraculous* by P D Ouspensky, p324
** *Beelzebub's Tales to His Grandson*, Chapter IV, p66

The Domain of Earth

> "In this formulation of his, Saint Venoma said further that everything when dropped into space, wherever it may be, tends to fall on one or another sun or on one or another planet, according to which sun or planet the given part of space belongs to, where the object is dropped, each sun or planet being for the given sphere the 'stability' or bottom."

This encourages us to equate the lowest square, "*Kernel*," with stability—indicating that it sounds the note *do* in the lateral octave that rises up to the Sun and the final *do* in the octave that descends from the Sun. Gurdjieff states rather mysteriously that "In these points, there is also concentrated the equilibrium which enables suns and planets to maintain their position." He is not simply describing the mass of a planet or sun, and he clearly does not suggest or imply that the equilibrium (or balance) of a planet within the solar system is due to gravity. He says that this is where "forces from all directions of the given part of the Universe definitely tend and where they are concentrated."

The Law of Falling applies everywhere, and thus the Earth is a "point of stability." Gurdjieff's choice of words is curious. He refers to the planets and suns as "the lowest points." This makes no sense as a mathematical reference—for where is the origin of measurement? "Lowest" must therefore refer to something else. Possibly it refers to the lowest substance (in respect of frequency of vibration) or the lowest class of being. More likely, it refers to the manifestation of "Holy the Firm."

The Forge

We are reasonably convinced that the *Kernel* acts as a forge for elements. It receives a multitude of influences from cosmic rays and particles that can penetrate deep into the Earth and also from Birkeland currents (streams of ions from the solar wind). All substances in this central sphere are under considerable pressure that may be impossible to replicate in a laboratory. The heavy *Hydrogens* H6144 that *Kernel* life "feeds on" can only be combinations of the heavy elements that can be found in the lowest two layers of the Periodic Table. It is also likely to be the case that *Hydrogens* H3072 are substances involving such elements. Below the element lead (Pb) on the sixth row of the table, all elements are radioactive.

"Radioactive" means that such elements are unstable on the surface of the Earth. It is possible that they are not unstable when subjected to the massive pressures deep below the surface of the Earth. While we cannot know what takes place there, in the lowest square of the Step Diagram, we suspect that it is a forge for elements that feed *Metals* life-forms, some of which are intended to rise to the surface of the planet.

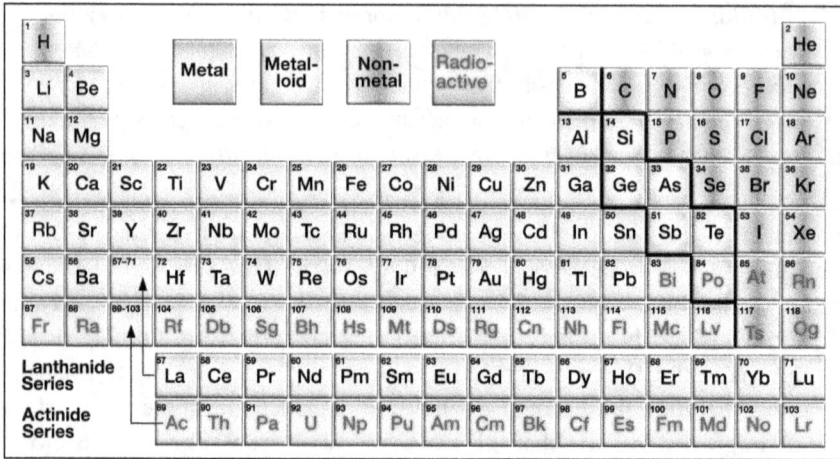

Figure 56. The Periodic Table

Metals Life

It is no easy matter to conceive of *Metals* as a category of being. We generally think of metals in the forms we encounter them; objects like knives, steel tools, gold jewelry, and so on. Man makes such objects from alloys, the ingredients of which are extracted from metallic ores. From our perspective, such metallic objects are akin to silk clothes or wooden cooking utensils—useful items created from material that was once part of a living thing.

Metals, in their natural unrefined state, are impure—they are ores that may contain many substances. Metals are usually oxidized or combined chemically with other elements. Even gold nuggets, almost an exception to the rule, are not pure gold. Most gold emerges to the surface as particles in quartz rock, which has to be pulverized to extract the gold.

Evidence suggests that man began mining metals about 4000 years ago. The Stone Age officially gave way to the Bronze Age in 3150 BC, by which time the use of metals was in full flow. The Iron Age followed in 1200 BC. Gold, silver, copper, iron, lead, and tin were mined and used as raw materials to manufacture useful items, mainly tools, and weapons.

The etymology of the word "metal" implies this. The Latin noun *metallum* meant "metal, mineral; mine, quarry," from the Greek *metallon*, which could mean "metal, ore" but was originally "mine, quarry-pit," probably a back-formation from *metalleuein* "to mine, to quarry."

In physics, from a theoretical perspective, a metal is regarded as any substance capable of conducting electricity at the theoretical temperature of ab-

The Domain of Earth

solute zero. It is not the only scientific definition. Commonly, a simple division of the periodic table, as shown in *Figure 56*, is used to sort the metallic elements from the metalloids and the non-metals.

However, we should not assume that the Step Diagram's *Metals* life-forms are comprised only of metals. As already noted, silicon is the dominant element in most rocks, and silicon is a metalloid, not a metal. Rock is thus a silicon-based being.

Other life-forms involve a wide range of substances of varying densities, so it seems logical to expect all life-forms to be that way. We find all the elements of the periodic table in rock, except for those that are unstable and never found in a natural state. Organic life makes no use of elements from the sixth and seventh row of the table.

A being is a complex system involving the organized interactions of many processes. If we examine man as an example, it is clear that this is so. A great deal of the activity that occurs within the human body depends upon the circulation of substances. Food circulates in a digestive process via the alimentary canal. When sufficiently digested food substances enter the bloodstream, fats (lipids) enter the lymphatic circulation. Air enters through the mouth and nose into the lungs and then the blood. Circulation of nervous impulses through the nervous system deliver impressions to the brain.

Scientists have assembled a wealth of information about the human body—far more than they have about any other life-form. Far less has been gathered about plants, which have distinctly different circulations and processes, but the data is rich even in that case. When it comes to rocks, the information is thin, and thus, in respect of trying to identify life processes, we can only theorize and make general statements.

Crystals

At the microscopic level, solid substances are arrays of crystals. Indeed, if there is something equivalent to a cell in the domain of EARTH, it is probably a crystal. By definition, crystals are a solid material whose constituents at a molecular level arrange themselves in a highly ordered structure, referred to as a crystal lattice. The lattice produces predictable geometric shapes. Because the lattice can extend in any direction, crystals can grow surprisingly large. For example, geologists have found gypsum crystals 12 meters long in caves in Naica, Mexico. Icebergs are crystals, and so are snowflakes.

Most inorganic solids are polycrystals—groups of fused microscopic crystals. Gemstones are crystals, some pure like diamond, and some polycrystals like opal. There are roughly 25 different kinds of gemstones, depending on how you classify them. Additionally, there are pearls, a highly valued variety of organic stones created by oysters create and less frequently by other mollusks. Fingernails and kidney stones are other examples of organically grown crystals. You can grow crystals of DNA. Scientists do that in the laboratory to study the structure of the molecule. There are also amorphous solids that are not crystals. Glass, wax, and many plastics are examples; they have no lattice.

Minerals Life

Because *Minerals* occupy a corner square in the Step Diagram, we discuss them from the perspective of the domain of EARTH here, and from the perspective of the domain of WATER in the next chapter. We repeat this approach for the other corner squares: *Invertebrates* and *Man*.

There are many different kinds of rock. Wikipedia lists 79 igneous rock varieties, 38 sedimentary rock varieties, and 28 metamorphic rock varieties. There are probably many more, including varieties that have never yet appeared on the Earth's surface. To this, we can add products made by man, from pottery to bricks and concrete.

If there is even a very rough parallel between the domain of EARTH and organic life, we should expect there to be many "species" of rock. Perhaps we ought to regard stones, from monoliths to pebbles, as life-forms of a kind. We only seem to encounter such stones on the surface, created by erosion. Beneath the surface, although there are definite layers, rock appears in large blocks and often appears to be continuous, if the sides of mountains are reliable indicators.

Earthquakes fracture rock. Perhaps when they do so, they create mineral beings. The National Earthquake Information Center in Rockland, MD, reports about 12,000-14,000 earthquakes each year. If you include very mild earthquakes that people might not notice, the estimate is about 100,000 per year. Major earthquakes, greater than magnitude seven on the Richter Scale, happen more than once per month on average—most commonly in sparsely populated areas. Very large earthquakes of magnitude eight and higher occur roughly once a year.

The world of EARTH is in motion, but its speed is ponderously slow. Tectonic plates move somewhere in the range of 0.6 to 10 cm/yr laterally. In places where mountains are rising, the Sierra Nevada in California, for example, the gain in height is minuscule—1 to 2 millimeters per year.

The speed of metamorphism is probably also very slow indeed. It is difficult to know, as it may be impossible to examine in laboratory conditions.

Volcanoes

The fastest processes within the world of EARTH are volcanic eruptions. More than 500 volcanoes are deemed active, and a roughly equal number are considered dormant. The measurement of underground activity determines whether seismologists classify a volcano as active or dormant. If there is little trace, the volcano is considered dormant, and if there has been no eruption in the last 10,000 years, a volcano is generally considered extinct.

Of the 500 or so active volcanoes, around ten are likely to be erupting on any given day. Such eruptions are usually small. Lava will usually flow at around 0.25 miles per hour unless gravity is on its side when it can achieve speeds of up to 40 mph.

The volcano naturally presents itself as a possible example of life in the domain of EARTH. The Institute of Geochemistry and Petrology of ETH Zurich created a computer-simulated subterranean volcanic behavior model in 2015. It provides the following picture.

A liquid magma chamber sits beneath the volcano. As the magma at the bottom of the chamber cools, it crystallizes into granite of some kind. Cooling in such an environment can occur from the intrusion of relatively cold groundwater. When this happens, metal-rich and salt-rich aqueous solutions are forced out of the chamber. They push their way upwards through volcanic vents of solidified igneous rock that has cracked. Once a particular height is attained, metals precipitate from the ascending fluids, creating ore deposits in a kind of veined area that will gradually solidify.

A cylindrical ascent zone is established above the magma chamber, which is dominated by these magmatic fluids. This zone merges into a cooler area where surface water circulates, and the rock is brittle. So the pressure and temperature change abruptly and drop dramatically over a height of about 200 meters. What began as a mechanical sieve (vents and brittle rock) becomes a chemical sieve, with the ascending fluid letting go of its metals.

None of this happens quickly—it is estimated to take 50,000 years. Although it is only a model, it has the virtue that the volcanic system behaves as an organized system, and metal ore accumulates as it appears to do in reality. If that's how volcanoes can behave internally, then maybe they are living beings of a kind.

Gurdjieff's Hydrogens: The Ray of Creation

Metamorphic Rock: A Feeding Process

It seems reasonable to classify the *Metals* square in the Step Diagram as comprising magma, igneous rock, and metamorphic rock. Metamorphism looks very much like a feeding process. Sedimentary rock, which we can classify as belonging to the *Minerals* square, can contain many different substances. While it can contain particles of igneous rock through erosion, it also contains the residue of all the layers above it: biological remnants that were not tidied up by scavengers and specialist bacteria.

Over time, a great deal of time, those substances sink deeper and deeper into the crust, within the layer of sedimentary rock to which they belong. Eventually, that layer of sedimentary rock encounters the temperatures and pressures where the process of metamorphism occurs. Then the constituents of the rock begin to be transformed. New minerals enter the equation, and different crystals are formed.

The destiny of the metamorphic rock is to melt into magma eventually and be further transformed or to rise back to the surface. If it rises, further changes may occur to it internally due to the reduction in pressure and heat.

Minerals and Organic Life

The *Minerals* square straddles the domain of EARTH and the domain of WATER. For that reason, we discuss it both in this chapter and the next. From the perspective of chemical composition, silicon dominates minerals on the planet's surface. Various processes gradually erode igneous and metamorphic rock—chemical reactions, chemical leaching, temperature variation, running water, rain, and a whole spectrum of biological processes.

If we set aside man's use of minerals and metals to create artifacts and consider how organic life uses minerals, it soon becomes clear that organic life needs these chemicals. Organic life feeds on minerals. Of the 118 elements found in the Periodic Table, 39 are ingested by biological life-forms and directly used.

That, by the way, excludes the bacterial use of elements. There are some bacteria, for example, that accumulate uranium. We know so little about the world of bacteria that as yet unexamined bacteria may make use of many other elements.

The list of elements known to be used by biological life is hydrogen, carbon, oxygen, nitrogen, sulfur, phosphorus, iron, magnesium, arsenic, boron, bromine, cadmium, calcium, cerium, chlorine, chromium, cobalt, copper, fluorine, gallium, germanium, iodine, lanthanum, lithium, manganese, molybdenum, nickel, potassium, silicon, sodium, strontium, thallium, tin, titanium, tungsten, vanadium, yttrium, zinc and zirconium.

The first six elements on this list are the true building blocks of life, as they enable protein and DNA creation. Of these, only sulfur and phosphorus come from mineral sources. The next two listed, iron and magnesium, are critical to life. Iron is necessary for hemoglobin, which transports oxygen around animals' bodies, and magnesium is needed for chlorophyll, which makes photosynthesis possible.

It is interesting to note that, after silicon, the next two major constituents of igneous rock are iron and magnesium. Higher life-forms can elevate minerals by consuming them. Minerals can also descend as sediment and sink within the Earth's crust. We also note here that stone plays an important role in the life of man in many contexts. We will discuss that in the next chapter.

EARTH in Overview

The domain of EARTH is slow-moving, and therefore we suspect its life-forms are long-lived. Our ability to study it is constrained by the fact that this domain is almost entirely subterranean—we see only its outer layer. The processes of crystallization and decrystallization, chemical reactions and chemical migration, heat and pressure, earthquakes, and eruptions are all largely invisible to us. Even some of the mineral processes on the surface are difficult to observe. Its processes are ponderously slow, and, we presume, the lifetimes of its beings are very long indeed.

What are the life-forms here?

We can say very little about the *Kernel* square, the core of the Earth. Is it one living thing or a vast population of living things? Whatever it is, it must feed the mantle in some way, providing it perhaps with the elements it requires. Beneath it lies the Holy Firm, which we can only think of as a mirror. Whatever descends to the lowest level must, of necessity, rise back up again, with the blessing of the Holy Firm.

The *Kernel* acts like a subterranean forge that transmutes elements. Heavy elements descend here from the layer above, and it may be that a slow process of atomic fusion occurs, causing those substances to descend further. But at the same time, when they touch the bottom, they are reflected from the Holy Firm; they transmute into lighter elements and rise. This may be what the inverted triangle Gurdjieff drew in the *Kernel* square signifies.

We can only speculate about the life-forms at the level of *Metals*: crystals, small stones, pebbles, rocks, boulders, hills, glaciers and icebergs, volcanoes, mountains, tectonic plates. Much more investigation into geological processes may be required before a coherent picture emerges.

The life-forms of the *Minerals* square are equally hard to discern. This will come into focus in the next chapter.

The Properties of EARTH

Each of the domains of the Four Elements is characterized not just by how it lives but also by how energy manifests within it. Energy can travel by vibrations within the medium that contains it. In EARTH's case, energy within that medium is propagated force and heat. The propagated force comes from tectonic plate movements and explosive eruptions.

Heat manifests as the vibration of atoms within a substance. We refer to its movement through the medium as conductance.

Take, for example, an iron bar. Heat one end of it and then remove the heat source. After a period of time, the heat spreads by conduction through the iron bar, and the whole bar will have a uniform temperature. Theoretically, conduction occurred because the atoms at the heated end vibrated faster and passed some of their energy of vibration to neighboring atoms and so on.

If the medium through which heat passes is a poor heat conductor, such as igneous rock, the heat moves very slowly. A trip to a volcanic area will convince you of that. Lava remains warm for a very long time. For example, the Parque Nacional de Timanfaya sits on the lava flows of extensive eruptions in 1720, 1736, and 1842. Although about 180 years have passed, the ground is still hot at the surface, and a few meters below the surface, temperatures range from 400°C to 600°C.

We know very little of the speed of the domain of EARTH.

The world of EARTH is in motion, but its speed is ponderously slow, except during earthquakes and volcanic eruptions. Land rises slowly; its growth measures in millimeters per year. Tectonic plates move somewhere in the range of 0.6 to 10 cm/yr laterally. The weathering, sedimentation, and chemical change of stones and metals measures in millimeters per year or slower. The growth or accretion of bony, horny, or woody substances (substances of H1536) that inhabit the boundary of the realm is a little faster—measured in a few centimeters per year. Activity in the *Kernel* of EARTH may be slower than we can even imagine.

Chapter 7

The Domain of Water

"How yielding is water. Yet back it comes to wear away the stone."

~ Lao Tzu

Water is a remarkable substance. Modern civilization has tamed it to the point where cities with millions of residents are well supplied with water, even watering their lawns and gardens. So we forget how much we, and the rest of organic life, are dependent on it.

We only need to consider what we know about Mars. It may be that Mars once had water in abundance. Certainly, there is photographic evidence from the Mars Rover Curiosity that there was once a significant amount of water there. The presence of conglomerates and sandstones similar to those found on Earth is only explicable by the presence of water in times past. NASA is fairly certain that liquid water flows on Mars, intermittently. It may prove to be the case that bacteria are present in various places on Mars. But there is no evidence of life at a more sophisticated level. The lack of a water cycle easily explains this. There are many volcanoes on Mars. The 27km tall Olympus Mons is the largest known volcano in the solar system. However, there is no evidence of recent eruption, and while there appears to be evidence of tectonic plates on Mars, there seems to be no tectonic activity.

Compared to other planets in the solar system, Earth is a very watery world. On no other planets is there evidence of lakes, seas, or oceans—just the presence of some water vapor or ice. Curiously, Jupiter's moons, Europa, Ganymede, and Callisto, are watery worlds, as is Saturn's Enceladus.

If all the water on Earth formed a sphere, it would measure about 1,385 km in diameter. That's massive, the size of a small moon—1.3 trillion cubic kilometers in volume—and 95% of it is saltwater.

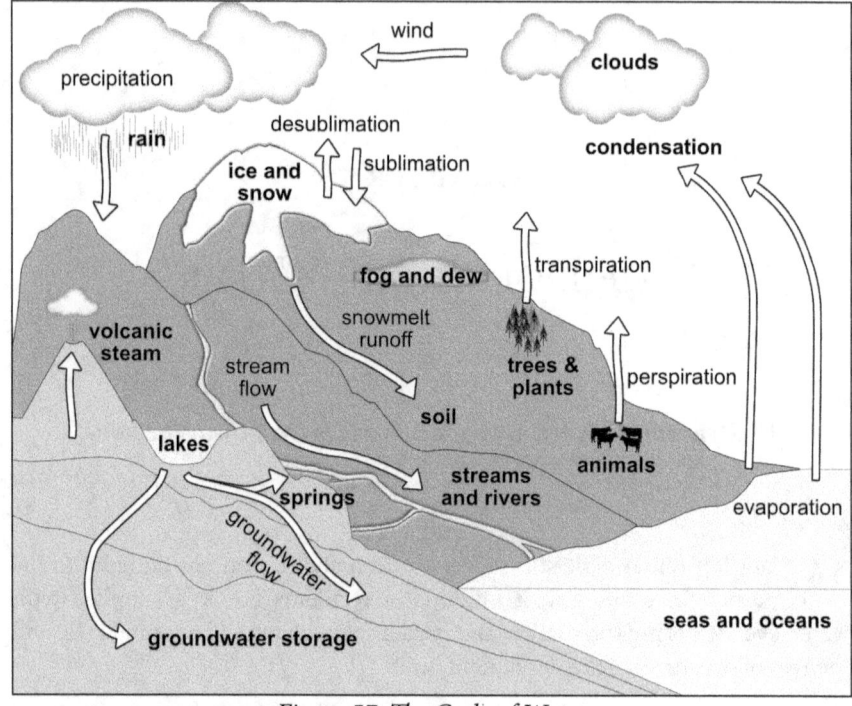

Figure 57. The Cycle of Water

The Earth's Water Cycle

Figure 57 above illustrates most of the processes in the cycle of water. Water constantly evaporates into the atmosphere from both land and sea. It also passes into the atmosphere from plants through the process of transpiration. Water passes from the roots to the leaves and then evaporates into the air. Animals usually drink it and pass some of it into the air by evaporation of water from the skin (perspiration) or breathed out (expiration).

As water rises through the atmosphere, it may form visible clouds. These mainly comprise ice (water crystals) or water droplets. Whether there are clouds or not, the air always contains water. The air is surprisingly wet. Low levels of humidity are uncommon. Technically, humidity is the measure of the capacity of air to absorb more water.

Thus at 100% humidity (approximately 3 grams of water per cubic meter of air), evaporation will not occur. Levels of humidity under 40% are considered low. Even desert air is not completely dry. For example, the average humidity of the Sahara desert is 25%. When the air temperature falls, water forms as dew on the ground; humidity may also become visible as fog and mist.

The Domain of Water

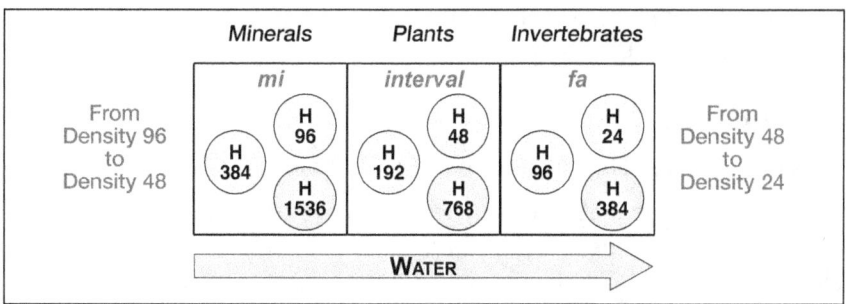

Figure 58. The WATER Triplet

Water descends from the clouds as rain, sleet, hail, or snow. Rain will get absorbed by the soil or run into streams and rivers, eventually making its way back into lakes or the sea. In very cold regions, snow falls and becomes compacted as ice. This may melt in warmer months to be absorbed by soil or to form streams. The surface of snow and ice can be subject to what is called "sublimation," where moisture in the air becomes ice directly, without ever becoming liquid, and also desublimation, where the ice transforms directly to moisture in the air.

As discussed in the previous chapter, water also makes its way into the Earth's crust. There, near the surface, it tends to flow through the courses it has eaten out over time, emerging in various places as springs. It may also form underground lakes and rivers that are not apparent on the surface. For example, Western Sahara sits on top of an underground river system that was discovered from space in 2015. Underground water also makes its way into volcano systems and may thus be returned to the atmosphere by volcanic action.

The Domain of Water

Within the WATER triplet, *Figure 58* above, we encounter life-forms that we recognize as life. According to the Step Diagram, *Minerals* life-forms are H384 (their being is WATER). These life-forms have a lower density than those from the domain of EARTH, which means that they are under fewer laws. We can get a sense of this through simple observation. *Plants* and *Invertebrates* have greater freedom of motion and seem to have greater freedom of decision. It is difficult to imagine a rock making a choice, but one can think of plants competing with other plants for a greater share of the available sunlight.

Once we enter the domain of WATER, we encounter the reality of evolution. We can think of organic evolution as intelligence (the intelligence of Planets) manifesting and mutating species of *Plants* and *Invertebrates*. We are indeed in a realm of fewer laws, and one of those orders of laws guides evolution.

Minerals Life

Erosion is a natural life process for *Metals* or *Minerals.*

Over time, rock on the Earth's surface is broken into fragments by a multitude of processes. When metamorphic or igneous rock rises to the surface, it may already be cracked in places, and tectonic movements may increase the extent of this. But once it reaches the land surface, it is subject to a whole series of "weathering" processes that break it down into smaller pieces. If it comes up beneath the sea, it experiences very little erosion.

Physical weathering

Physical weathering comprises processes that cause the disintegration of rocks without there being any chemical change. Such processes include:

- **Abrasion weathering:** This is the main physical weathering process. The main agents of this are water, ice, and wind loaded with sediment. Consider the fact that Niagara Falls retreat at the rate of three feet of rock per year, and you appreciate the power of water abrasion. In glacial areas, the moving masses of ice, loaded with sediment, grind down the rock beneath. Aeolian erosion (erosion by wind) is most visible in desert areas where sandstorms have eroded the lower layers of rock formations. The sand is denser close to the ground in a sand storm.

- **Thermal stress weathering:** This is the expansion and contraction of rock due to temperature changes—not just from sunlight but also forest fires. As some minerals expand more than others, temperature changes may cause rock to crack. This process can also lead to a peeling away of the warmer outer surface of rock. Such weathering can occur relatively fast in deserts where the temperature difference is high between day and night. The presence of moisture within the rock will accelerate the process.

- **Frost weathering:** There are several rock weathering processes involving ice. The expansion of ice in cracks within rock causes frost shattering. It can produce piles of fragments called "scree" that tend to settle at the foot of mountains. Frost-wedging is the result of the freeze-thaw cycle. Each time water freezes, it forces the rock apart, making space for more water in the next cycle to force the rock further apart. Capillary action is often a factor in bringing water into newly created spaces in the rock. Some rock types, chalk is an example, are particularly susceptible.

- **Pressure release weathering:** When overlying materials are removed, perhaps by erosion or in recent times by glacial melting, the pressure on the rocks beneath reduces. When this occurs with intrusive igneous rocks, the rocks' outer parts expand, setting up stresses that cause fractures parallel to the rock surface. Over time, sheets of rock break away from the exposed rocks along the fractures.

Chemical weathering

Chemical weathering refers to processes whereby atmospheric chemicals or biologically produced chemicals cause the breakdown of rocks or soils. Chemical weathering is most intense where the climate is wet and hot. Physical weathering plays a role in this, exposing more areas to chemical action, reducing particle size, and increasing the surface area. Chemical weathering includes:

- **Salt weathering:** This is similar in process to frost weathering. Saline solutions seep into cracks and spaces in rocks. The solution evaporates, leaving salt crystals behind. These expand with temperature, fracturing the rock. That can be just a physical effect. However, salt crystallization can also occur when chemical solutions decompose rocks (such as limestone and chalk). They form salt solutions of sodium sulfate or sodium carbonate, which fracture the rock when they crystalize. Some salts can expand by a factor of three.

- **Dissolution and carbonation weathering:** Rainfall dissolves a certain amount of carbon dioxide to produce weak carbonic acid. This affects rocks such as limestone and chalk that contain calcium carbonate, dissolving them to some degree. Nowadays, there's also acid rain from sulfur dioxide and nitrogen oxides, some from volcanic eruptions but mostly from fossil fuels. These acids also dissolve limestone. Some minerals, due to their natural solubility, oxidation propensity, or simply chemical instability in air and water, naturally dissolve.

- **Oxidation weathering:** The chemical oxidation of metals occurs when rock is exposed to the elements. The oxidation of iron in rocks commonly occurs, providing it with a reddish-brown coloration and weakening it, so it becomes crumbly. The same process of oxidation occurs in other mineral deposits, such as copper ore.

Biological Weathering

Plants feed on *Minerals*. In doing so, they cause weathering in both physical and chemical ways.

- **Root wedge weathering by plants:** Plant roots enter cracks in rocks and pry them apart. This could be small plants or tree roots. They seem to be capable of bringing hydraulic forces to bear. Lichens and mosses grow on bare rock surfaces to create a humid chemical microenvironment.
- **Chemical weathering by plants:** Frequently, both physical force and chemical capability is brought into play. The most common chemical process involves chelating compounds that free up the chemicals the plant needs. *Plants* produce such digestive juices anyway to break down compounds in the soils. They can as easily release them into rock and they do.

Man also has an extensive weathering effect on *Minerals*. It involves mining, quarrying, and manufacturing both bricks and cement. Bricks and cement are made from sandstone, shale, clay, limestone, flash, and aggregates. In such activity, man is introducing *Minerals* into his world, the anthroposphere. Think of Manhattan. Is it not an anthill made of stone?

Pebbles, stones, and dolmen are probably *Minerals* life-forms rather than substances. All minerals comprise a mixture of substances of EARTH, WATER, AIR, and FIRE at the microscopic level, and psychic substances may attach to them. Many of us who have spent time experiencing stone circles or old churches and cathedrals sense an atmosphere associated with the stones.

There are well-recorded apparitions that seem to attach to places. Thomas Charles Lethbridge,* an archaeologist who focused on paranormal research, hypothesized that apparitions are imprints on invisible fields around bodies of water, forests, and mountains. He speculated that these fields are recharged by ions in the air and could be accumulative. Places that are notorious for suicides might, as a result, be conducive to suicide.

Various theorists have tried to propose a mechanism by which such things occur, with little success. Such traces are plasma traces, and science does not yet have sufficient knowledge of plasma to carry out a physical analysis. The phenomenon of psychometry, where sensitives can deduce information about someone by holding an object the other has held onto, provides a clue to what this may be.

---

* Books by T C Lethbridge include *Ghost and Ghoul*, and *Ghost and Divining Rod*.

Gurdjieff taught that* when a man moves, particles of his atmosphere are sloughed off. They leave a trail, by which he can be traced for a while. They disperse but may also stay in one place and settle on his clothes and objects in his possession. Psychometrists interact with this material. Haunting and apparition phenomena are common, so the plasma explanation that some emotional trace is wired into stones or even trees is credible.

Soil

It seems reasonable to compare the process by which rock is broken down on the surface of the Earth to create soil with the way that food is broken down in the digestive process. A set of physical processes break the food/rock into smaller fragments that are easier to digest/weather. But we also have to apply a variety of chemical processes to transform the food/minerals. At the end of this process, in man's case, we get chyle—a mix of nutrients that can enter the bloodstream. In the case of rock, we get soil, on which the domain of *Plants* firmly sits.

Typically, surface soils are 45% minerals, 25% soil solution, 25% soil air, and 5% organic matter. The term "soil solution" is used here instead of "water" because it contains hundreds of dissolved mineral and organic materials. The term "soil air" indicates that its composition is different from atmospheric air—the amount of carbon dioxide is about ten times greater. Thus, earthworms breathe a different air to us and drink a different water to us.

The minerals in soil are generally a mix of sand, silt, and clay in a rough ratio of 2:2:1—a ratio that can vary considerably, of course. Sand is typically small fragments of quartz but can include grains of other minerals of a similar size. It is the coarsest of soil particles and useful structurally because it discourages compaction and helps water flow, but it has little food to offer to plant life. Silt, a combination of quartz and rock crystals, has smaller particles than sand and is richer in nutrients than sand. Clay is the richest in respect of soil minerals and the smallest in respect of particle size.

Soil is probably the highest form of mineral life. We can think of no reason not to regard soil as a silicon-based life-form, alive with and assisted by microbiological life.

Roughly a fifth of the organic matter that constitutes 5% of soil is microbiological life, mainly bacteria. It amounts to about a billion bacteria per gram of soil, of which we can probably identify just a small number. However, al-

---

* *Views from The Real World by G I Gurdjieff, lecture Essentuki, 1918.*

though we may not know the individual functions of many bacteria species, we can group them according to their roles, under the following designations:

- **Decomposer bacteria.** These bacteria's primary role is to decompose organic waste that enters the soil converting it into forms helpful to the soil ecosystem. Some participate in the nitrogen cycle by converting proteins and urea into ammonia. Some break down pesticides and pollutants. Some even provide storage for chemicals, particularly nitrogens, which are not immediately required by the ecosystem.

- **Mutualist bacteria.** The most important subgroup of bacteria in this category is the one involved in the nitrogen cycle, without which plants could not live. This includes nitrifying bacteria that convert ammonia to nitrite and nitrate salts and the nitrogen-fixing bacteria that directly turn atmospheric nitrogen into nitrogen compounds.

- **Deconstructive bacteria.** Some bacteria appear to be unhelpful to plants but are, in fact, helpful to the soil ecosystem. One example is the denitrifying bacteria that turn nitrogen compounds back into nitrogen gas—the exact opposite of nitrogen-fixing bacteria. They are anaerobic and thus thrive in situations where there is a low oxygen level in the soil.

- **Actinobacteria.** These can be terrestrial or aquatic, and they grow like fungi, forming long filaments that stretch through the soil. A common form of actinobacteria, Actinomycetes can decompose a wide array of substrates. They are particularly important in degrading hard-to-decompose compounds, such as chitin and cellulose. Fungi also do this but operate at low pH levels, whereas Actinomycetes operate better at high pH levels.

- **Lithotrophic bacteria.** The term "lithotrophic" means rock-eating, which is how these bacteria live. They acquire electrons (energy) from inorganic compounds enabling ATP synthesis via a process of oxidation. (We will discuss ATP in more depth later—all you need to know here is that it is the internal energy source of living cells.) There are a huge variety of these bacteria, and they are ubiquitous, to be found high in the atmosphere, in soil, in wastewater, seawater, and freshwater, as well as deep in mines, in volcanoes, and in deep-sea ocean vents.

- **Pathogenic bacteria.** In every environment, pathogens cause disease in some life-forms. The Erwinia bacterium genus, which is found in soil, contains many pathogenic species. Some infect woody plants, causing a blight on fruit and roses. Others infect herbaceous vines and gourds—

Agrobacterium species that cause tumors in plants (also known as gall formation).

- **Healthful bacteria.** The action of pathogenic bacteria is opposed by other bacteria that form an ad-hoc immune system for the soil and its ecosystem. Some strains of Pseudomonas fluorescens have anti-fungal properties that inhibit some plant pathogens. The Pseudomonas and Xanthomonas species increase plant growth. It is thought that this is a dual-action of discouraging pathogens and producing growth compounds.

There are seventeen elements essential for plant growth and reproduction. In addition to the obvious ones: hydrogen, carbon, oxygen, nitrogen, phosphorus, sulfur, and magnesium, plants also require potassium, calcium, iron, boron, manganese, copper, zinc, molybdenum, nickel, and chlorine. Aside from the first six, most metals need to be extracted from minerals in the soil.

Erosion

Aside from weathering, which breaks down rock in many ways, we also need to draw attention to erosion. Erosion refers to the processes that distribute rock, soil, or dissolved material from one location to another. The agents of erosion are water (surface runoff, streams, rivers), snow and ice (glaciers), air (wind), plus the biological ones, (plants, animals, and humans).

The action can be mechanical such as a river eroding rock or the river depositing eroded material into the sea. It can be chemical as, for example, water simply dissolving various salts. It happens most quickly with fast-moving water or dust blown on the wind. It happens with all catastrophic processes: hurricanes, landslides, tsunamis, floods, forest fires. It happens at coastlines where the sea erodes the land. *Plants* participate by providing food to animals or birds, which move small amounts of material from where they eat to where they deposit the residues.

Humans have been dramatic in their influence in recent decades, moving material hundreds or thousands of miles and farming thousands of square miles of land previously unfarmed. This is anomalous, of course. There were irrigation schemes and a good deal of building in previous eras, but nothing as dramatic as has occurred in the last two centuries.

Erosion is a natural process that has in recent years been interfered with unnaturally.

Plants Life

Figure 59 illustrates the *Plants* square of the Step Diagram. This square moves us from silicon life-forms to carbon life-forms, from what is conventionally deemed inanimate to what is indisputably animate.

It is also the place of the interval in the lateral octave from the Sun. Thus, it is reasonable to suggest that this dramatic shift in the class of creature occurs because this is where the interval occurs in the lateral octave. Something new appears that crosses the interval from *mi* to *fa*. And this new piece of the puzzle is not *Plants* per se, but cellular life. *Plants* are a consequence of cellular life. Under fewer laws (48) they are not constrained by the sluggish speed of *Minerals*.

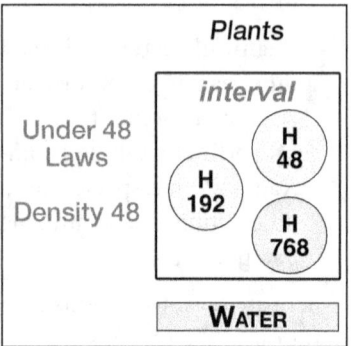

Figure 59. Plants

To think of them in respect of being under fewer laws is easy. They can interact with the Sun, they can breathe the air, and water can course through their capillaries. From the perspective of the domain of EARTH, this is freedom indeed.

Bacterial Origins

Contemporary scientists believe that the origin of life occurred on an Earth devoid of oxygen but rich in methane. The earliest life-forms were microbes (estimated at 3.5 billion years ago). They formed sticky undersea mats that bound sediment into layers called stromatolites. Scientists study living stromatolite reefs to try to understand the past. The next significant event on the time line was the evolution of cyanobacteria (dated to 2.4 billion years ago). These bacteria were capable of photosynthesis, creating oxygen in the gradual development of an oxygen-rich atmosphere. This period is named the Great Oxidation Event.

If that is indeed what happened, then the cyanobacteria are the ancestors of modern plant life. It not surprising that single-celled life evolved to produce larger and more varied life-forms.

Gurdjieff says the following in *The Tales*:

> "So, my boy, owing to all the aforesaid, there first arose on this planet Earth also, as there should, what are called 'Similitudes-of-the-Whole,' or

as they are also called 'Microcosmoses,' and further, there were formed from these 'Microcosmoses,' what are called Oduristelnian' and 'Polormedekhtic' vegetations.

"Still further, as also usually occurs, from the same 'Microcosmoses' there also began to be grouped various forms of what are called 'Tetartocosmoses' of all three brain-systems. *

Gurdjieff uses the term "Microcosmos" to refer to single-celled life. Thus these two paragraphs imply that such microcosmoses formed two types of vegetation, and these, in turn, formed one, two, and three-brained beings. The phrase "Similitudes-of-the-Whole," emphasizes the idea that the microcosmos and the megalocosmos are the same systems, just different in scale.

Over time, bacteria came to support all these life-forms that they preceded. They are found deep in the Earth's rocky crust scavenging for chemicals and even high in the atmosphere. They are abundant in soil, managing the soil environment for plant communities. They play a role in the life of every higher organic creature, primarily aiding the digestion of food.

When rock emerges to the Earth's surface, and fresh sediment is formed, bacteria populate it first, starting with bacteria capable of photosynthesis. They fix nitrogen and carbon, produce organic matter, and kick start the nitrogen cycle. When that is complete, the plants arrive.

Photosynthesis

If we adhere to the definitions of modern science, we will need to exclude some algae and plankton from this square in the Step Diagram and the whole world of fungi. However, we choose not to. We take the defining characteristic of life-forms in this square to be immobility. Some algae float and thus go where the water takes them—there are even some algae that float in the air. Fungi cannot shift location, but neither can they photosynthesize, feeding instead on other life-forms. The typical plant does not resemble these at all. Whether as tall as a redwood or as small as a violet, it has leaves and roots, reproduces by flower and seed, and feeds by photosynthesis.

Plants that inhabit the sea rather than land, algae, seaweed (which is a multicellular form of algae), and seagrasses live in shallow water and feed on the sunlight that penetrates the water.

Photosynthesis is endothermic (it absorbs heat). Chlorophyll makes it possible. This sophisticated organic compound comprises a single magnesium

---

* The Tales, Ch IX, The Cause of the Genesis of the Moon, p86 atom connected to a complex of hydrogen, carbon, nitrogen, and oxygen atoms. It participates in a process that converts carbon dioxide and water into sugars, exhaling oxygen in the process and thus renewing the atmosphere.

Remarkably, *Plants* are the only class of creatures capable of photosynthesis and thus the only ones that transmit the light of the Sun directly to our planet.

Morphology and anatomy

Despite their variety, plants have a relatively unsophisticated structure. They have roots, stems, leaves, flowers, and fruit. They reproduce either sexually via pollen and ovule or asexually via propagation (cloning) of roots, stems, or leaves. Asexual reproduction is the only possibility for plants that do not flower, but some plants can do either.

Many plants have a regeneration capability that higher life-forms lack. Take a cutting of such a plant and simply bury part of its stem and, if there are adequate water and food in the soil, it will grow a root system. Put it in a glass of water with a sprinkling of plant food, and you can watch the process. Similarly, when you amputate a plant's growing point, it simply develops another one or even several.

The circulation of water, nutrients, and enzymes is, of course, an up-and-down affair. If we examine the anatomy of wood, we observe that it consists of multiple different layers, as follows:

- **Bark.** The bark is the equivalent of skin and hair or fur, cells that are no longer alive but serve to protect the tree from the outside world. It insulates against cold and heat, wards off pests, and regulates moisture. As it wears away, it renews from within.
- **Phloem.** This is the inner side of bark and is alive. It is the pipeline, the vascular system through which the products of photosynthesis (sugars and hormones) pass both up and down. For example, sugars pass from the leaves of the tree to the roots. Phloem cells are short-lived but continually renewed. The dying outer part of it turns to cork and becomes part of the bark.
- **Xylem.** This is the other transport system within a tree or plant. It is unidirectional, taking water and soil nutrients from the roots to the rest of the plant, using a combination of capillary action and hydraulic pressure. It is complementary to the phloem in its circulation function and also in how it combines with the phloem to create pressure differentials.

The Domain of Water

- **Cambium.** This layer sits between the phloem and the xylem. It is the growing part of the trunk. Looking at a cross-section of a trunk, you see annular rings. The outer one is the most recent, and all the interior rings are no longer alive. The cambium responds to hormones that accompany sugars passing up or down through the phloem. The hormones stimulate growth in cells, both within the tree and in producing leaf buds and flowers.
- **Heartwood.** Heartwood comprises all the tree rings that are not alive. It is xylem that no longer has a function other than as a supporting pillar. It can be remarkably strong, depending on tree variety.

Among the remarkable things about plant life is that the tallest trees (there are sequoia over 350 feet (110 meters) tall in Redwood National Park in California) are capable of circulating water and nutrients (their lifeblood) to such great heights without any kind of pump. The mechanism used is capillary action within very small bore tubes in the xylem. Water evaporates from leaves, which naturally draws water up hundreds of feet from the roots creating a negative pressure in the ground, which draws water into the roots via osmosis. These trees consume about 150 gallons of water each day.

Plant Awareness

Gurdjieff taught* that plants have relations between themselves and also that *Man* relates to plants. Modern science has confirmed the first assertion. Regarding the second, it has long been known that some people have green fingers.

In recent years botanists have accepted that plants have a kind of awareness. They have no discernible nervous system, so the mechanism is unclear, but various plant behaviors indicate that it is so. They can detect touch. This is easily observable in the mimosa pudica plant, which responds to the slightest touch by quickly folding its leaves. The "mouth" of the Venus flytrap snaps shut when it senses an insect's touch—a magnetic field change is detected when this happens.

In addition to touch, plants can react to chemicals, infections, parasites, gravity, moisture, temperature, oxygen and carbon dioxide concentrations, sound, and obviously light. Clearly, plants have to respond to gravity so that roots work downwards and growth goes upwards. A mechanism for this is found in the root tip, where amyloplasts contain starch granules that fall in the direction of gravity. This activates secondary receptors, which pass the in-

---

* Views from The Real World by G I Gurdjieff, Prieuré lecture May 24, 1923.

formation on via hormones. One example of this is HY5, which is known to control rates of photosynthesis and travels with sugars through the phloem.

A plant's response to light is bound up with an inner circadian clock that enables time-based activity so that daily, seasonal and annual behavior is possible: when to produce flowers, when flowers should open, when to die back in the autumn, and so on. Plants also sense the quality of light, producing more photosynthetic pigments if the light is consistently poor. They detect when a cloud or a fellow plant shades them, and some species also know whether that fellow plant is related to them. They recognize their genetic kin and rearrange their bodies to avoid competing with siblings.

Plants have communal defense mechanisms. For example, wounded tomatoes produce methyl jasmonate. It diffuses through the air and neighboring plants that detect it prepare for an attack, perhaps by producing a chemical defense against insects or by producing something that will attract insect predators.

Plants can also deliver electrical signals. Some plant cells are electrically excitable and can display rapid electrical responses. These can influence movement, wound responses, respiration, photosynthesis, and flowering. They can also initiate the synthesis of numerous organic molecules, including some that prove to be neuroactive substances in other organisms.

The strong symbiotic relationship between plants and mycorrhizal fungi is rarely discussed, but it is fundamental to the plant world. About 95% of plants have symbiotic relationships with such fungi, of which there are thousands of species. The relationship is complex. The fungi trade salts for carbohydrates, taking as much as 30% of the carbohydrates a plant (or tree) produces.* Some plant species, such as orchids, would die out without the fungi.

The network of mycorrhizal hyphae (fungus roots) is more extensive than the network of plant roots, forming a network, not just for transporting foods but also for communications. Research done by Dr. Alexander Volkov, and Dr. Yuri Shtessel demonstrates that tomatoes communicate electrically with each other through mycorrhizal networks, with the fungal hyphae acting as wires.** It's possible, although unproven, that cross-species communication takes place. It seems likely.

Other research suggests that trees of the same species form communities and often form alliances with trees of other species. Communication happens

---

* http://www.anbg.gov.au/fungi/mycorrhiza.html
** http://www.tomatonews.com/en/research-tomato-plants-communicate-by-electric-signals_2_1088.html#

through the roots and mycorrhizal networks. Trees send chemical, hormonal and electrical signals. Edward Farmer at the University of Lausanne in Switzerland, who has been studying the electrical pulses, reports that the voltage-based signaling system appears very similar to the activity of animal nervous systems. The main topic of conversation seems to be anything dangerous that's going on.

Invertebrates

About 97% of animal species are *Invertebrates*. They vary widely in size, from microscopic rotifers to giant squid (10 m/33 ft). They have no backbone, but most are capable of motion and changing their location. Like *Plants*, some can reproduce asexually, but the vast majority reproduce sexually. They are usually classified as follows:

- Sponges (Porifera)
- Comb jellies (Ctenophora)
- Hydras, jellyfishes, sea anemones, and corals (Cnidaria)
- Starfishes, sea urchins, sea cucumbers (Echinodermata)
- Flatworms (Platyhelminthes)
- Round or threadworms (Nematoda)
- Earthworms and leeches (Annelida)
- Insects, arachnids, crustaceans, and myriapods (Arthropoda)
- Chitons, snails, bivalves, squids, and octopuses (Mollusca)

Contemporary science is no longer convinced of the division of the Animal Kingdom into *Vertebrates* and *Invertebrates*, believing it to be artificial and, perhaps, a reflection of human bias. Objective science inherited the designations *Vertebrates* and *Invertebrates* from the science of Gurdjieff's time, but sees the actual distinction as one of brain system. The lower of the two classes, *Invertebrates*, has a single brain system governing moving/instinctive/sexual functions. The higher, *Vertebrates*, also have an emotional brain.

Almost all *Invertebrates* have a brain of a kind and a nervous system that serves the brain. (There are exceptions. Sponges, for example, have very primitive nervous systems with no synaptic junctions.) While they do not possess skeletons, most have some rigid structures that assist in their motion. They are diverse in that respect. Jellyfish and worms have fluid-filled, hydrostatic skeletons. Insects and crustaceans have hard exoskeletons that they shed regularly. Others like oysters, sea urchins, and snails have shells. Thus some of these life-forms make use of silicon in their anatomy, but many do not. The exoskeleton of insects and crustaceans is made from chitin, which turns out to be a long-chain polymer derived from glucose.

They have breathing systems that vary widely and digestive systems that also vary widely. Because they are capable of motion, they are also capable of social behavior. It goes far beyond the sexual function and is observed in many species, including cockroaches, termites, aphids, ants, bees, and spiders. Bees, ants and other species communicate.

WATER In Summary

Water is a remarkable substance. The Earth is the only planet in the solar system on which it is clearly abundant. It supports all life and it is the dominant constituent of all biological life-forms. Animals are typically between 50% to 75% water. With *Plants*, it is more like 80% to 90%. And yet water provides life with no calories or organic nutrients.

It is a very effective solvent, sometimes (incorrectly) described as universal. It is a vehicle for transporting substances. When observed playing a role in some creature's life process, it is usually carrying substances from one place to another. It is versatile, the only common substance we naturally encounter on the Earth's surface in solid, liquid, and gaseous forms.

Water's transition from liquid to solid feels like a cleverly designed-in feature because it expands when it freezes. As a rule, most liquids do not. Incidentally, it is not the only substance that does this. Silicon does too—a property that likely exerts an influence on processes beneath the Earth's surface. Water's property of expansion ensures that icebergs float rather than sink, that water will split rocks when it seeps into cracks and freezes, and that lakes freeze from the surface down. If ice sank, then lakes could freeze solid, destroying the fish.

Even a cursory study of the water cycle leads one towards the view that it is the "blood of the planet." It covers 82% of the Earth's surface, 71% as seas or oceans, 10% as ice, and 1% as rivers and lakes. It permeates the domain of EARTH as groundwater within the crust and forms 25% of soil. At any time, about 1% by volume is within the bodies of plants or animals. It is rarely fresh water in any of these contexts. It becomes fresh when it evaporates, but when it falls as rain, it normally carries gases (carbon dioxide, sulfur dioxide, etc.) with it.

Gurdjieff makes this point in the following extract from *The Tales*:*

---

* *The Tales*, Ch XXI, The First Visit of Beelzebub to India, p230

The Domain of Water

> "The said pearl is formed in one-brained beings which breed in the 'Saliakooriap' of your planet Earth, that is to say, in that part of it which is called 'Hentralispana,' or, as your favorites might express it, the blood of the planet, which is present in the common presence of every planet and which serves the actualizing of the process of the Most Great common-cosmic Trogoautoegocrat; and there on your planet this part is called 'water.'

Water also plays a huge role in man's life, supplied directly to almost all houses for drinking, cooking and washing. It is used extensively for agriculture and in many industrial processes, exploited for hydroelectricity, and for transporting goods by boat via canals, rivers, lakes, and seas.

An area becomes a desert when plant life cannot survive there, and rain does not fall. By contrast, rainforests experience consistently heavy rain. The Amazon rainforest has formed its own water recycling environment. What evaporates from there returns to the same ground.

The Properties of Water

Liquids, aside from the fact that they flow, are distinctive from solids in several ways:

- **Surface tension**. Liquids are not crystallized into lattices, but the cohesion that exists between their molecules creates a skin or surface. We refer to the phenomenon as surface tension. Because of this, a volume of liquid tends to shrink to the minimum surface area possible, becoming a sphere when no other forces are present. Water molecules are strong dipoles (magnets). Consequently, the surface tension of water is relatively strong, and water is a powerful solvent for other polar molecules.

- **Convection**. This is the effect of hotter liquid rising upwards and colder liquid descending. Conduction of heat also occurs in liquids but not so quickly as in solids because the molecules are further apart. Convection occurs because hotter liquid is simply less dense and is thus displaced by denser liquid.

- **Diffusion**. Diffusion is the net movement of a substance (its atoms or molecules) to expand from a region of higher concentration to a region of lower concentration without the assistance of any kind of flow. Osmosis is related to diffusion and also important. (Osmosis is the net movement of a liquid substance through a semipermeable membrane from higher concentration to lower.)

Gurdjieff's Hydrogens: The Ray of Creation

Water is subject to tidal gravitational force from both the Sun and the Moon. The Earth's crust is also subject to the same influence. The so-called Earth tide is estimated to have an amplitude of about 1 meter. Ocean tides on the open ocean are estimated at merely 0.6 meters. This is, of course, differential. You need to add the ocean tide to the Earth tide to get the impact of both. However, ocean tides are variable and because of currents and wind can be as large as 13 meters (40 ft).

Wave motions pass energy through water, as briefly discussed in Chapter 5. The speed of tsunami waves is very fast, as fast as 725km/hr (450mph). That's the upper limit to wave speeds, which usually travel at a leisurely speed of around 8km/hr (5 mph). The tsunami waves travel fast because they have a very long wavelength, in the range of 100km. The wavelength compresses when the wave comes close to land.

Other speeds from the domain of WATER that are notable are the speed of plant growth, which is in the range of 1 to 100m per yr. (At the higher end of this range is the growth of some bamboos, which can approach 30 cm per day. The speed of sap flow through plants can be as high as 56 cm/hr, which is similar to the speed of rising and falling tides and also (not surprisingly) the speed of liquid passing through cellular walls by osmosis.

When we consider the speed of motion of *Invertebrates*, they rarely go faster than 100meters/hr. However, flying insects can go faster. A dragonfly can achieve speeds above 60km/hr (36mph).

Chapter 8

The Domain of Air

"A human being is only breath and shadow."

~Sophocles

The creatures that populate the domain of Air are breathing creatures. They live mainly in the medium of air, but many are also found living in the medium of water, in lakes, rivers, and seas. Although plants breathe, absorbing and releasing oxygen and carbon dioxide, they do not have a complex respiratory system. All animals do.

Underwater creatures have gills. Insects breathe through spiracles, small holes in their body that conduct oxygen through trachea to individual cells. They are simply different kinds of lungs. Most vertebrates have lungs, which absorb oxygen from the atmosphere into the bloodstream.

The main constituents of the atmosphere are nitrogen (78%) and oxygen (21%). The remaining 1% comprises argon, water vapor, carbon dioxide, and trace amounts of hydrogen, krypton, methane, helium, and neon. Surprisingly, given that all life requires nitrogen, only *Plants* can absorb nitrogen from the air. The animal world absorbs it from food.

The distinction between warm-blooded and cold-blooded creatures is an important one. For sensible reasons, modern science prefers not to emphasize the term "warm-blooded" as there are gradations and variations in how body temperature is regulated. The term homeothermic describes species that maintain a stable body temperature. The only known homeotherms are birds and mammals.

Because of air-conditioning and central heating, we pay far less attention to the passing seasons than we once did. Most life-forms are "designed" to accommodate the Earth's seasons. The Earth's tilt, at $23.4°$, is sufficient to ensure that the seasons have a huge impact on the activity of life. However, the effect is less towards the equator, where temperature and sunlight vary less.

Consequently, there is a Trogoautoegocratic chain of cause and effect. Less sunlight and warmth means less plant and tree growth. So *Plants* have become seasonal, putting a strain on the species that feed on them. From spring to autumn food is abundant, but after autumn it becomes scarce. This impacts the whole food chain. Cold-blooded creatures tend to hibernate, remaining buried throughout the winter. Fish do not hibernate, but their metabolism slows if the water cools. Even some warm-blooded creatures hibernate, bears being the obvious example.

Cold-blooded creatures face the problem of body heat on a daily basis. They require heat from the Sun for their metabolism to function. Some species, referred to as bradymetabolic, can significantly lower their metabolic rate at night, almost hibernating, only to burst into activity once they wake.

The Cycle of AIR

Figure 60 illustrates the cycle of AIR. For those less familiar with chemical symbols, the table below provides a key:

SYMBOL	DESCRIPTION	SYMBOL	DESCRIPTION
Ar	Argon atom	CO_2	Carbon dioxide molecule
CH_4	Methane molecule	N_2	Nitrogen molecule
NH_3	Ammonia	NH_4^-	Ammonium
NO_2-	Nitrite ion	NO_3^-	Nitrate ion
O_2	Oxygen molecule	SO_2	Sulfur Dioxide

Table 5. Key to Figure 60, The Cycle of AIR

Argon is an inert gas that does not even pair with its own atoms. Despite making up almost 1% of the atmosphere, it doesn't participate in any known way in biological processes. Aside from volcanic activity, there are two distinct cycles involving the air.

The first is the nitrogen cycle, shown in the foreground of the illustration. On land, it proceeds as follows. Nitrogen enters the soil from the atmosphere by diffusion. Nitrogen-fixing bacteria (cyanobacteria) act on it, transforming it into ammonia (NH_3). Decomposer bacteria transform decaying plant waste and animal waste into ammonium (NH_4^-). Ammonium and ammonia are further transformed into nitrite and nitrate salts (or ions), which plants can ab-

The Domain of Air

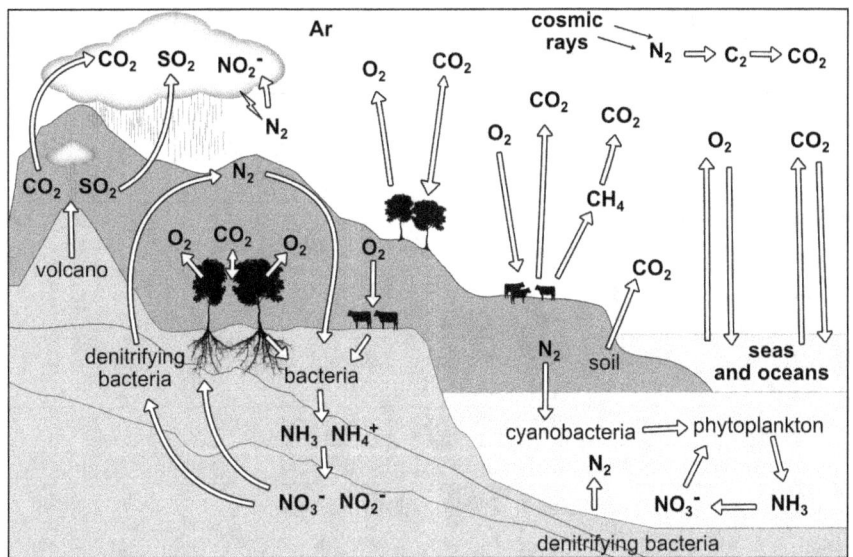

Figure 60. The Cycle of AIR

sorb directly. Surplus nitrite and nitrate salts are converted back to nitrogen by denitrifying bacteria in the soil to be released back into the air.

Lightning also creates nitrite ions. Usually, the rain from the clouds will absorb this, and it will fall to the ground as nitric acid to form a nitrate salt that plant life can feed on. The rain will also absorb sulfur dioxide and carbon dioxide, both of which fall to the ground as acid and become part of the soil or the sea. Rain has a cleansing effect on the atmosphere.

A similar process occurs in the seas and oceans with nitrogen fixation by cyanobacteria and phytoplankton absorbing and also producing ammonia. Bacteria transform ammonia into nitrate salts, which feed undersea plant life. However they may also be converted back to nitrogen by denitrifying bacteria.

The carbon/oxygen cycle is simpler. Both on land and sea, in the process of photosynthesis, plant life consumes carbon dioxide and creates oxygen. Animal life, both on land and under the sea, converts the oxygen back to carbon dioxide. Somewhere between 50-80% of the Earth's oxygen comes from phytoplankton on the ocean's surface. They oxygenate the water. Simultaneously, oxygen from the atmosphere also diffuses into the water with the assistance of wind and waves.

As you sink deeper below the surface, the oxygen in the water diminishes. Nevertheless, fish exist at great depths. Deep-sea fish live in the Mariana

Planet	Atmos. Pr.	Atmospheric Gases
Mercury	10^{-15}	O_2 42%, Na 29%, H_2 22%, He 6%, K 0.5%
Venus	92	CO_2 96.5%, N_2 3.5%, SO_2 0.015%
Earth	1	N_2 78.08%, O_2 20.95%, O2, Ar 0.934%, CO_2 0.038%, H_2O variable <1.0%
Mars	0.004-0.009	CO_2 95.32%, N_2 2.7%, Ar 1.6%, O_2 0.13%, CO 0.08%, H_2O 0.021%, NO 0.01%
Jupiter	1000+	H_2 89.8%, He 10.2%, CH_4 0.3%, NH_3 0.026%
Saturn	1000+	H_2 96.3%, He 3.25%, CH_4 0.45%, NH_3 0.0125%.
Uranus	1000+	H_2 82.5%, He 15.2%, CH_4 2.3%
Neptune	1000+	H_2 80.0%, He 19.0%, CH_4 1.5%

Table 6. Planetary Atmospheres

Trench at a depth of 26,200 feet below sea level. Such species can survive with less oxygen. Some can survive with one-fortieth or less of the oxygen that is available near the surface.

Consider now *Table 6* above, which shows the make-up of the atmospheres of other planets in our solar system. Mercury is like the Moon, having no significant atmosphere and minuscule atmospheric pressure. The gas giants, Jupiter, Saturn, Uranus, and Neptune, are much like each other, perhaps surprisingly so. The dominance of hydrogen and helium is reminiscent of the Sun's atmosphere, which is 91.2% hydrogen and 8.7% helium. However, methane and ammonia are also present, and so are clouds of ammonia ice, methane ice, water ice, and ammonium hydrosulfide. Their atmospheres provide no indication of carbon-based life.

One of the notable features of Venus' atmosphere is the lack of hydrogen. Venus is unusual in not having its own magnetosphere. (It has a derived magnetosphere caused by the solar wind.) Some astronomers have suggested that Venus once had much more hydrogen, but it was stripped away by the solar wind. Because it has a dense atmosphere with abundant carbon dioxide, in the future it might become more conducive to organic life, through the bacterial deconstruction of carbon dioxide. That is, of course, speculation.

We have more data about Mars than any other planet aside from our own. The Mars Rover has assembled abundant evidence that there were considerable flows of water on its surface. It may have had atmospheric oxygen too, so maybe it once supported life. Currently, the Sun is gradually stripping it of its atmosphere and like Venus, it has no magnetosphere of its own with which to shield itself.

The Domain of AIR

This triplet gives us the full range of animal life from mollusk to man, or better put, from one-brained being to three-brained being.

Within this triplet, we encounter lifeforms that we more easily recognize as life and better understand. Beings, no matter which of the three squares they belong to, have nervous systems and brain functions that we recognize. They eat, breathe and digest impressions in similar ways to ourselves. They do not digest sunlight, as *Plants* do. Their breathing is a metabolic process for burning rather than creating sugars. They have sense organs similar to ours, and we imagine they process the impressions they receive in a similar way.

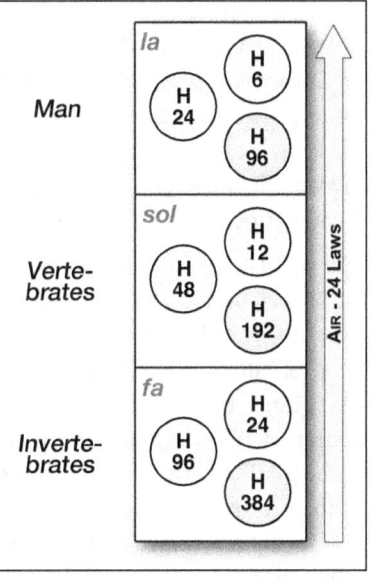

Figure 61. The Triplet of AIR

We observe that the domain of AIR is the three notes, *fa*, *sol*, *la*, in the lateral octave from the Sun that Gurdjieff describes in *In Search of the Miraculous*. In that context, we should probably include *Plants*, which fills the *mi-fa* interval in the lateral octave. The question is: how exactly does the domain of AIR and its myriad creatures fill that interval?

Invertebrates

The earliest animal fossils are thought to be a species of sponge. They are dated at 665 million years old and found in the Trezona Formation at Trezona Bore, South Australia. For perspective, using the same radiometric dating, the Earth is estimated to be 4.54 billion years old (from dating Earth rocks), and the Moon is estimated to be a similar age. Using meteoroid dating, the Solar System is estimated to be a little older, at 4.6 billion years.

Where geology proves far more valuable is in its classification of the fossil record. Geologists apply impeccable logic in the ordering they give to the fossil record and in ensuring its consistency. Fossils are particularly valuable, not only because they are the only record of life that survives for such an immense time but also because of the detail preserved.

Typically bacteria break down cell structures. So most of an organism is consumed before it is thoroughly covered by sediment. The bonier details of the fossil are preserved. However, if the organism is covered by sediment immediately, a process called per-mineralization can occur, as follows. When the organism is buried, the internal spaces occupied by liquid or gas fill with mineral-rich groundwater. Mineral precipitates then begin to fill such spaces. This petrification can occur at a fine grain level, even within cell walls, producing very detailed fossils. This process doesn't require millions of years to occur; it can happen in a few decades.

All soft-bodied organisms (roundworms, earthworms, jellyfish, squids, etc.) are invertebrates and, understandably, rare in the fossil record because bacteria destroy the whole body, leaving no trace. Thus the first multicellular invertebrate life-form cannot be known for sure. A good candidate is Grypania spiralis, found in fossils dated as 2 billion years old. They are flat discs with scalloped edges and radial slits, measuring almost 5 inches across. According to the fossil record, life never got much more complex than that until the onset of the Cambrian period, about 540 million years ago. There were fungi later, and sexual reproduction in multicellular life-forms seems to have begun in the billion years that followed. It may be that most of life then was soft-bodied.

The explosion of evolution during the Cambrian included various invertebrates, mollusks, arthropods, and echinoderms, many with body forms similar to modern invertebrates. All of it was underwater. And it includes some vertebrates (very primitive fish). There is no fossil record from that time of land-based life, not even plants.

The subsequent geological period, the Ordovician, began 485 million years ago. Arthropods (invertebrates with exoskeletons, segmented bodies, and legs) established themselves on land along with plants and mycorrhizal fungi. The evolutionary foundation for life on land and sea was firmly set in place.

Anatomy and Physiology

Plants are immobile. All *Invertebrates* can move by their own volition during at least some stage of their life—they have muscles. Some, like the jellyfish for example, have no skeletal structure but achieve motion without it. Others, like insects and crustaceans, have a very definite skeletal structure to which muscles attach to enable movement. Most such species have exoskeletons like crabs or clams, but a few, like starfish, have endoskeletons. Those with exoskeletons regularly shed the exoskeleton to regrow a larger one. Others simply enlarge the exoskeleton as their bodies grow.

Like *Vertebrates* and *Man*, they consume food, breathe oxygen and take in impressions. Naturally, their digestive apparatus is adapted to the food they consume. However, it has the familiar components: mouth, esophagus, stomach, caecum, intestine, rectum, anus, whether it's a mollusk or an insect. The pattern is similar across all animals, irrespective of brain system.

With invertebrate breathing systems, lungs are a rarity. Underwater invertebrates usually have gills. Insects have spiracles in each body segment (holes to the outside world), which feed oxygen through trachea to the bloodstream. They all have cardiovascular systems to circulate nutrients and oxygen, and they have endocrine glands that feed the blood stream with hormones.

Brain and Nervous System

Invertebrates are one-brained. Their senses can range from simple groups of cells to more complex organs depending on species. Ocean dwelling invertebrates sense changes in pressure, as do all fish. Whether on land or in the sea, they have chemoreceptors that provide a sense of smell and/or taste. They distinguish the chemical environment (acids, alkali, salts), detect prey and mates (via pheromones), and some can measure humidity. They sense gravity and temperature. They sense sound; most have vision and all sense touch.

The eye is a particularly interesting evolutionary development that is necessary for a mobile creature. *Plants* need to sense the Sun, but most animals need far more specialized senses than that—to sense their environment in a comprehensive way to search for food and avoid peril.

While most invertebrates have an identifiable brain that exerts central control over the body via the nervous system, the lower invertebrates simply have a decision-making network of neurons distributed throughout the body. All such invertebrates are sea animals: jellyfish, corals, starfish, sea urchins, and quite a few others.

From the Work's perspective, *Invertebrates* possess a single brain composed of three complementary centers. Each is responsible for different organic functions: the instinctive center, the moving center, and the sex center.

The Instinctive Center

The instinctive center manages all the organism's inner functions—the senses, the heart, breathing, blood circulation, growth, digestion, excretion. If we consider the inner workings to be like a factory, it quickly becomes apparent how difficult this activity is. It never stops, so there can be no rest.

A malfunction in any of the organism's critical functions could be fatal. It is not just true of eating, breathing, and sensing, but for every organ of the body. Things not only have to work, they have to coordinate, and the quantities have to be correct. Eating too much or too little is a mistake. The circulation and the breath must be in harmony. The energy demands of the moving center must be catered to.

Food must be examined for poison and pathogens and rejected if it is unsuitable. There are digestive enzymes to be created. Vitamins and fats must be stored for later use. The immune system must go into action at a cellular level if there is any threat. Damage to the organism must be addressed immediately. The organism needs to keep track of time, at least for the sake of sleep or, in the case of invertebrates that do not sleep, times of rest.

This task cannot be much less difficult for a mollusk than a man. The human system is more sophisticated, but not significantly so. We can get some sense of this simply by pondering what we know and observe of our own instinctive center.

The question naturally presents itself as to whether the instinctive center of an *Invertebrate* employs the same *Hydrogen* as the instinctive center of *Man* or *Vertebrate*. Logic suggests that it must do. It is simply a matter of speed and intelligence. The substance that has the appropriate intelligence and speed is H24.

The instinctive center works with H24. It has its own memory, imagination, (instinctive) emotions, and will. It has two halves, affirming and denying. It "denies" pain and responds to it accordingly, it denies unsuitable food, it recognizes external threats and prompts the moving center to deal with them. It is reasonable to assume that all of this skill and capability is somehow encoded in DNA and activated by a fertilized egg from the moment it begins to grow.

The Moving Center

Instinctive functions exist in *Plants*. A slower *Hydrogen* most likely governs them: H48 rather than H24. *Plants* have no moving center—their limited movements are managed by some kind of "instinctive" brain. In contrast, *Invertebrates* move and have a moving center.

In *Man*, the only external functions controlled by the instinctive center are reflexes. But this is not the case in *Invertebrates* and *Vertebrates*. An obvious example is a newly born giraffe's ability to stand up and walk a few hours after

The Domain of Air

it is born. It is not something it has to learn. Human babies take at least nine months to learn this. The baby giraffe's innate ability to stand and walk is instinctive. Examples of the instinctive center kick-starting the newly born can be observed in many species of invertebrates and vertebrates.

All invertebrates that reproduce sexually lay eggs. Some species hatch directly into adults, while others go through a larvae stage. Some invertebrates parent the offspring, but there are many examples where they find a suitable location to lay eggs but then leave the eggs to their fate. When parenting takes place, the parent serves as a model for the offspring's moving center to copy.

The moving center learns by imitation and by experience and learns most if not all of its capability. Some scientific experiments have focused on the learning capabilities of crayfish, aplysia, slugs, bees, and other insects. For example, in a crayfish experiment, scientists fooled crayfish into an escape response when there was no real threat. After multiple repetitions, the crayfish realized there was no real threat and began to ignore the stimulus. This suggests, unsurprisingly, that conditioning works on crayfish.

For *Invertebrates*, the moving center provides their only means for learning. They have neither an emotional nor intellectual center that can be involved in that activity. The moving center has memory and a degree of will (all centers do).

Scientific experimentation on macaque monkeys led to the identification of "mirror neurons." Scientists observed that when a macaque monkey saw an action by another monkey or a human, such as picking up food, a specific set of neurons fired. The same neurons fired when they themselves picked up food.

Mirror neurons are found in the brain areas that receive impressions and are involved in movement (the primary somatosensory cortex, the inferior parietal cortex, the premotor cortex, and the supplementary motor area). The presence of such neurons in invertebrates has yet to be established, but they are probably there.

The moving center controls dreaming. There is no direct evidence that invertebrates dream—they have not been observed to experience REM sleep as some vertebrates have. It is possible that their dreaming does not involve rapid eye movement. Dreaming is a learning activity in the sense that it involves past experience.

There is no direct evidence that invertebrates yawn or laugh. However, both of these are moving center functions, and they are contagious behaviors. So it is likely that they do in some way. Laughter is caused by an impression falling on both the affirming and denying parts of a center. It has the effect of raising the energy of the impression. Yawning, the filling of small accumulators with energies for a center, is also a moving centered activity. Both probably occur in invertebrates—both have been observed in some vertebrates.

The Sex Center

Invertebrate reproduction practices vary with species. Asexual reproduction is fairly common, but sexual reproduction is more typical, most likely because it is better for the individual variability of the species—the diversity of the gene pool is important.

Zoologists estimate that about 5% of animal species are hermaphrodites, and most of those are invertebrates. In many cases, the fertilization of eggs occurs after the female has laid them, so it does not necessarily involve bodily contact or a persistent (or even monogamous) relationship between parents.

Reproduction drives the survival of the species, which contributes to the survival of the ecosystem—ecosystems always involve a balanced co-existence between multiple diverse species. The survival of the ecosystem contributes to Nature's health, which, in turn, serves the Earth and the universal Trogoautoegocrat. The sex center takes its place in this chain.

It is inevitably the dominant force in the invertebrate's life and exerts a determining influence on behavior. According to Gurdjieff,* the three centers of the lower story work together as follows. The sex center is neutralizing in relation to the other two lower centers. *Invertebrates* are easy to understand in this respect as they have no intellectual and emotional center to interfere with the activity of the sex center.

Either the instinctive center is active and the moving center is resting passively, for example, when the organism is digesting food. Alternatively the moving center is active, for example hunting for food, and the instinctive center is passively supporting it. Both centers have positive (affirming) and negative (denying) parts. Typically, incoming impressions will fall on both centers, and an action or a change of activity will be triggered by whichever center is energized most by the impression. Either instinctive emotions (H24) or moving center emotions (H24) will dominate.

The sex center becomes activated only when alerted to the possibility of reproductive activity, which is triggered seasonally in *Invertebrates*. It has no

---

* In Search of the Miraculous by P D Ouspensky, p115

positive and negative side; it is activated by pleasant sensations or feelings, or it is indifferent.

Vertebrates

Vertebrates are generally larger than *Invertebrates*—the largest being the blue whale. The elephant is the largest land animal. *Vertebrates'* skeletal structure gives them superior strength and mobility. Their internal organs can be more complex than *Invertebrates*, but the general pattern is the same. Depending on species, they may feed on *Plants*, *Invertebrates*, or other *Vertebrates*, and large predators may even feed on *Man*.

If we consider the order of evolution suggested by the fossil record, the earliest indication of life is estimated at 3 to 3.5 billion years ago. These initial cells may have been RNA rather than DNA-based. However, DNA soon became the vehicle for cell replication and organic evolution. It can be thought of as sophisticated biological software. It is, in essence, a single molecule forming a single crystal, composed of about 200 billion atoms.

It is fragile and rarely preserved as a fossil. Nevertheless, scientists have discovered intact DNA in ancient salt deposits from the Michigan Basin. Estimated to be 419 million years old, it proved to be similar to bacteria from our era. It is probably little different to the Earth's initial bacteria.

Bacterial DNA is simpler than human DNA, which is tightly coiled and more complex. However, the genome size (the total amount of DNA in a single complete species genome) has no direct relationship to an organism's complexity.

Some single-celled organisms have much more DNA than humans, and *Plants* tend to have larger genomes than animals. Currently, no-one has a good theory as to why.

From the perspective of the Work, DNA is important. It imposes some very specific laws. Neither men nor monkeys can grow wings and fly; neither can we breathe underwater like fish. It is not in our physical program. Nevertheless, it may be the case that a genetic program to enable such behavior is stored somewhere in our genome.

The Putative Origins of Life on Earth

Fred Hoyle, the English astronomer, once said that "the evolution of complex life-forms by natural selection was as probable as a tornado blowing through a junkyard and assembling a 747 jet aircraft." Religious creationists, wishing to discredit evolutionary biologists often quote this argument. Atheistic biologists oppose the suitability of the analogy. The two sides differ only in the dogma to which they wish to glue their arguments.

The origin of DNA itself (or RNA, which probably came first) is a matter of theory. In 1953, Stanley Miller and Harold Urey conducted an attention-grabbing experiment. They recreated the presumed conditions of the early Earth in a flask containing an atmosphere of simple gases (water, methane, ammonia) and some "lightning" in the form of electric sparks. Resulting samples contained five different amino acids. Since amino acids are the building blocks of proteins, scientists concluded that the process revealed how life on Earth began.

The scenario of the sun, rock pools, methane, and ammonia is easy to imagine. And that happily coincided with an idea from no lesser light than Charles Darwin. He had expressed the notion that life began in rock pools. However, the plot thickened when, in the mid-1990s, Robert Hazen conducted pressure-based experiments that produced amino acids from a cocktail of chemicals, including rock dust, in the complete absence of electrical sparks.

He had been investigating how life might have arisen near mid-ocean vents, which, it had been discovered, are surrounded by an organic ecosystem. The ecosystem seems to have developed locally and includes both invertebrates (tube worms, clams, crabs, and shrimp) and vertebrates (fish). These vents are so deep beneath the ocean that light from the Sun never penetrates. R Hazen, an American mineralogist, and astrobiologist subjected the substances surrounding deep ocean vents to considerable pressure in a laboratory experiment. When he did so, amino acids formed. Hazen concluded, reasonably, that life might form anywhere.

From the Work perspective, we expect Nature to emerge if a planet has a moon in a particular stage of gestation. It may develop under other circumstances, but if so, we do not know what they are.

These deep ocean vents provide us with particularly interesting information.

In the presence of the lowest three notes of the lateral octave from the Sun, *do* (*Kernel*), *re* (*Metals*), and *mi* (*Minerals*), and sufficient deep water, it appears that life can evolve. It apparently happens in the **complete absence of any plant life**. *Invertebrates* and *Vertebrates* appear to evolve from *Metals*, *Minerals*, and bacteria alone.

Single Cell Symbiosis

The fossil record suggests that evolution began with cyanobacteria, a bacterium that could create oxygen by photosynthesis. While oxygen is toxic to

The Domain of Air

some bacteria, it energizes the metabolism of those that can use it. The oxygenation of the atmosphere was required for the evolution of complex eukaryotic cells.

The emergence of eukaryotes appears as a dramatic evolutionary event. Surprisingly, eukaryotes appear to be a symbiotic union—a merger— of two life-forms. Eukaryote cells have DNA located sensibly in the nucleus and also have organelles called mitochondria within their cell walls.

Mitochondria are the engine within the cell that produces ATP, which is the cell's primary energy source. Mitochondria have their own DNA.

Plant-based eukaryotic cells contain organelles called chloroplasts, which carry out photosynthesis within their cell walls. These appear to be the consequences of a different merger because chloroplasts also have their own DNA.

There is a theory named endosymbiosis, which suggests that, back in the day, some large cells engulfed some small cells, and instead of feeding on them, they decided to allow them to persist as organelles. Whether or not this is so, this symbiotic combination made multicellular life possible. Almost all multicellular organisms appear to have evolved from eukaryotes. Since eukaryotic cells are capable of sexual reproduction, this development launched sexual reproduction into multicellular life.

Multicellular Life

Multicellular life may have emerged as far back as 3.5 billion years ago. The earliest fungi fossils date back 2.5 billion years. The earliest plant-like fossils formed a billion years later. *Plants* gradually enriched the Earth's atmosphere with oxygen, enabling more sophisticated oxygen breathers to evolve. By 665 million years ago, we saw the emergence of invertebrate sponges, and then the Cambrian (submarine) explosion occurred around 541 million years ago.

The Cambrian period was alive with varieties of invertebrates, including an estimated 17,000 species of trilobite, as well as varieties of crustaceans and mollusks. It saw the evolution of the first vertebrates. Fossils of haikouichthys, the first known fish, date back 518 million years. They were jawless, with jawed fish evolving much later.

Colonizing The Land

The next giant leap for organic life was its evolution into forms that could live on land. When you consider the details, it is clear that major evolutionary adjustments are required to step from the domain of water into the domain of Air. Aside from anything else, land-based life needs a reliable water supply.

Gurdjieff's Hydrogens: The Ray of Creation

Early land life needed to carve out a living without enjoying the luxury of soil. It had to withstand gravity—a force that is far less of a challenge in the domain of WATER. The respiration system had to change, and the reproductive system too. And as for food, it was an entirely different ecosystem populated by different organisms.

The colonization of land could not occur in a single leap. It began with the emergence of amphibious plants and animals. The fossil evidence of land plants suggests that this colonization was in progress by 460 million years ago. Evidence of scorpions dates back 420 million years, and the first evidence of four-legged animals dates back 395 million years.

Flowering Plants

Flowering plants and insects tend to be symbiotic, the flower providing nectar in exchange for the insect's pollination service. Early plants, mosses, and ferns, for example, use spores to reproduce. They tend to exist in moist soil where the spores can swim to find female ovules to fertilize.

Flowering plants are estimated to have evolved, along with many insect varieties, in the period between 256 million years ago and 149 million years ago. By that time, plant ecosystems had formed, which resemble what we have today: rich soil, plants, and forests. However, dinosaurs populated them.

The Warm-Blooded

Mammals and birds are warm-blooded. The first mammals dated to 225 million years ago, and the first birds (as opposed to flying reptiles) date back 100 million years. Given that mammals are the dominant vertebrates, you might expect that keeping the body at a constant temperature is obviously advantageous. There are some advantages, but they are not obvious.

It's an observable fact that warm-blooded animals adapt better to colder climates. If we ignore marine life, about 100 species of birds and 20 mammals live in the arctic area. These include caribou, arctic fox, dall sheep, wolverine, and lemmings along with polar bears and penguins and skuas near the south pole.

Maybe that's the point. Nevertheless, that extra body heat comes at a price. Mammals and birds need to eat more to maintain the temperature. Only a few mammals hibernate—the bear is the only large one that does. The rest are stuck with the problem of finding food in the winter and finding more than they would need if they weren't warm-blooded.

Warm-blooded mammals and birds perch at the top of the empire. And in the water, whales and dolphins rule supreme.

Evolutionary Considerations

From the perspective of objective science, the current theory of evolution is awkwardly flawed. Nevertheless, the research activity that surrounded it has generated a great deal of valuable data.

Objective science and the evolutionary theory of modern science are irreconcilable. This can be explained simply. Evolutionary theory suggests that higher life-forms self-evolve without any interaction with a higher level. This cannot be reconciled with the Law of Three. To create a substance of middle intelligence, a substance of lower intelligence must be mixed with a substance of higher intelligence. There is no debate to be had here; there is simply a parting of ways.

There are questions we need to ponder about evolution. The evidence suggests that life has evolved from a point far in the past where the Earth was host only to bacteria and archaea. We have described most of the major milestones involved. The question is: How? What are the processes?

We can establish some markers here.

At some point, probably billions of years in the past, bacteria combined with archaea to form eukaryotes. They may have done so in various ways, but all we know of it is what survived. Strains of eukaryotes, built to metabolize oxygen effectively, survived to became the building blocks for animal life. Photosynthetic eukaryotes survived and became the building blocks for plant life.

This is remarkable. Just as it is remarkable that two of the major elemental constituents of igneous rock are iron (useful for oxygen transport) and magnesium (useful for photosynthesis). It may be that the only possible combinations of bacteria and archaea that would ever work are the two that occurred. That may or may not be the case, but the curious thing is that such biological combinations are no longer happening. The Earth is awash with untold species of bacteria and archaea, but there's no evidence that they're forming eukaryotes now. Why?

We can only presume, circumstantially, that billions of years back in the day, the conditions were exactly right for this merger of revolutionary proportions, and that's all she wrote. If this merger had not occurred, then life would never have become multicellular.

The Taxonomic Tree

The taxonomic tree in *Figure 62* (next page) illustrates the presumed evolutionary progression. Ordinary bacteria begat eukaryotes, which begat the multicellular kingdom, and so on. It continued over millions of years to arrive at the modern era. Note how many differently named levels there are. If you dive into the detail, you also discover that individual levels can have sub-levels. For example, *Vertebrates* do not qualify as a Phylum but only a subphylum of Chordata.

Looked at this way, it all seems well ordered. We can easily imagine a path of evolution stepping forward from eukaryotes to marine plants and invertebrates, then to amphibians and then land plants and animals. Finally, we arrive at familiar ecosystems: plains, forests, and tropical jungles, swarming with flora and fauna that we recognize. But the fossil record provides us with very, very little information, and the taxonomy we can map out is almost certainly missing entries at every level below the top one. We simply do not know who all our ancestors were, and perhaps we never will. This is partly because of extinction events.

	Life
Domain:	Bacteria - Eukaryota - Archaea
Kingdom:	Animals - Plants - Fungi - Protista
Phylum:	Chordata - Hemichordata...
	Vertebrates - Lancelets - Tunicates
Class:	Mammals - Birds - Amphibians...
Order:	Carnivores - Ungulates - Primates...
Family:	Cats (Felidae) - Dogs - Bears...
Genus:	Panthera - Leopardus - Catopuma...
Species:	Lion - Leopard - Jaguar - Tiger...

Figure 62. The Taxonomy of Life

Extinction Events

If the first evolutionary step from bacteria and archaea to eukaryotes seems incredibly well planned, many subsequent events that can be found in the fossil record look deeply unfortunate. The fossil record provides abundant evidence of extinction events—events when large numbers of species were annihilated. They must have changed the whole character of the biological ecosystem. The five most frequently mentioned are:

- **The Ordovician–Silurian extinction event(s).** These are two distinct events dated between 450 and 440 million years ago. These events are estimated to have eliminated 60-70% of all (multicellular) species.
- **The Late Devonian extinction.** This is dated to between 375–360 million years ago and eliminated at least 70% of all species
- **The Permian-Triassic extinction.** This is dated as 252 million years ago and eliminated 90-96% of all species.

- **The Triassic–Jurassic extinction.** This dates back to about 201 million years ago and eliminated 70% to 75% of all species.
- **Cretaceous–Paleogene extinction.** This is dated to 66 million years ago and eliminated 75% of all species.

There are theories as to the cause of these extinction events: swift changes in the chemical make-up of the atmosphere, chains of massive volcano eruptions, large asteroids colliding with Earth, nearby supernovas, and so on. The geological record is inconclusive.

But think of this. Historically, the year 536AD is often characterized as "the worst year to have been alive." There was a volcanic eruption in late 535AD (probably in Iceland), which injected large amounts of sulfur and ash into the atmosphere. Summer temperatures in North America, Asia, and Europe dropped somewhere between 1.6 to 2.5°C, and there were crop failures and famine. They were desperate years for man, and most likely for the world of nature, but that wasn't even close to being a minor extinction event.

The question that naturally presents itself is:

What is the purpose of these extinction events?

The five detailed above are not the only such events; they are just the biggest. Since about 542 million years ago, geologists have recorded 24 such events—an average of one every 22.5 million years. Evolution appears to be a very hazardous process. Not to put too fine a point on it, the Permian-Triassic extinction came quite close to sending Nature back to square one. And for all we know, it could happen again.

And if Nature is vulnerable, if those notes *fa-sol-la* are vulnerable, and it looks like they are, there is no guarantee that the *mi-fa* interval in the Ray of Creation will remain filled in this solar system. These extinctions may be a danger to the life of the Moon itself.

Nature—the Great Survivor

Nature's defense against cataclysm is diversity. She has colonized all the territory available and bet on every possibility. She has expanded in every dimension available on every scale. There are estimated to be about 8.7 million eukaryotic (multicellular) species, and there are only about 300 primate species. If some extinction event eliminated the primates and a few other orders of mammals, Nature would have to climb back up her tree and come back down with an alternative to *Man*. Nature would adapt.

Gurdjieff's Hydrogens: The Ray of Creation

Consider the fact that the Earth tilts $23.44°$ on its axis. Planets in our solar system spin on their axes with very different tilts. Mars tilts at $25.19°$, Saturn at $26.73°$, and Neptune at $28.32°$. But then Jupiter is at $3.13°$, Venus at $2.64°$, Uranus is at $82.23°$, and Pluto at $57.47°$. Venus is perhaps the oddest because it spins in the opposite direction to the other planets. And to cap it all, even the Sun tilts—about $6°$ on its axis relative to the ecliptic.

For Nature, the consequences of the Earth's tilt are severe. It imposes seasons: spring, summer, autumn, winter. It means a cycle through the year from inactive to active and back for plants and cold-blooded animals. There are periods when food is plentiful and when it's scarce, and reproduction must be scheduled accordingly. There is no sense in producing offspring if they cannot eat.

It stresses the lives of warm-blooded animals. How do they survive? Some learned to migrate, some, like squirrels, learned to store food, and one or two knew how to hibernate. Nature found different ways to adapt with different species.

Indeed Nature's ability to adapt seems miraculous. For example, take flowers that bloom in desert regions—in places where rainfall happens only once in several years. After rainfall, the desert bursts into bloom. Seeds sprout, plants grow and flower. They are pollinated, form seeds and scatter them, and then they die. It lasts for a few weeks, and in that time, a whole generation of plants grows and dies.

Drilling by International Ocean Drilling Programme under the seabed at Wilkes Land in East Antarctica has revealed sediments containing the pollen of plants that only thrive in the tropics. Similar biological traces can be found in the Arctic. The implication is that the Earth was once warmer, or that the continents drifted around like unmoored boats, or that the Earth's tilt was different. Whatever may have happened, Nature happily colonized the available territory and retreated when conditions deteriorated.

Nature's Eccentricities

Nature may be imbued with an iron will to survive at all costs, but in some instances, it acts in ways that are apparently suboptimal, if not downright foolish.

Consider flightless birds. New Zealand is famous for them. Its islands were without mammals for an estimated 60-65 million years, except for the odd species of bat. The birds dominated, and many species happily evolved away from their aerial skills.

The Domain of Air

The arrival of humans and the mammals that came with them wrought havoc in the flightless bird population. It isn't known for sure how many species bit the dust—at least 43, but possibly more. So now there are only 16 flightless species left, and some of those have to be protected.

The question that presents itself is: "Was it necessary for Nature to create species of flightless birds in New Zealand?" The answer is probably "yes."

Nature had land to colonize, no mammals turned up, and only a few reptiles, so birds were its best bet. Thus, the kiwi bird fed on grubs and bugs and chewed on worms in leaf litter. It filled the role that moles and badgers fill in other lands. That appears to be how Nature works; she evolves species to maintain the Trogoautoegocrat. In Australia, there were no other mammals, so Nature made do with marsupials.

Consider flying fish. Why do they even exist?

Sure, you can invent an explanation. Nature wanted them is as good an explanation as any. Remarkably, they can achieve a speed of about 40 mph as they approach the surface. Then they launch themselves to glide through the air. Predators chasing them are left floundering as the flying fish glides as far as 200 meters away. Extraordinary. But is there some point, or is it just Nature announcing "and for my next trick?"

Let's talk about the salamander. This talented creature can regrow not only tails but arms, legs, and other body parts should it lose them. It's a neat trick, one that Nature might have had reason to pass on to many other species, but it didn't. Well, that's not entirely true. Human beings and other mammals are capable of liver regeneration—regrowing the whole liver when about 75% of it is removed. But only the liver. Why?

Another act of Nature that doesn't seem to make much sense in the Darwinian perspective is the evolution of co-dependent symbiotes. Symbiosis makes obvious sense; you scratch my back, and I'll scratch yours. But when it becomes a two-way dependency, it seems like deliberate stupidity.

Consider, for example, the yucca plant and the yucca moth. The yucca moth depends entirely on yucca plant blossoms as a place to lay its eggs, and the yucca depends entirely on the moth for pollination. If one of these species pops its clogs, then so does the other. There are many examples of this; the fig tree and the minute fig wasp, hummingbirds with beaks that fit only particular flowers, which in turn can only be pollinated by that hummingbird.

The truth is that, in many of her activities, we don't understand why Great Nature does what she does.

Gurdjieff's Hydrogens: The Ray of Creation

A Reluctance to Redesign

Nature is capable of great evolutionary variety, but she seems always to work within a particular pattern determined by common ancestry. The arms of a man, the feet and legs of turtles and horses, the wings of birds and bats are all derived from the same pattern that is probably hundreds of millions of years old.

This fundamental pattern that can evolve to an excellent skeletal design for hands with an opposable thumb also works out well if you don't want opposable thumbs. And it works fine if you want to develop it into the front hooves of a horse or the cloven hooves of sheep and goats. Oh, you want wings, well sure, this is fine for wings, and by that, I mean bird wings or bat wings, take your pick.

Nature is not keen on redesign.

Fecundity

An oak tree's life expectancy is somewhere in the range of 150 to 300 years. After about 30 years, it begins producing about 10,000 acorns per year. If, over one hundred years, one or two of those acorns become trees, the tree will have replaced itself. A large bluefin tuna can produce 10 million eggs in a single year. Only a few need to survive for it to replace itself.

Human beings are reproductively prolific too. A female fetus can have as many as 6 million eggs. The number reduces to about 400,000 by the time puberty comes round. The numbers continue to reduce, and only three or four hundred will ever have a chance of being fertilized in practice. On the male side, each ejaculation produces a population of somewhere between 300 million and a billion sperm. The evidence suggests that the sperm-egg combination that created you, me, and everyone else was random, or at least utterly unpredictable.

Nature plays the numbers game, in many ways, all the time. If you think you're important, you're deluded. As a human being, you are one in over 7 billion, going on 8 billion. You do not matter much to Nature, even if you're in a position of genuine power. And even if we accept that humanity as a whole matters to Nature, it's unlikely that you and I do. Nature has already lived for billions of years. Our lives are gone in a blink of its eyes.

Ecosystem Evolution

Most likely, the force of evolution does not act on an individual species but on an ecosystem. By ecosystem, we mean a geographical area where many species

The Domain of Air

co-exist in dynamic harmony. In the life of an ecosystem, predators and prey are symbiotes rather than enemies. If one particular species grows too dominant, then other species rein it in.

The enemy of an ecosystem is any species that is truly disruptive and cannot be reined in. The introduction of the cane toad into Australia provides a good example of this. It was brought from Hawaii by Australia's Bureau of Sugar Experiment Stations. The hope was to control the native population of grey-backed cane beetle, which were detrimental to sugar crops. It was a forlorn hope. Following their release, the toads quickly went feral and multiplied. Now, 80 years later, their population is estimated to number over 200 million. They have damaged or devastated natural ecosystems throughout Queensland. They turned out to be a kind of pathogen.

Nature seems unable to respond to such an event quickly—otherwise it would already constrain the cane toad. The opposite appears to be happening. The cane toad has developed (or evolved) larger legs and is spreading more rapidly at 60 km per year rather than 40 km per year.

Does Nature even care?

We cannot know for sure, but she probably cares about her ecosystems. The reason for thinking this emerges when we begin to examine individual species not simply as a species but as an ecosystem unto itself. In recent years the evidence has piled up that pretty much every multicellular species is an ecosystem. So is *Man*.

The latest estimates suggest that we possess something in the region of 30 trillion human cells and about 39 trillion microbial cells. The microbes dominate us, but not by much if you just count the population. But if you count the population of genes, it is wildly lopsided. Our cells are armed with about 25,000 genes, but our microbiome, with its 10,000 or so microbial species, wields many more—about 12,500,000 genes.

And aside from the pathogens (of which there are relatively few), they are all playing on our team.

What do they do?

The full extent of how they serve us can only be guessed. We know they play a major role in digesting food, extracting vitamins, storing fats, and supporting our immune system. We probably wouldn't survive without them. That's definitely the case for grazing mammals, like cows, sheep, and deer. They depend entirely on gut microbes to break down the fibers in the plants they eat.

And, if you think we all have similar microbiomes, it ain't necessarily so. Microbiologists who assembled an inventory of different human microbiomes discovered that there wasn't a specific "core" microbiome that everyone shared. Some bacterial species were common, but none seemed to be ubiquitous. There seems to be a core of functions that the microbiome helps with, like extracting certain nutrients from food, but it's not always the same microbe.

Ecosystem Isolation

The evolutionists of contemporary science have been stuck with a problem ever since Darwin first described and popularized the idea of species evolution. Natural selection is masterful at changing species characteristics (bigger, smaller, more fur, less fur, and so on). Dog breeders learned this long ago.

But no biologist has ever been able to provoke a new species into existence by experimental means. There were fruit fly experiments that combined breeding with irradiation of the flies that could cause mutations in the fruit fly's DNA. It created mutant four-winged flies. No genetic changes provided a "positive advantage," and no new species emerged. Nowadays, evolutionists simply blur the distinction between the changes in characteristics caused by selection and new species, which they have never demonstrated, by suggesting that a new species will "happen eventually."

Evolutionists are well aware that isolated populations evolve. It was the uniqueness of flora and fauna of the Galapagos islands that set Darwin alight. The same evolutionary diversity occurs in the Hawaiian Islands, the Canary Islands, the Caribbean Islands, Madagascar, New Zealand, Australia, and other places.

Unfortunately, this diversity just provokes awkward questions to which biologists have no good answers. Why, for example, are there no marsupials in Asia, Africa, or Europe. There is just one in North America; there are 120 different species in South America and about 250 in Australia. Why only in those places?

Why did the horse, elephant, and camel die out in North America and South America? It seems to have happened around 10,000 years ago. It seems insane. Those are exactly the animals that it would have been good for man to domesticate. While humans of 10,000 years ago may not have been capable of that (who knows), horses and camels are good survivors. And the idea that they were hunted to death by unwise indigenous folk is laughable.

The Domain of Air

The likely solution to this and a thousand other such questions is that ecosystems rather than species evolve. That seems to be what happens at the microbial level, so most likely, it happens at every scale. Species become extinct in specific contexts rather than globally when the ecosystem fails them.

If ecosystems achieve a stable harmony, they are likely to persist for very long periods. The historical record seems to suggest that even man, the greatest of all disruptors, seems not to have inflicted much damage on any ecosystems until the age of exploration began. We will revisit the topic of evolution in the next chapter.

The Emotional Center

It helps to remind oneself that *Vertebrates* have no intellectual center. One consequence of this is that they do not deliberate; they decide quickly and act. They do not think; they feel. They do not construct moralities or mental models. They are incapable of measuring, comparing, and formulating in the way that men do.

Vertebrates know things in a different way from *Man*, with their highest level of knowing coming through the emotional center. All centers learn and acquire knowledge. Consider how birds make nests. The ability is critical to their reproductive survival, and they are born with it. The bird's instinctive center knows how to select a site for a nest, what to gather to build it, how to build it, and what size —and it must coordinate with the moving center to make it happen. And yet each new generation of birds must learn how to fly by moving center imitation. Oddly, this is something that flying insects do not need to be taught.

The British milk bottle phenomenon provides a well-documented example of avian learning. Around 1918, British dairies began to seal milk bottles for doorstep delivery with aluminum foil. In 1921, in a suburb of Southampton, several blue tits learned how to peck through the milk bottle tops to feed on the cream beneath. Within 30 years, the entire British blue tit population had acquired this knowledge and subsequently passed it on to Europe.

It spread by imitation. However, as the years passed, the knowledge was eventually lost, probably due to various factors, from milkmen covering the bottles to doorstep delivery becoming less fashionable. Nowadays, although doorstep milk delivery still occurs, the blue tits have forgotten what they once knew. At some point, the knowledge lost its importance, and a generation of blue tits grew up that simply never knew.

Another interesting avian phenomenon is the dawn chorus. It's a behavior that lacks a convincing scientific explanation. Most likely, it is simply an expression of joy at the rising sun, especially as it is louder in spring. If so, it is an example of *Vertebrates* expressing emotions.

In *Man* and in *Vertebrates* too, the enteric nervous system (ENS) is the substrate for the emotional center, or at least the lower part. The biology here is complex. *Invertebrates* have what is called a peripheral nervous system. It divides into three separate subsystems, the somatic system, which serves the moving center; the autonomic system, which serves the instinctive center; and the ENS, which governs the organs of digestion and serves the instinctive center.

In *Vertebrates*, the brain and spinal cord are more developed, and both the moving and instinctive centers rely upon and to some degree inhabit the spinal cord. The ENS is part of the peripheral nervous system and has the same organic functions for *Invertebrates*. The ENS appears to be present in all animals, from small invertebrates like the hydra, to vertebrates with complex central nervous systems, like the elephant.

In vertebrate embryo development, the nervous system divides into three parts at a very early stage, the ENS, the spinal cord, and the head brain. In *Man*, these three parts are the substrate for the three brains. In *Vertebrates*, the head brain does not form a separate brain—it becomes the center for the collection of impressions.

Organically, the emotional center appears to be an evolutionary child of the instinctive center. It seems to make logical sense when we consider medical traditions that associate emotions with particular organs. We need to be careful here not to infect ideas about *Vertebrates*' emotions with *Man*'s emotional defects. With that in mind, perhaps the following associations have some validity:

- **The heart**: affirming, expresses love or joy; denying, rejection.
- **The lungs**: affirming, pride; denying, sorrow.
- **The liver**: affirming, generosity; denying, melancholy.
- **The spleen**: affirming, trust; denying, anxiety.
- **The small intestine**: affirming, patience; denying, impatience.
- **The large intestine**: affirming, confidence; denying, nervousness.
- **The kidneys**: affirming, courage; denying fear.

Here we are not focussing on *Man* with all his specialized negative emotions that do not exist in the *Vertebrates* world. With *Man*, negative emotions are based initially on instinctive emotions that have intellectual postures linked to them. They are learned by imitation as a child, from family and peer groups.

In *Vertebrates*, emotions have the function of amplifying or economizing the creature's energy during specific activities. Domesticated animals clearly express emotions, acting pleased when they please their owners. Those who have experience with horses, perhaps the most visibly emotional of all domesticated animals, know full well that the horse's power comes through its emotions.

The domestication of animals probably occurred through conditioning linked directly to feeding—to the instinctive center. Those who train dogs know well that emotional interactions with the animal are part of the training. The emotional center is subject to conditioning in animals as well as humans. *Vertebrates* clearly have a formatory apparatus that can be conditioned. It just happens to be wordless.

The Work teaches that emotions have cognitive value. Knowing something includes an emotional state. Consider, for example, what a horse knows when it moves like the wind.

The Body Kesdjan

The word "animal" comes from the Latin noun *animale* meaning "living being, or being which breathes." Both it and the adjective *animalis* are derived from the noun *anima* meaning "breath, soul; a current of air."

That a "soul" is associated with the air (with the breath) is a common ancient belief that spread beyond the Roman world. It is in the pre-Christian Greek tradition and the Hindu and Buddhist traditions (as prana). It's also there in the Hebrew tradition. When God created Adam, he exhaled into Adam's nostrils, giving him life.

In the Work, the second body or body Kesdjan relates to the octave of breathing. *Figure 63* provides a simple definitive overview of the four (possible) bodies of man. Gurdjieff said that having a body Kesdjan was not a necessity for man. It was a luxury that a normal man could easily live without. The physical body possesses all the functions man needs.

According to this diagram, the center of gravity of the emotional center lies in the body Kesdjan rather than the physical body, although in neither *Vertebrates* nor *Man* is the body Kesdjan fully formed (i.e., crystallized). It depends upon the physical body. With *Vertebrates*, the air octave cannot be helped as much by the octave of impressions as it can in *Man*. Nevertheless, a small

1st Body	2nd Body	3rd Body	4th Body
Carnal body or Physical body	Natural body or Astral body or body Kesdjan	Spiritual body or Mental body	Divine body or Causal body
Carriage (body)	Horse (feelings, desires)	Driver (mind)	Master (I, consciousness, will)

Figure 63. The Four Bodies of Man

amount of H12 is created for use by the higher emotional center. For a vertebrate, the emotional center spans both the lower and higher manifestations of emotion.

The lower emotional activity is strongly associated with the instinctive center, which works with the same energy (H24) as the lower part of the emotional center. The lower emotional center will normally resonate with the affirmations and denials of the instinctive center.

The body Kesdjan, or astral body, is composed of fine matter and resides within and around the physical body, providing it with an atmosphere. Under normal circumstances it has the form of the physical body. It remains within the physical body until death, when the physical body begins to decay, and the astral body separates from it. If this body crystallizes in *Man* before death, it survives the death of the physical body and is indestructible within the confines of the Earth.* However, it can die all the same. It must learn to survive if it is to avoid a second death.

With *Vertebrates*' astral bodies, there is no possibility of crystallization, so at death, the astral body either disintegrates or immediately attaches itself to another life, with no memory of its previous existence. In *Vertebrates*, the body Kesdjan breathes in harmony with the physical body, and its heart beats in harmony with the physical heart.

The Higher Emotional Center

Evidence suggests that the higher emotional center functions in animals to some degree. It enables some animals to sense events, particularly disastrous events, before they happen. Animals have been observed to become anxious before strong earthquakes and volcanic eruptions. It may be the case that they detect various vibrations that human beings do not, but that does not explain their rational behavior. The tsunami of December 26, 2004, provided a

---

* In Search of the Miraculous by P D Ouspensky, p41

surprising example. It was caused by an earthquake along the floor of the Indian Ocean. It claimed thousands of lives in Asia and East Africa, including more than 35,000 people in Sri Lanka. Sri Lanka's Yala National Park lies on the South-Eastern tip of Sri Lanka. The tsunami impacted the whole of Sri Lanka's East coast, however, wildlife officials from the park reported no mass animal deaths among the elephants, leopards, monkeys, or other mammals or reptiles in the park. Most of the animals had moved inland before the tsunami came ashore. In some way, they knew and also knew what to do (either individually or collectively).

Man

> "What a piece of work is a man, how noble in reason,
> how infinite in faculties, in form and moving how
> express and admirable, in action how like an angel!
> in apprehension how like a god! the beauty of the
> world, the paragon of animals—and yet, to me,
> what is this quintessence of dust? Man delights not me—
> nor woman neither, . . ."

Shakespeare puts these words in Hamlet's mouth. Expressing in his masterly manner man's apparent virtues and at once decrying him. Given the record of history and the pageant of our times, these words speak to us across the centuries.

Occupying the highest square in the domain of AIR, *Man* is master of the animal kingdom. His intellectual center marks him out among the rest of Nature's creatures. The advantage it confers on him is unique. By comparison, other species are imprisoned by their biological capabilities and awkwardly limited by their ecosystem.

Man has the power to cast off such constraints. It is almost as if the Earth were his playground. He digs deep beneath the Earth's surface to acquire metal ores that he forges into tools and weapons. He siphons vast volumes of oil from underground chambers. He carves stone and fabricates bricks for his dwelling places. He works clay into pottery. Trees serve him with material for his furniture. He turns forests into farmland and wild plants into crops. He domesticates animals to serve his purposes; dogs, donkeys, horses, camels, and elephants are his slaves.

Like Prometheus, he has stolen fire from the Gods, and he uses it in a thousand ways: for cooking, expanding the broad array of substances that can serve him as food, for automation of his factories, for moving himself over the land

and the sea, and through the air, and beneath the oceans. He has subjugated geothermal forces, dammed rivers for hydroelectric power, and harnessed the wind and the waves. His electric grid spreads across the planet, and so do his communication networks.

The intellectual center's potentialities are indeed remarkable, and *Man* has explored and exploited most of them. Nevertheless, he is a child of Nature. And although he knows this well enough, he pays her scant regard. As a species, he appears to suffer from a fundamental flaw. He has the possibility of being Nature's true ally, the gardener of the planet and the guardian of harmony—but, in this day and age, few people are interested in that possibility.

Air in Summary

Like liquids, gaseous substances exhibit convection and diffusion, but the process occurs more rapidly because gases are far less dense. They flow more easily and do not exhibit surface tension. Whereas water will happily dissolve a huge variety of salts and other substances, the Earth's atmosphere will accept only a few gases, and most of those will be washed out of the air by rain. So the atmosphere is composed almost entirely of nitrogen and oxygen (about 99%). The rest is water vapor, carbon dioxide, methane, and noble gases (argon, helium, krypton, etc.).

However, the air acts as a distribution mechanism for substances that rise above the Earth: sands from sandstorms and volcanic ash are among the material it carries far and wide. A sufficiently large volcanic explosion can dim the atmosphere for a year or two, to the point where famines result from the failure of harvests. Under the right conditions hurricanes and tornadoes form that can achieve air speeds up to 190 mph. Air is not an electrical conductor, but it is a medium through which plasma can and does flow. Hence the air is host to both gentle clouds and violent thunderstorms.

The speed of life in the domain of Air is far faster than in the domain of Water. The cheetah can achieve speeds of 75 mph (120 km/h) and the Peregrine falcon 242 mph (389 km/h). At ground level, wind speeds can reach over 250 mph (408 km/h). Man can outdo all of that with trains that go faster than 350 mph (550km/h) and airplanes that go as fast or even faster than bullets.

Walking speed for man (and many animals) is about a meter per second, which is also the speed of blood circulation. The speed of information along nerve pathways is 35 to 120 m/sec (268 mph or 430 km/h). Human reaction times to stimuli are in the region of one-tenth of a second.

The Domain of Air

Planetary Influence, Astrology

Gases are generally transparent, so visible light easily passes through. Nevertheless, the Earth's atmosphere is quite selective in respect of the EMR it allows in. Most EMR does not get through. The upper atmosphere blocks a good deal of short wave radiation: gamma rays, X-rays, and ultraviolet light from the Sun and elsewhere. The ozone layer is instrumental in this.

The greenhouse gases, carbon dioxide, water vapor, and methane serve to regulate the Earth's surface temperature. Too great an abundance and the surface heats up; too little, it cools. Another cause of cooling is volcanic eruptions that throw dust into the atmosphere. The dust blocks EMR that would otherwise heat the Earth. So it cools until the dust clears.

Earth's ionosphere blocks EMR with wavelengths longer than 10m, but shorter wavelengths in the range of 5cm to 10m make it through as, of course, does visible light. Because of this, until we were able to use high altitude balloons and satellites for observation, our knowledge of the cosmos was limited to analyzing a limited range of EMR.

What the atmosphere blocks and what it admits is important. The planets can directly affect organic life on Earth by two mechanisms, by the EMR they transmit to Earth and by the impact of their electric and magnetic fields. Organic life on Earth is a receiving station for planetary influences, and some of that comes through EMR.

Gurdjieff described the planetary influence on life, and thus man, as being similar to a big wheel in the sky surrounding the Earth with nine spotlights of different colors fixed to it.* Different lights dominate life on Earth according to how physically close or distant a planet is. Each planet influences the Earth by emanations (magnetic influence) and radiations.

Taken together, two planets have a joint magnetic effect; when they approach one another, it becomes stronger. The influence they have at a particular time on a particular location on the Earth's surface is proportional directly to the cosine of the angle made at that point by the two planets.

The Planets affect mankind in general and men on an individual level. The light (EMR) shining from the planets has a direct influence at the time of conception, determining a man's type. The typing that Gurdjieff mentions probably corresponds with the human planetary typing described by Rodney Collin as types of essence,** with types related to the light of the planets.

---

* *Views From The Real World* by G I Gurdjieff, Lecture Feb 24, 1924, p191
** *The Theory of Celestial Influence* by Rodney Collin, p221

Gurdjieff's Hydrogens: The Ray of Creation

A newborn baby's first breath exposes it to the atmosphere for the first time. It experiences the plasma state of the air in its vicinity which is affected by the the emanations of Planets. This determines the *do* of the infant's breathing octave, which has a direct impact on essence.

Gurdjieff said that the planetary influence was not as modern astrologers suggest in the sense of planets determining an individual's fate. Most men are not under the law of fate, so their fate passes them by. In *The Tales*** Gurdjieff mentions genuine astrologers existing long ago and using horoscopes both to assist in treating individual illnesses and for selecting suitable marriage partners. That same tradition still exists in some areas of South India.

** *The Tales* by G I Gurdjieff, Ch XXIII, p287

Chapter 9

The Domain of Fire

—∿—

"We cannot steal the fire. We must enter it."

~ Sufi Proverb

In *The Tales*, Gurdjieff explains that two different principles can determine a three-brained being's duration of existence. One principle, he calls Fulasnitamnian and the other Itoklanoz. He writes:*

> "The first kind or first 'principle' of being-existence, which is called 'Fulasnitamnian,' is proper to the existence of all three-brained beings arising on any planet of our Great Universe, and the fundamental aim and sense of the existence of these beings is that there must proceed through them the transmutation of cosmic substances necessary for what is called the 'common-cosmic Trogoautoegocratic-process.'
>
> "And it is according to the second principle of being-existence that all one-brained and two-brained beings in general exist wherever they may arise...
>
> "And the sense and aim of the existence of these beings, also, consist in this, that there are transmuted through them the cosmic substances required not for purposes of a common-cosmic character, but only for that solar system or even only for that planet alone, in which and upon which these one-brained and two-brained beings arise."

Gurdjieff clearly states that *Man* has a specific role in respect of the Trogoautoegocratic process regarding the substances he creates. It is his natural role, his intended position in Nature, and that this role is potentially far-reaching. The substances he creates result from how he digests food, air, and impressions—how he lives his life.

---

* *The Tales* by G I Gurdjieff, Ch XVI, The Relative Understanding of Time, p130

Gurdjieff's Hydrogens: The Ray of Creation

He also writes:*

> ...in the beginning, after the organ Kundabuffer with all its properties, had been removed from their presences, the duration of their existence was according to the 'Fulasnitamnian' principle, that is to say, they were obliged to exist until there was coated in them and completely perfected by reason what is called the 'body-Kesdjan,' or, as they themselves later began to name this being-part of theirs—of which, by the way, contemporary beings know only by hearsay—the 'Astral-body.'

He says that, at one point, it was natural to perfect the body-Kesdjan within a single lifetime, mentioning in passing that the process of coating and perfecting is carried out by reason. However, because of man's failings, Nature adapted and imposed the second principle, Itoklanoz, on *Man*. He writes:*

> "...they began to exist already excessively abnormally, that is to say, quite unbecomingly for three-brained beings, and when in consequence of this they had, on the one hand, ceased to emanate the vibrations required by Nature for the maintenance of the separated fragments of their planet, and, on the other hand, had begun, owing to the chief peculiarity of their strange psyche, to destroy beings of other forms of their planet, thereby gradually diminishing the number of sources required for this purpose, then Nature Herself was compelled gradually to actualize the presences of these three-brained beings according to the second principle, namely, the principle 'Itoklanoz,' that is, to actualize them in the same way in which She actualizes one-brained and two-brained beings in order that the equilibrium of the vibrations required according to quality and quantity should be attained.

We can think of the domain of AIR, organic life on Earth, as obeying the Itoklanoz principle. It applies to most of humanity, particularly those who show no interest in self-perfection. The domain of FIRE is under the influence of the Fulasnitamnian principle, and those who seek to perfect themselves seek to enter the fire.

The Plasma Cycle

Figure 64 illustrates the Earth's plasma cycle. The Earth is negatively charged relative to the Sun, as are all the Planets. The surface of the Earth is negatively charged relative to the ionosphere. The ionosphere spans the region from about 50 miles (80 km) to 600 miles (1000 km) above the surface. The iono-

---

* The Tales, Ch XVI, The Relative Understanding of Time, p131

The Domain of Fire

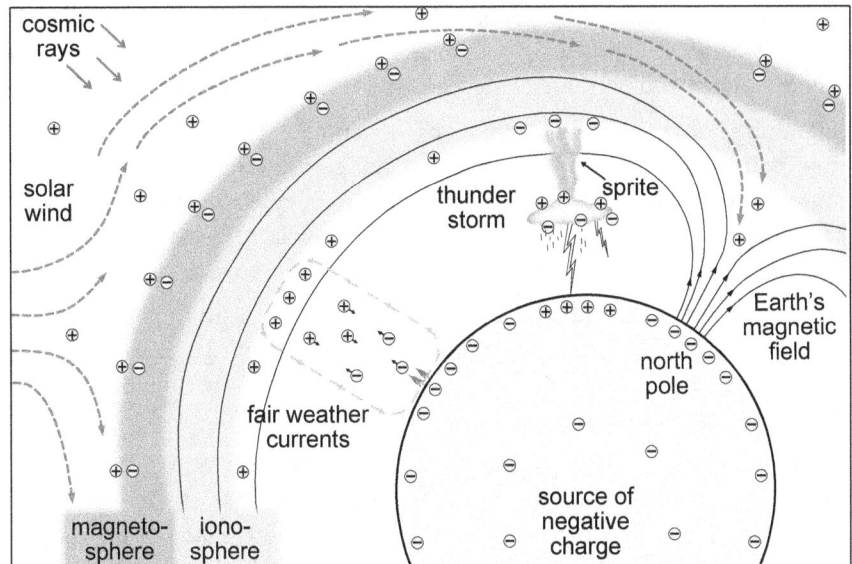

Figure 64. The Plasma Cycle

sphere is almost all plasma, a mix of both positively and negatively charged ions. The voltage, the potential difference, between the surface and the ionosphere is in the range of 250-300 kV.

As a consequence of this potential difference, positive ions tend to drift down toward the surface, and negative ions tend to rise into the atmosphere. It causes a "fair-weather current" in the atmosphere, which gradually reduces the voltage. The alternative scenario is a thunderstorm when the gentle, fair-weather convection is absent.

Thunderstorms are not yet understood in fine detail. It is generally agreed that a strong updraft of air provokes them. The main area of electrical activity is in the center of the thundercloud. The rising moist air cools rapidly to temperatures in the range −15 to −25°C (5 to −13°F). A mixing of supercooled cloud droplets, small ice crystals, and graupel (soft hail) follow. As the rising ice crystals collide with graupel, the rising ice crystals acquire a positive charge and the graupel a negative charge.

It creates the situation illustrated in *Figure 64*, with the lower half of the thundercloud having a negative charge. Positive charge is attracted to the ground beneath, setting up the situation where a current (a lightning bolt) can flow between the two regions. The lightning reduces that potential difference. However, a potential difference also forms between the top of the cloud and the ionosphere, where negative charge gathers, just as positive charge gathered on the ground. This potential difference resolves by a discharge that is often visible. Such phenomena are named sprites, red sprites, blue jets, and elves.

Lightning bolts can produce "fountains" of gamma rays which travel upwards from points quite low in thunderclouds. These are so intense they can blind sensors on satellites hundreds of miles away. They also create anti-matter particles. The gamma-ray bursts precede the associated lightning discharge, just as gamma-ray bursts precede the visible light in a nova or supernova.

The upper atmosphere discharges occur concurrently with the lightning. So in fair weather, the voltage between the ionosphere and the Earth's surface gradually reduces but with thunderstorms, it increases. A harmonious balance exists between these distinct movements of plasma.

However, there are other plasma effects in play. The Sun emits a continuous stream of positive ions, mainly protons, referred to as the solar wind. The solar wind emanating from the Sun's equatorial belt into the plane of the ecliptic moves relatively slowly, varying between 300–500 km/s. The solar wind emerging from close to the Sun's poles typically has a velocity of 750 km/s and a higher temperature.

With its double layer of plasma, the Earth's magnetosphere guides the solar wind around and away. Nevertheless, some of this stream of positive ions breaks through the magnetosphere, coming down to Earth at the poles. Attracted by the Earth's negatively charged surface, it descends in parallel with the Earth's magnetic field at the poles.

Thus the Earth constantly receives large amounts of positive charge, sometimes boosted by energetic solar flares. There are no other major electrical effects in play here. Energetic EMR arriving from space—gamma rays, X-rays, and ultraviolet—is not deflected by magnetism. Most of it is absorbed by the atmosphere before it gets to the surface, ionizing the upper atmosphere in an electrically neutral way.

As *Figure 64* indicates, there is a source of negative charge at the center of the Earth. Contemporary scientists tend to wiseacre in their efforts to explain how that can be. From an objective science perspective, one possible theory is this. The Absolute, as the Holy Immortal, is a ubiquitous source of positive charge. This current flows through to galaxies and is distributed by galaxies to suns and by suns to planets. This flow is complemented to some degree by the Absolute, as the Holy Firm, which is a localized source of negative charge at points of stability (the centers of moons, planets, and suns) scattered throughout the universe.

The Sun directly feeds the Earth with positively charged plasma at its polar regions. It is attracted to the negative charge of the Earth, which seems to continually refresh itself. The resulting current, flowing through the Earth, contributes to all activities on the Earth.

Domain of FIRE

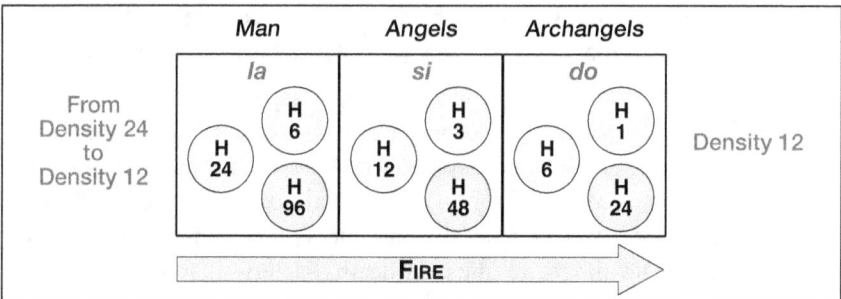

Figure 65. The FIRE Triple

It is important to realize that normal men do not live within the domain of FIRE. *Man* appears in this domain only as an evolving man (man number 4 or higher). Normal man may step into this realm in some of his higher moments. He can remain permanently when his body Kesdjan has been coated and crystallized.

FIRE's density ranges from 24 (under the law of fate) to 12 (no longer under the law of fate). At the lower end of this domain, *Man*, under the law of fate, is subject to the Planets. Nevertheless, he has even greater possibilities that can raise him to the level of the Sun. To achieve this, he must develop a third body: the spiritual or mental body.

When we consider the three squares in the domain of FIRE, we are beset by a shortage of reliable data and a plethora of hearsay. At levels below this, we can harvest data from science, as long as we carefully separate scientific theories form the associated data.

We can acquire some data about *Man* from psychological and other scientific sources. Additionally, we can glean concepts and data from the Work, some of which we may verify personally. But when it comes to the squares titled *Angels* and *Archangels*, there are few reliable data sources, so we need to use the Law of Seven deductively.

Gurdjieff's Hydrogens: The Ray of Creation

Man as a Cosmos

From the perspective of objective science, biological evolution is a force exerted by the Planets at the level of an ecosystem. The planets need organic life as a medium through which to communicate with the Earth, and they influence the structure of that medium according to geographical location. The idea of a species evolving independently makes no sense. A successful species requires an ecosystem in which it has a necessary role, where it can eat and be eaten. The fossil record indicates (but doesn't prove) that ecosystems evolve in a kind of mutual harmony. Geographically isolated ecosystems (Madagascar, New Zealand, Australia, etc.) provide circumstantial evidence for this.

It is, however, a complex topic with multiple aspects. It would be much simpler if a single organism, a sheep or a cow, say, was an independent creature, but it is not. It is itself an ecosystem that hosts a large array of microbes. And so, of course, is *Man*. Microbes are found almost everywhere within the human body. They have colonized the hair and skin and the whole of the digestive system from mouth to anus. At death, bacteria that were previously harmless symbiotes turn into demolition agents and gradually consume the lifeless body.

Some microbes are well known for causing disease. Thrush, the common cold, AIDs, cancers, tuberculosis, and many other diseased conditions are caused or aided by various microbial fungi, bacteria, or viruses. And incidentally, the responsible microbes are often found in many other human hosts, causing no disease at all. And if we ever doubted the ability of a microbe to have a dramatic world-changing impact on humanity, Covid-19 has provided an example.

The human microbiome is far more influential than we might suspect. Cell for cell, we are outnumbered about 3 to 1 by the microbiome. We host around 1000 different species of microbe in the gut alone. And each human microbiome is thought to be unique.

The microbial population of the gut is composed predominantly of bacteria rather than viruses or fungi. About a third of the strains are common to most adults, while the rest vary according to genetics, diet, age, state of health, and, of course, geographical location. The primary symbiotic role of gut microbes is to break down food that we cannot digest and extract vitamins, notably vitamin K and most B vitamins, that would otherwise simply pass through.

The microbiome interacts with the immune system, although the relationship between the two is as yet unclear. Experiments with bacteria-free mice

The Domain of Fire

demonstrated that the immune system becomes underdeveloped in the absence of gut bacteria. Gut bacteria are implicated (sometimes positively and sometimes negatively) in obesity, diabetes, inflammatory bowel disease, various forms of cancer, asthma, heart disease, and even depression.

Infants are born with a starter pack of gut microbes. The fetus may ingest microbes from amniotic fluid. Colonization occurs rapidly during birth (from bacteria in the birth canal) and after birth by direct contact with the mother and through breastfeeding. By the age of about three, the composition of a child's microbiome resembles that of an adult.

The Solar System Parallel

There is a parallel between the microbiome, organic life in *Man*, and organic life on Earth. We show this in *Table 7*.

Note	Solar System	Step Diagram	Man
sol	Sun	*Archangels*	Heart
fa	Planets	*Angels*	Organs
interval	Nature (Organic Life)	*Plants, Invertebrates, Vertebrates, Man*	Microbiome
mi	Earth	*Minerals*	Gut
re	Moon	*Metals*	Womb (Fetus)

Table 7. Parallels Between Man and The Solar System

To be clear, the parallel is between the lateral octave from the Sun and the equivalent lateral octave in *Man*. Note that when we include the Solar System, as we have done in the second column of the table, there appears to be a natural correspondence between *Archangels*, the Sun, and the Heart in *Man*. Similarly, there is a correspondence between the Planets and *Angels*, which Ouspensky noticed.* *Archangels* are solar "gods" and *Angels* planetary "gods."

The parallel is quite startling in the sense that neither *Man* nor the bacteria that serve us so well is aware of their role and their contribution. It was only relatively recently (in the 17th century) that we became aware that bacteria even existed. The Sun, however, is probably well aware that we exist, although we have no idea of how it thinks of us.

Planetary Influence and the Endocrine System

The endocrine system is a bodily communication system based on releasing hormones into the bloodstream by specific (endocrine) glands in various

---
* *In Search of the Miraculous* by P D Ouspensky p324

locations throughout the body. In plants every cell can produce and circulate hormones, whereas, in animals, specific glands are responsible for doing so.

Collectively the hormones regulate behavior from birth to death, with responsibility for long-term activity such as the body's growth, and immediate activity such as the fight or flight response to a threat. The hormones are purpose-designed molecules that lock into receptors on the organs they communicate with. The arrival of a hormone can set off a cascade of activity. Insulin, for example, provokes the rapid uptake of glucose into muscle cells. If you think of the body as a family of organs that needs to operate in harmony, the endocrine system is a mechanism for preserving the harmony.

In *The Theory of Celestial Influence*,* Rodney Collin suggests correspondences between the Planets and the endocrine glands—linking perhaps to Gurdjieff's comments we reported earlier. *Table 8* on the next page expands on this idea, listing the hormones produced by each gland and the organic roles they play.

The functions of the endocrine glands govern necessary life processes. It is unlikely that planetary influence triggers activity. It most likely modifies activity, amplifying the influence of particular glands at particular times and thus increasing the secretion of hormones. We note that the thymus gland kick-starts part of the immune system and then atrophies once the adolescent years are over. Curiously, both the anterior and posterior pituitary glands have a controlling role in respect to some of the other endocrine glands. For example, the anterior pituitary gland can trigger the adrenals.

We are not aware of anyone gathering direct data on the psychological influence of the endocrine glands, although personal observation can confirm that human types are a genuine phenomenon: saturnine types are generally tall intellectuals, martial types are generally physical and confrontative, and so on. The hypothalamus is regarded as the bridge between the central nervous system and the endocrine system. It releases hormones to the pituitary glands, which are stimulated to release hormones into the bloodstream. Those hormones cause specific behaviors by the organs they target.

These hormones are H96, the energy of the immune system, which employs such substances. Vitamins and the proteins of the complement system are substances at this level. Most likely, the Law of Three acts in the endocrine system by mixing H48 (nerve impulses from the senses) with H192 (blood or compounds in blood absorbed by the endocrine gland) to produce H96, hormones

---

* *The Theory of Celestial Influence* by Rodney Collin, p221

The Domain of Fire

Planet	Endocrine Gland	Hormone	Function
Sun	Thymus	Thymosin	Stimulates T-cells
Moon	Pancreas	Insulin	Energy management, digestion
Mercury	Thyroid	Triiodothyronine, Thyroxine	Regulates metabolism, respiration
Venus	Parathyroid	Parathormone	Calcium (bone) management, Vitamin D synthesis and phosphate management in kidneys
Mars	Adrenals	Adrenaline, Aldosterone, Cortisol	Fight or flight response, increases blood flow, blood sugar, sodium conservation
Jupiter	Posterior Pituitary	Oxytocin, vasopressin	Social behavior, reproduction, childbirth, milk production, blood pressure management, fluid control
Saturn	Anterior Pituitary	Somatotropin, Corticotropin, Thyrotropin, Lutropin, Lactotropin	Promotion of growth, management of liver, activation of adrenal gland, activation of thyroid gland, activation of gonads
Uranus	Gonads/Testicles	Testosterone, Androstenedione	Sexual function, management of male characteristics
	Gonads/Ovaries	Estrogen, Progesterone	Sexual functions, management of female characteristics
Neptune	Pineal	Melatonin	Higher intellectual functions

Table 8. *The Planets and the Endocrine Glands*

that are secreted into the bloodstream. Hormones in the body carry the neutralizing force.

The Law of Fate

The general theory here is that the endocrine glands receive influences from the Planets. If no other influences intervened, then, according to the time and place of their conception and birth, an individual's life would proceed entirely under planetary influence, that is, under the law of fate. But this rarely happens because most individuals live under the law of accident (at the level of the Moon) or the law of cause and effect (at the level of the Earth).

Gurdjieff's Hydrogens: The Ray of Creation

For someone living under the law of accident, life events are accidental according to the following process. At birth, a child's mind is a blank slate. As life events gradually proceed, the contents of the slate, memories, habits, and attitudes accumulate. It occurs through interactions with a series of individuals: parents, relatives, siblings, peer groups, and educators. Added to that is the media's influence—in this age, primarily television, computer games, and the Internet.

This process is the formation of personality, which forms a shield for essence but in most cases forms a prison that eventually suffocates the essence. From a certain point in childhood, the essence no longer participates in responding to events. The person responds using only habit. The ability and tendency to imitate are part of this. Most adolescents build complex personalities through the imitation of their friends, peer groups, and media stars.

For most, the essence never recovers and ceases to participate in the immediate responses to the events of their life. They only respond mechanically. Even if the response appears clever, it will be mechanical. It will arise by association linked to some stored memory linked to a specific behavior pattern. If we think of mankind as a living being, then we should expect the life of most of its cells to be predictable in this way, just as the behavior of the cells in a man's body are predictable.

Planetary influence can trigger the behavior of the masses by acting on just a few influential people. When they act, a cascade of imitative behavior proceeds. You can witness this in large events that affect the masses—wars, riots, disasters, and so on. You can also witness this in the tide of fashion, whether it is fashion in clothes, or art, or music, or science, or politics. An idea triggered in even a single individual can provoke a meaningless fashion. It is the nature of man under the law of accident; his mechanicality is his destiny.

Those who are under just 48 laws at the level of the Earth are better placed. They have grown up wise enough to avoid falling precisely in line with the mass of humanity. In general, they are decent people, good householders. They live their lives under the law of cause and effect. When they incur debts, they pay them.

The law of fate is a term used to describe the influence of the Planets on organic life, including their influence on humanity. It may seem paradoxical that it does not apply to many individuals, but it is not. If you want someone to do something, you ask them. You do not ask the permission of every cell in their body. Relatively few cells in their body will be involved in deciding how to respond to your request. Those cells correspond by analogy to those few men and women who are under the Law of Fate. Such individuals may not be perfected; however, they are at a higher level under fewer laws.

The Mechanism of Evolution

The fossil record indicates that a specific order of events took place in the evolution of biological life. *Table 9* shows this in the context of the Step Diagram. It begins with the presence of bacteria. We suspect that bacteria exist in most planetary environments. There is some evidence for this hypothesis. A NASA scientist reported the detection of fossilized bacteria on three meteorites that proved not to be identifiable as a strain of Earth bacteria.* It seems likely that DNA and life at the bacterial level form quite easily everywhere.

The more intriguing question is what happened next and why.

Objective science asserts that life on Earth has two specific functions: transmitting influences from the planets to Earth and substances from biological life on Earth to the Moon. The form that life takes should, in theory, conform to such a purpose. Thus the evolution of Eukaryotes appears to be a surprisingly intelligent way to initiate evolution. It begins with a strand of life that creates oxygen from sunlight and a strand of life that consumes oxygen for the sake of physical energy. It seems unlikely that this was a happy accident.

It is at least possible that the Sun and its family of planets, including the Earth and the Moon, intended it. No other planetary atmospheres are dominated by nitrogen and oxygen, the two gases that support fundamental life processes. The natural question is, "what agency was capable of provoking the development of Eukaryotes?" What is at work here is an intelligence beyond the intellect of *Man*. Rather than assign a label like "archangel," it may be better to simply think in terms of something with the level of intelligence of the Sun.

The Eukaryotes fill the *mi-fa* interval in the lateral octave from the Sun. They are the microscopic seeds of all plant and animal life. And they are also the agency by which the atmosphere of the Earth was able to completely change its character through microbial activity that clearly spanned a very long time frame.

As far as we know from space probes, there is no evidence of biological life beyond bacterial life on any other planets in our system. If that's the case, then for the other planets in our solar system, the *mi-fa* interval in the Ray of Creation is filled in some other way. Logically, if the Earth (or possibly the Sun) intended the birth of an infant planet (the Moon), there was a specific process

---

* https://www.reuters.com/article/us-meteorites-life-idUSTRE7252KQ20110307

Square	Octave	Biological Life	Geological Period	Est. Date
Kernel	do	-		
Metals	re	-	Archaean	4,600 m.yrs
Minerals	mi	Bacteria	Archaean	3,500 m.yrs
	int.	Eukaryotes	Proterozoic	2,000 m.yrs
Plants	int.	(Sea) Plants and invertebrates	Proterozoic	600 m.yrs
Invertebrates	fa	(Land) Plants and invertebrates	Paleozoic	488 m.yrs
Vertebrates	sol	Vertebrates	Paleozoic	397 m.yrs
Man	la	Man	Holocene	Recent
Angels	si	-		
Archangels	do	-		

Table 9. The Speed of Evolution

by which it happened involving the development and evolution of biological life.

There is abundant circumstantial evidence, particularly in the area of symbiotic species, that ecosystems rather than species evolve, although the mechanism is unclear. Since the new species can only result from changes to specific genes, the same change to DNA needs to occur many times. There needs to be many more than one pair (male and female) of the new species created in the same mating season. The gene pool needs to be as broad as possible for the sake of in-species diversity. So a significant population of the prior species needs to participate.

Species from the microbial level up to the largest species and across all classes of creature must maintain a balance. A dynamic harmony must emerge between predator and prey, between parasite and host and between all the symbiotic species: the mutualists, the commensalists, and the endosymbiotes. These interspecies relationships, which appear critical to an ecosystem, may need to be preserved or possibly reforged for a changed ecosystem to emerge. The Trogoautoegocrat requires a level of harmony.

Nature must perform such a bio-engineering task whenever needed. The evidence that Nature is capable of this is the world we live in. From the fossil record, we know there have been major extinction events that destroyed and disrupted ecosystems world-wide, and Nature recovered. No doubt it can recover from less disruptive events.

The Domain of Fire

Nature's aim for organic life included the evolution of a three-brained being. The fossil record suggests that *Plants* and *Invertebrates* developed first in the sea and later on land. By 397 million years ago, two-brained beings (*Vertebrates*) had developed. The subsequent development of *Man* took considerable time.

Such a development may be relatively hazardous. Gurdjieff said that Nature had tried to people the planet five times and that *Man* was the fifth (and perhaps even the final) attempt. The modern scientific view is that homo sapiens first appeared in Africa about 200,000 years ago.* This perspective is predicated on the idea that the human intellect (the thinking center) somehow just invented itself.

Gurdjieff said otherwise. When asked whether man evolved from animals, he replied, "No—man is a different formula."** If so, there must have been some kind of agency that transformed a hominid two-brained being into a three-brained being.

It is an open question as to how much Nature influences *Man*'s activity and how much comes from *Man*'s initiative. Gurdjieff insisted, for example, that the building of skyscrapers was Nature's intention rather than *Man*'s, as was the growth in human population, the latter being a case of Nature substituting quantity for quality.

Man, Organic Life and Plasma

We may think of cells as microscopic creatures with the consistency of gelatin that absorb water and breathe. They span all four elements, including plasma, which plays a fundamental role in cell activity. The cell wall is a barrier that denies entry to most molecular substances. Some molecules, particularly water, oxygen, and carbon dioxide, pass through the cell wall by diffusion. If the cell needs oxygen and it is plentiful outside the cell, it simply diffuses in. If the cell has too much carbon dioxide, it diffuses out. However, heavier molecules like glucose cannot gain entry that way.

Cells' walls are punctured by porins—protein channels that guide required substances into the cell. Every human cell has porins for glucose. Inside the cell, glucose combines with oxygen in a biochemical reaction to create ATP (adenosine triphosphate). ATP rather than glucose is the true high-energy fuel of the cell.

---

* https://en.wikipedia.org/wiki/Human_evolution
** Ladies of The Rope p 28

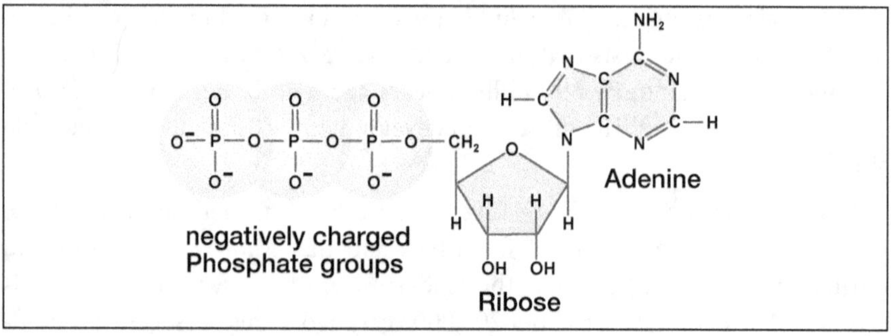

Figure 66. The Ionic Structure of ATP

The conversion process has three stages. The first is glycolysis which occurs in the cell cytoplasm and breaks glucose down into pyruvate molecules. From this point on, the mitochondria in the cell complete the process. A mitochondrion is a cell within a cell. Mitochondria have cell walls, cytoplasm and porins, and their own DNA. They are the engines that create ATP from pyruvates.

The next two stages in the process of ATP creation, the Krebs Cycle and the Electron Transport Chain, are a complex series of chemical and electrical interactions involving the manipulation of both protons and electrons across the inner membrane of the mitochondria. ATP created in the mitochondria then passes into the outer cell.

As illustrated in *Figure 66*, ATP molecules have three negatively charged phosphate groups attached to a combination of Ribose and Adenine. They are ionic plasma. The phosphate groups can be used one at a time to provide energy, and the Adenine and Ribose can be recycled back into a mitochondrian for the re-addition of phosphate groups. It is a repetitive cycle that consumes glucose and oxygen and generates carbon dioxide.

ATP works as follows. The cell is negatively charged in relation to its external environment, just as the Earth and other planets are negatively charged in relation to their external environment. There is thus a voltage (a potential difference) between the cell and its external environment.

The porins conduct water molecules to ATP, causing hydrolysis, which removes one of the phosphate groups. The energy thus released can be used, for example, to transport positive ions (of magnesium, calcium, or whatever) into the cell. Alternatively, the negative charge can transmit an electrical signal, as happens when neurons communicate.

ATP is also instrumental in the process of photosynthesis. It is one of the fundamental pillars of biological life, and it is plasma.

Intelligence

Electricity is fundamental to life in the cell, and it is instrumental in many bodily functions. Every muscle contraction is initiated by an electrical signal and made possible by the depolarization of cellular membranes, followed by influxes of ionized minerals into the muscle cells. The nervous system sends signals throughout the body and to the brain, using electricity to relay information—impressions.

Everything in the realm of intelligence: feelings, language, deductive capability, insight, and so on, occurs in plasma. We naturally apply sophisticated pattern recognition to the signals we receive from the retina, the eardrum, the taste buds, the nerve endings, and so on. Interestingly, rough imitations of neural activity by computers (so-called neural networks) can mimic the level of sophistication of biological nervous systems.

The physical structure, the hardware that computers use, is very different. It is silicon-based, whereas life is carbon-based. Silicon sits directly below carbon in the Periodic Table and hence has some similar atomic properties, but it is denser. The software techniques that computers implement may well be almost identical to human intelligence, since man built them. Artificial intelligence (AI) is a creation of human intelligence.

While it has become so sophisticated that it can outperform man in many fields, much of that is the brute force application of silicon. Computers are incapable of pondering or experiencing revelation. The science-fiction-inspired idea that the machines might become "aware" can be set aside for lack of evidence. Their highest achievements are roughly equivalent to the lower parts of the lower centers—exceptional mechanical intelligence.

Plasma and Organic Healing

For many years, medical science denied the possibility that electricity had any role to play in healing. It naturally led to the emergence of fringe plasma-related activities that claimed to have a positive impact on health: ionizers, electric blood purifiers, earthing, which can include tree-hugging and walking barefoot on grass, acupressure, acupuncture, Reiki, and therapeutic approaches. The evidence or the lack of it for any of these techniques tends to be anecdotal.

However, there has been one interesting and deeply surprising series of experiments carried out by Bob Becker M. D., which he documents in detail in *The Body Electric*.* Becker spent almost thirty years carrying out experiments on salamanders—lizards capable of regenerating amputated parts of their body.

What he demonstrated, among many other important results, was that when a salamander's leg is amputated, and it begins to grow a new one, the growing point, which is called the blastema, is negatively charged. If the negative charge increases, regrowth is faster, and if it is opposed, then regrowth slows down or stops.

Wherever biological repair is in progress, it is enhanced by a flow of negative charge to the damaged area and suppressed by an opposite flow (in human or animal bodies). This book, incidentally, was first published in 1985. Since then, there has been less skepticism about electrically-based therapies, but they have not become mainstream.

Being Bodies and Gradations of Reason

In *The Tales*, Gurdjieff uses one of his invented words, "Rascooarno," to denote death. Etymologically the first part of this word is from the Russian prefix *rask*, which means "to break apart." In respect of the various possible bodies of man, death is a process of breaking apart. The physical body feeds something on Earth, whether consumed by bacteria, worms or incinerated to ashes.

Its higher parts, its instinctive center, and everything within the realm of personality become food for the Moon. These are the plasma components of the physical body—the personality and the instinctive center. The body Kesdjan separates, and so does the mental body. What subsequently happens to them depends upon their level of development. Both may reincarnate in some way or simply dissipate after a time.

In *The Tales*, Beelzebub frequently refers to man's normal reason as "bobtailed," implying that it is diminished. He describes the various gradations of the Reason that are possible for man, with these words:**

> "Data for these three kinds of being-Reason are crystallized in the presence of each three-brained being depending upon how much—by means of the 'being-Partkdolg-duty'—the corresponding higher-being-parts are coated and perfected in them, which should without fail compose their common presences as a whole.

---

* *The Body Electric* by Robert O Becker and Gary Selden
** *The Tales*, Ch XXXIX, The Holy Planet "Purgatory", p770

The Domain of Fire

> "The first highest kind of being-Reason is the 'pure' or objective Reason which is proper only to the presence of a higher being-body or to the common presences of the bodies themselves of those three-brained beings in whom this higher part has already arisen and perfected itself, and then only when it is the, what is called, 'center-of-gravity-initiator-of-the-individual-functioning' of the whole presence of the being.
>
> "The second being-Reason, which is named 'Okiartaaitokhsa,' can be in the presences of those three-brained beings, in whom their 'second-being-body-Kesdjan' is already completely coated and functions independently.
>
> "As regards the third kind of being-Reason, this is nothing else but only the action of the automatic functioning which proceeds in the common presences of all beings in general and also in the presences of all surplanetary definite formations, thanks to repeated shocks coming from outside, which evoke habitual reactions from the data crystallized in them corresponding to previous accidentally perceived impressions."

The third kind of being-reason described here is man's normal reason, which operates entirely as automatic responses to external impressions based on associations. The second kind is possible only when the body Kesdjan is perfected. The first kind reflects the situation where the third body, the mental body, has been perfected.

In describing a three-brained being's existence according to the Fulasnitamnian principle, Beelzebub mentions the perfection of the body Kesdjan by Reason up to the sacred "Ischmetch."*

In describing the levels of Reason in *The Tales*, Beelzebub refers to the following gradations, ordered from lowest to highest: Martfotai (self-individuality), Degindad, Ternoonald, Podkoolad, and Anklad. He also says this, in respect of an archangel: *

> But the Reason of the sacred Podkoolad, to which Beelzebub had already perfected himself, is also very rare in the universe, hence even the venerable archangel prostrated himself before Beelzebub because his own degree of Reason was as yet only that of the sacred Degindad, i.e., wanting three degrees to the Reason of the sacred Anklad.

It suggests that in his possible evolutionary path, *Man* can rise beyond the level of the *Archangels*. It seems likely that *Angels* have a level of being that corresponds to a perfected body Kesdjan. They may also have a corresponding

---

* The Tales, Ch XLVII, The Inevitable Result of Impartial Mentation, p1177

mental body, perfected at least to the level of self-individuality. Something similar will also be true of *Archangels*.

Angels and *Archangels*

The life-forms above the level of *Man* have a higher "*Hydrogen* of being" than H24. *Angels* are H12; *Archangels* are H6. They are invisible to our normal senses, so it naturally follows that they are not protein-based creatures grown from DNA and RNA. Most likely, their bodily structures are invisible plasma structures based, perhaps, on hydrogen.

Square	Hydrogen	Function
Archangels	H6	Higher Intellectual Center activity
Angels	H12	Higher Emotional Center activity
Man	H24	Moving Center or Lower Emotional activity
Vertebrates	H48	Impressions, Lower parts of centers, Formatory apparatus
Invertebrates	H96	Animal magnetism, Vitamins, Hormones
Plants	H192	Breath—Air, Oxygen, Nitrogen, Carbon Dioxide
Minerals	H384	Water, circulation and transport
Metals	H768	Elasticity, Muscle, Food for Man, Fats, Carbohydrates
Kernel	H1536	Wood, Bone

Table 10. The "Average" Hydrogens

Table 10 lists the Step Diagram squares and the *Hydrogen* that represents them (their average *Hydrogen*). As discussed in Chapter 5, the average *Hydrogen* is the middle *Hydrogen* of the psyche of the class of creature. For example, in *Plants*, it is H192, so the three *Hydrogens* H384, H192, and H96 represent automatism, emotion, and reason (or adaptive intelligence). So for *Plants*, H384 (water) governs salt and hormone circulation, not just through a plant's body into its environment but also by communicating with other plants through its root system. Its emotionality is H192 (air), which is the heart of its life—it facilitates the breathing of all organic life. Its reason is H96, the basic plasma of life, inspired by the Sun.

For *Angels*, the three *Hydrogens* are H24, H12, H6. Their automatic manifestation, H24, corresponds to the normal (identified) feelings for *Man*, although probably oriented to above. Their emotionality is H12, which is

The Domain of Fire

impartial—the energy of higher emotion and the organ of conscience. Gurdjieff told Martin Benson, "The angels are pure, and there is no place for them to go. We on this Earth are fallen angels, but we have a place to strive for, objectively and actively to come to."*

Angels are asexual, as is everything higher than *Man*—sex is a capability of organic bodies. The reason of *Angels*, at H6, is beyond the level of Man's normal understanding. It seems logical to suggest that *Angels* feed on air and impressions, in a similar way to *Man's* body Kesdjan. Consequently, *Angels'* bodies must have corresponding "organs" to digest such substances. We can say very little about *Archangels*, beyond assuming that their impartiality is automatic, as they are even further above our level. In theory, higher beings form a hierarchy above *Angels* and *Archangels*. Below we provide an edited summary of the Wikipedia entry about this, which derives from Biblical sources.

- **Seraphim**: Highest in the hierarchy. They are said to surround and maintain the divine throne. Literally translated, the word means "burning ones." They have six wings, of which two cover their faces, two cover their feet, and two are for flying.
- **Cherubim**: Second in importance, Cherubim are said to have four faces: an ox, a lion, an eagle, and a man. Alternatively, they have the legs of an ox, the body of a lion, the wings of an eagle, and the head of a man. They guard the way to the tree of life in the Garden of Eden and the throne of God.
- **Thrones**: Also referred to as Ophanim, these are throne angels, symbols of absolute justice and authority. They are said to express and relay the wisdom of God. The Hebrew word "ophanim" means "wheels." The Hebrew text of Ezekiel's description of God's chariot uses the word. The Ophanim are thus sometimes represented as fiery wheels.
- **Dominions**: Fourth in the hierarchy, they are also referred to as Lordships or Angelic Lords. They are said to preside over the duties of lesser or lower angels and hence rarely appear physically before men. They are said to resemble angels, as usually depicted, with wings and divinely beautiful faces. They carry swords and scepters, embellished with orbs of light.
- **Virtues**: Fifth in the hierarchy, the virtue angels are the ones through which signs and miracles appear in the world. They are the holy source of virtue, perfectly attuned to the Absolute, the source of virtue itself.

---
* *Gurdjieff: Mysticism, Contemplation, and Exercises*, Joseph Azize, p243

- **Powers**: Sixth in the hierarchy, and also called Authorities, the primary duty of these angels is to supervise the movements of the heavenly bodies to ensure that the cosmos remains in harmony. They oppose the work of demons and evil spirits. Artists' depictions represent them as soldiers wearing full armor, with weaponry such as shields and spears.
- **Principalities**: Also referred to as Rulers, these are seventh in the hierarchy. Their task is to guide and protect nations and tribes and also the institution of the Church. They are middle management, obeying orders from above and presiding over the ordinary angels below them. Artists depict them wearing a crown and carrying a scepter. They are the educators and guardians of the Earth, the inspirers of art and science.
- **Archangels**: Tradition asserts that there are seven archangels: Chamuael, Michael, Raphael, Uriel, Gabriel, Azrael, Jophael (names may vary). They are usually assigned specific roles. For example, Raphael is concerned with healing, and Michael opposes Satan. The word "archangel" means "chief angel," from Greek. Artists portray them as regular angels with a pair of wings.
- **Angels**: The word "angel" (from the Greek angelos) means "messenger." Angels are the lowest order of celestial being, supposedly the most concerned with the affairs of men. Guardian angels fall into this class; it is commonly believed that everybody is assigned one.

To complicate matters, sources within Judaism, Islam, and Christianity do not agree precisely, although it is clear that they relate to each other. Other traditions, Zoroastrianism, Hinduism, and Buddhism, present yet other descriptions of a hierarchy of intelligent beings beyond man's level. The common thread is that such a hierarchy exists, and some of its members have contact with humanity.

In most traditions, a hierarchy of demons also exists with different roles with respect to organic life. It is over-simplistic to divide the invisible world between "good guys and bad guys." In an ecosystem, such an approach makes no sense. There are not good species and bad species; a whole range of species comprise the ecosystem.

The Trogoautoegocrat

The Trogoautoegocrat is an octave of ecosystems within an octave of ecosystems, within an octave of ecosystems, descending through 7 levels of scale. The Earth can be thought of as the note *mi* in a Ray of Creation or as a living being which is also a whole octave composed of other notes. Mankind is one of those notes, the note *la* in the lateral octave from the Sun. But mankind itself is an octave, and so on.

Gurdjieff defined the three cosmic traits of a creature (or species) as what a creature eats, what it breathes, and the medium in which it lives.* We can view these traits as a different perspective on three kinds of food. Every creature occupies a specific place in a food octave. Its role in the Trogoautoegocrat is to consume specific food, breath specific air, and take in impressions from the specific medium in which it lives. Balancing this, it will, by what it excretes and by its death, produce foods for other creatures in other octaves.

For example, earthworms consume decaying vegetable matter. They breathe the air within soil through their moist skin, which behaves as a kind of external lung. Their perception is limited to the experience of light (it has no eyes), sensations, taste, and smell. It is food for a host of other creatures (birds, hedgehogs, badgers, reptiles).

Its metabolic waste includes nitrogenous material, such as ammonia, urea, uric acid, and salts. It thus contributes to the nitrogen cycle for the benefit of plants. It breathes oxygen and produces carbon dioxide. But what about its psychic life, whatever that is? It absorbs impressions (H48), and this must generate some psychic activity. But what happens to the substances produced in that way? What feeds on them?

Table 11 on the following page shows the range of *Hydrogens* in man from H768 to H6 alongside the substances H1536 to H12. These are the eliminated residues from the normal use of those *Hydrogens*. These residues naturally become food for other creatures within the Trogoautoegocrat.

In Row 8 of this table, the digestion of food (H768) produces the unusable residues we excrete. The excreted materials serve as food for bacteria and the *Plants* world. We discard some organic H1536, nails, and hair while alive. Other H1536 teeth and bones become available to the ecosystem on death.

On Row 7, H384, chyme, and water are both used by the body. We discard H768, dead skin, urine, and sweat. Dead skin is food for dust mites and similar small creatures. Urine and sweat feed the soil and ultimately plants or bacteria. In Row 6, inhaled oxygen, H192, leads to the exhaling of carbon dioxide (H384). The oxygen is first circulated to the body's cells which create the carbon dioxide that is then returned to the bloodstream to be exhaled by the lungs along with water vapor (H384). The excreted carbon dioxide serves plant life, whereas water serves all life.

---

* *In Search of the Miraculous* by P D Ouspensky, p320

Gurdjieff's Hydrogens: The Ray of Creation

Row	Food Octave	Substance	Elimination	Substance
1	H6	Objective consciousness	H12	Higher emotions
2	H12	Self-remembering, self-observation, sex	H24	Considered behavior
3	H24	Representation, negative emotions	H48	Animated behavior, negative expression
4	H48	Nerve impulses, impressions	H96	Automatic physical activity, behavior
5	H96	ATP, Vitamins, Hormones	H192	Sweat/heat, protein fragments, lost aura substance
6	H192	Oxygen	H384	Carbon dioxide, water vapor
7	H384	Chyme, Water	H768	Urine, sweat, dead skin
8	H768	Proteins, Fats, Carbohydrates	H1536	Excreta, nails, teeth, bones

Table 11. Man's Food and Elimination

With Row 5, we come to substances that are plasma (ionic) in their usage. Substances such as vitamins and, particularly, ATP (H96) play a dominant role in the moving center's activity and the instinctive center. All muscles function because of the electrical force of ATP. H96 dominates the immune system. In their battles with pathogens, lymphocytes make extensive use of it. Physical exercise and immune system activity generate heat. A local increase in temperature (inflammation) occurs whenever the immune system goes into action in a big way. Most, but not all diseases, raise body temperature.

Exothermic reactions dominate muscular activity—we burn glucose and fats. The same is true of immune system activity. So the body has its mechanisms for excreting heat, chief of which is sweating. The beneficial effect of steam baths and saunas goes beyond the cleansing of the skin. They enable the excretion of waste materials that our bodies no longer require, and they relieve the immune system of having to deal with the pathogens that feed on such waste matter.

Internally specialized cells, phagocytes run an internal garbage disposal system. They hoover up the remains of dead cells, including our own, and dead viruses and other pathogens. Fragments of protein (ribose, guanine, and many other compounds) find their way into urine via the kidneys. They are food for bacteria once excreted.

The Domain of Fire

Hormones

We denote Hormones as H96, partly because of how Gurdjieff explained them in one of his lectures. He said:*

> *The work of human centers, the speed of which is so different and which are so easily influenced by extraneous matters introduced into the organism, is controlled and governed in the organism itself, under normal conditions of existence, not only by currents passing through the nerves to the brain but also by a certain chemical action inside the organism.*
>
> *The theory of hormones in modern physiology is a fairly close illustration of the state of affairs in our organism. It was thought for a long time that psychic centers communicate with one another and with external organs by means of nerve-ducts. This is to a certain extent true, but it does not exhaust everything that can be said about the relationship of centers to one another and to the external organs of perception, as well as to the periphery of the body in general.*
>
> *The theory of communication through nerves failed to explain many facts, among them the extraordinary speed of communications inside the organism, because transmission by means of nervous ducts everywhere requires a certain time, however short. And a transmission in all directions and a total subjugation of the whole organism to some one emotion, some one feeling, would require a certain length of time, easily recorded and calculated, if the period of time was as long as seconds. Observations show, however, that these transmissions and subjugations take place instantaneously, without any possibility of establishing the interval of time between the impact and the result. This is the result of the activity of hormones.*
>
> *Hormones are clouds of fine matter, finer than the gaseous matter known to us, which is given off by various organs of our body. These clouds permeate our whole organism with incredible rapidity and, intermingling, are the cause of the state in which the organism finds itself at a given moment. Moreover, they also constitute the atmosphere of emanations which envelops a human organism for a certain distance and which under certain conditions can even be seen. These emanations or radiations of the organism connect it with the fine atmosphere which surrounds it and which penetrates the atmosphere in which we move and breathe.*
>
> *The radiations of the human body, or rather, the network of radiations which forms the emanations, is of two kinds: First, the absorption, the*

---
* Gurdjieff's Early Talks 1914 - 1931, p45

sucking into the organism of certain substances from the surrounding atmosphere, and second, the throwing out of certain matters from the organism. If the activity of radiations of the second kind is too intense, the organism uselessly loses its energy. If the activity of radiations of the first kind predominates, the organism gets stronger and healthier. Certain forms of nervous diseases and disorders, for instance, contusions, falls, and bruises-when there is no definite traumatic injury-depend on the violation of the right radiations. A strong shock may break off radiations, but such a breaking off is possible only with a very strong and quick shock. The slow movement taking place around us does not break off radiations because the vibrations of radiations are so quick that slow movement cannot affect them.

As regards people influencing one another, there is a great deal that has remained unknown to Western European science. But experimental investigations in esoteric schools have established the fact that there are people who, by their emanations, have a good or a bad influence on others. There are people who give to others and take from them and thus, as it were, counterbalance one another. But there are other people who give too much and take nothing in return, and yet others who seem to take energy from other people. One or another state of the emanations of our body determines the state of our health. Right and healthy emanations make a man completely or almost immune from infectious diseases, for many microbes perish in the atmosphere of man's healthy emanations.

Both the science of Gurdjieff's era and modern science depict hormones as little more than chemical messengers. However, modern science has greater knowledge of how they act and has a longer list of such substances. The hormone molecule may be the transport for a smaller ion delivered to a cell receptor, in the same way that hemoglobin acts as transport for oxygen and carbon dioxide. As Gurdjieff indicates, the hormones in a person's atmosphere can be acquired by others or from others. We also secrete pheromones (also called ectohormones) from our skin into the air.

The Plasma Substances

With *Row 4* in *Table 11*, we depart from what we usually think of as substances and encounter plasma. We have many biological receptors for impressions: the skin, the tongue, the nose, and olfactory system, the eardrums, the semicircular canals (for balance), and the eyes. Each receives external impressions and translates them into nerve impulses, H48, which collectively arrive at the part of the brain called the amygdala, where all inputs are merged and processed.

H48 moves in a wave through the nervous system from neuron to neuron, carrying information. As the content of the impulse varies, it may be that the 'waves' carry information in a similar way to how radio waves carry sound and images. The operation of the nervous system as a whole is not yet well understood, although the neural network clearly filters signals at the level of H48.

The amygdala is the physical location of the formatory apparatus, which acts as a central data distribution point to the centers. Typically people respond to external impressions through mechanical behavior in the second state (waking sleep). They consume impressions and excrete behavior. The excretion of an automatic physical response is clearly H96, muscular activity fueled by ATP. The impression data is stored but not evaluated. Instead, the formatory apparatus responds automatically.

The waste matter of the formatory apparatus frequently leads to endless repetitive chatter that proceeds without the participation of the thinking center or the emotional center. Alternatively, we mindlessly and passively watch television almost without interest.

Row 3 is where impressions are digested, where centers are working with H24. The moving/instinctive center's ability to present information comes into play. An image or a moving image is one kind of (plasma) structure; sounds are another, taste and touch are others. These structures can be thought of as similar but distinct according to the sense involved. *Man* experiences and responds to the blended result. However, it is composed of pieces, just as a movie is composed of individual frames.

Man notices and processes differences. We observe this when, for example, a small gnat appears in our peripheral vision. We notice it immediately even though it is a small fraction of our visual input, which is itself only part of all the data our senses are feeding us. Also, in receiving information, we bring our associations (memories) to the party. These are also plasma structures.

At this level, man generally responds more animatedly than in simple waking sleep, although it may be driven by identification and largely automatic. Most of what we regard as our life experiences, our imaginary fears and hopes, our negative responses, our "intelligent" conversation, dreams, and much more is the manifestation of H24. As with all other points in the food octave, the H24 can ascend or descend. If it is the expression of negative emotion, then it may infect whoever receives it. If a man works to suppress negative emotions—a definite inner struggle—the H24 may rise to become H12.

If there is no inner struggle, behavior at the level of H48 will be the result, whatever the impressions ingested. Such behavior in another is readable. A man's facial expressions and micro-expressions, of which he is usually ignorant, can be accurately interpreted by skilled observers.

On Row 2, we arrive at the substance of self-remembering, self-observation, and sex. The emanations from this level of activity are subtle. In many contexts, they are simply not observed and are only likely to be observable by someone in a similar state. Certain kinds of affirmation and even certain kinds of laughter come from this level. We can think of this in general as considered behavior. For example, a person who is consciously acting a part will consider their actions in line with their attention. The sex act is one where there is at least the opportunity of presence.

On Row 1 of the table, we encounter very rare states. In almost everyone's life, they are momentary. For most of us, objective consciousness, which depends on H6, is a theoretical state that can only be fully understood by those who experience it. Sexual ecstasy is the only likely point of entry for the average person. But in the fraction of a second where this might be possible, men and women descend into oblivion rather than ascend into an extraordinary awareness. And in the union of man and woman, new life, which requires the full gamut of *Hydrogens*, is created.

The Second Lateral Octave

From an intellectual perspective, the question as to whether there are higher life-forms than man, including a "heavenly hierarchy," begins with the question:

Is there form in plasma?

This is now known to be the case, although scientific knowledge in this area is relatively recent. The natural subsequent question is:

Are there life-forms composed entirely of plasma?

While objective science asserts that this is so, and logically must be so, it is difficult, if not impossible, to establish by employing modern scientific methods. To do so requires an apparatus that can examine plasma structures in fine detail. MRIs and other plasma-sensitive devices provide images only in the human context. But even if someone invented such an apparatus, it is unlikely that plasma beings would volunteer to become lab rats. Thus, we are left with the situation that the only apparatus able to interact with plasma life-forms is man himself. So man can be the crucible, but whatever he discovers will not be accepted by mundane science.

The Domain of Fire

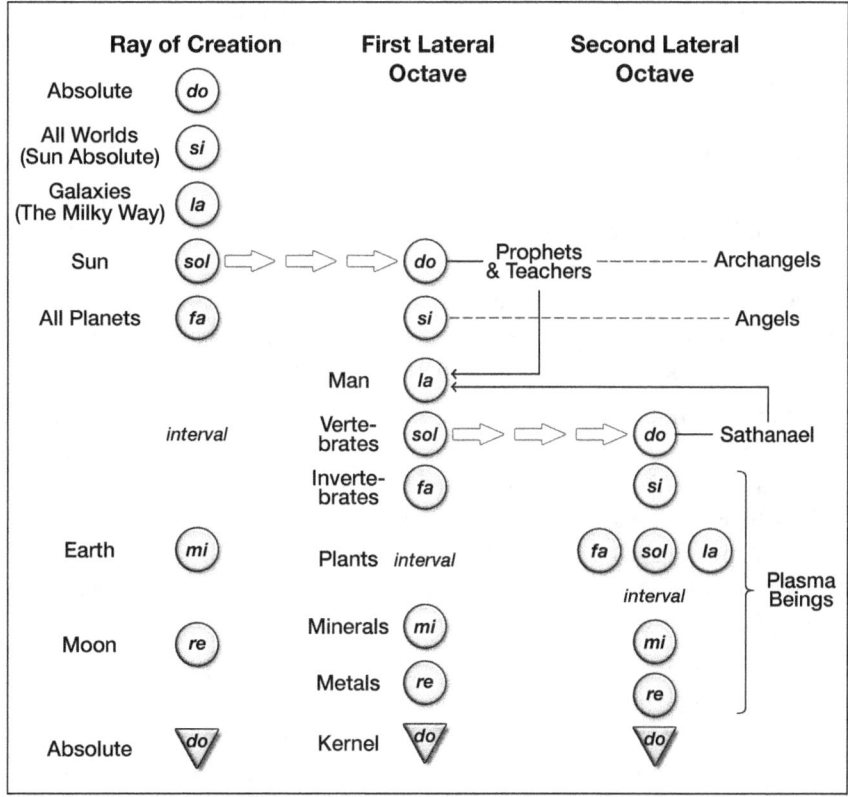

Figure 67. The Ray and the Two Lateral Octaves

The Ray of Creation can help establish an idea of the landscape of plasma beings. *Figure 67* illustrates an approach to this. It depicts two lateral octaves rather than just one. The note *sol* in the first lateral octave initiates the second lateral octave. This is the note corresponding to *Vertebrates*. At the level of *do* in the first lateral octave, we have the archangels: Chamuael, Michael, Raphael, Uriel, Gabriel, Azrael, and Jophael. This is also the level of the Sun, from which "messengers from above" are said to descend to teach and assist *Man*.

Traditions vary in their description of the nature of such messengers (Krishna, Moses, Buddha, Lao Tzu, Christ, Mohammed, and others). Sometimes scripture suggests that such individuals evolved by their own efforts. Alternatively, the holy record may assert that they were born perfected.

Whatever the truth about such men, each left an indelible mark on *Man*, helping many individuals pursue self-perfection.

Gurdjieff's Hydrogens: The Ray of Creation

If we now look to the second lateral octave, initiated by the note *sol* of the first lateral octave, we have assigned Sathanael to the note *do*. We chose to use the name Sathanael with care. This name is usually shortened to Satan,* possibly to indicate that he was cast down from the level of the *Archangels* or to conceal his origin at that level.

Sathanael is the representative of the Moon on Earth. He is the Lord of the world for life-forms at the level of *Vertebrates* and below. So he is depicted as hairy, cloven-hoofed, sexually lascivious, with goat's horns and a tail. It is also the way that the Greek god Pan is depicted. Pan is the Greek Nature god, clearly associated with sex, the companion of nymphs and the god of shepherds and flocks, and rustic music. That is a depiction of Sathanael.

Both the Catholic Church and most Protestant sects depict God and Satan as adversarial, with Satan as a demonic antagonist who opposes God's plan for man's salvation. This idea has remained fashionable for centuries. Irrespective of how he is depicted, Christian, Islamic and Judaic sources regard the Devil as 'a real personage,' which is how Gurdjieff described him. However, the Devil's true role is not to wage war with heaven; it is to preside over organic life and represent the interests of the Moon.

The notes of the second lateral octave are most probably plasma beings of various orders. We can refer to these as elementals. They most likely feed on the psychic excretions (dreams, wishes and prayers, imagination, hopes and fears, anger, hatred, and other negative excretions) of *Man*. Some may feed on the psychic excretions of beings at lower levels, from *Metals* to *Vertebrates*. There are likely many such plasma creatures at every level, contributing to the Trogoautoegocrat in their own way. Possibly, they too serve and feed the Moon.

Folk traditions claim that there are 'deities' or spirits associated with rocks, standing stones, volcanoes, mountains, streams, and rivers, plants, trees, groves, and animals of all kinds. They are common in myth and folklore, going by various names: nymphs, devas, elementals, gnomes, udines, sylphs, fairies, etc. None of these 'Nature spirits' appear to play a significant role in the life of *Man*. However, Sathanael does. He represents and oversees Man's animal nature, creating in him an internal arena for struggle. It is not in Sathanael's interests for many men to succeed in the struggle.

On the one hand, *Man* has a physical body subject to the same biological needs and tendencies as any vertebrate, including the need to procreate. The

---

* In Hebrew Sathanael means "adversary of God," the 'el' ending means "of God."

The Domain of Fire

physical body is thus under the sway of Sathanael. On the other hand, Man's aspirations, for the evolution of his body Kesdjan and mental body are the concern of *Archangels* and "messengers from above."

In the struggle between those two forces, a man's psychic efforts are the third force that determines which direction he will choose. He either squanders his finest energies on the force of sexual attraction and the growth of personality, with all its "glittering prizes," or he pursues B influences and becomes a seeker after truth. The Devil prefers him to takes the former path.

Kundabuffer

In early lectures given years before writing *The Tales*, Gurdjieff refers to the organ Kundabuffer directly. These are his words:*

> *...At the same time Nature has given him the possibility of changing, but this does not mean that any change will necessarily take place...*
>
> *The chief reason is in ourselves, and it is the Kundabuffer as it is called.*
>
> *To understand clearly what this new thing is, we must stop here and go into further details.*
>
> *Nature, in her foresight, has given to man's machine a certain property, which protects the man from feeling and sensing reality.*
>
> *That is the Kundabuffer.*
>
> *Let us take a real fact. All men are mortal and every man can die at any moment. I can imagine that Mr. Smith comes out from the theater, and crossing the street, he falls under an automobile, which crushes him to death. Or a signboard is torn off and falls just on the head of Mr. Jones and kills him on the spot. Or Mr. Brown eats crayfish, poisons himself, and dies the next day without anyone being able to save him.*
>
> *All this, everybody can easily imagine. But, we ask, can anyone imagine that he himself this moment or tomorrow, or in one year or in ten years will also die? Really if we think of this carefully-death is a terror. What is more terrible than death? What would happen if he really imagined this terror, his own death? Can you imagine the terror? You cannot imagine your own death, but you can imagine the death of another.*

Thus, one of the consequences of Kundabuffer is that *Man* finds it almost impossible to be aware of his own death. And yet, an awareness of his death is vital to him. In a different lecture, Gurdjieff explains why. In answer to a question about the Moon, he says:**

---
* Gurdjieff's Early Talks 1914 - 1931, p36
** Gurdjieff's Early Talks 1914 - 1931, p367 (see also p331, p388)

Gurdjieff's Hydrogens: The Ray of Creation

The moon is man's big enemy. We serve the moon. Last time you heard about kundabuffer. Kundabuffer is the moon's representative on earth. We are like the moon's sheep, which it cleans, feeds and shears, and keeps for its own purposes. But when it is hungry, it kills a lot of them. All organic life works for the moon. Passive man serves involution; and active man, evolution. You must choose. But there is a principle: in one service, you can hope for a career; in the other, you receive much but without a career. In both cases, we are slaves, for in both cases, we have a master. Inside us we also have a moon, a sun and so on. We are a whole system. If you know what your moon is and does, you can understand the cosmos.

Kundabuffer serves Sathanael. In *The Tales*, Gurdjieff describes the introduction of the organ Kundabuffer into *Man* with the following words:*

> "You must know that by the time of this second descent of the Most High Commission, there had already gradually been engendered in them—as is proper to three-brained beings—what is called 'mechanical instinct.'
>
> "The sacred members of this Most High Commission then reasoned that if the said mechanical instinct in these biped three-brained beings of that planet should develop towards the attainment of Objective Reason—as usually occurs everywhere among three-brained beings—then it might quite possibly happen that they would prematurely comprehend the real cause of their arising and existence and make a great deal of trouble; it might happen that having understood the reason for their arising, namely, that by their existence they should maintain the detached fragments of their planet, and being convinced of this their slavery to circumstances utterly foreign to them, they would be unwilling to continue their existence and would on principle destroy themselves.
>
> "So, my boy, in view of this the Most High Commission then decided among other things provisionally to implant into the common presences of the three-brained beings there a special organ with a property such that, first, they should perceive reality topsy-turvy and, secondly, that every repeated impression from outside should crystallize in them data which would engender factors for evoking in them sensations of 'pleasure' and 'enjoyment.'
>
> "And then, in fact, with the help of the Chief-Common-Universal-Arch-Chemist-Physicist Angel Looisos, who was also among the members of this Most High Commission, they caused to grow in the three-brained beings there, in a special way, at the base of their spinal column, at the root of their

---

* The Tales, Ch X, Why "Men" Are Not Men, p88

tail—which they also, at that time, still had, and which part of their common presences furthermore still had its normal exterior expressing the, so to say, 'fullness-of-its-inner-significance'—a 'something' which assisted the arising of the said properties in them.

"And this 'something' they then first called the 'organ Kundabuffer.'"

We interpret this excerpt from *The Tales* and other related passages also to indicate that the organ Kundabuffer grows in us while we are in the womb, when men had tails. The embryonic tail usually develops about four or five weeks after conception and is reabsorbed into the embryo after a further three weeks. The text seems to suggest that this introjection happened to "our ancestors." If the development and action of this organ occur as a natural part of the human embryo's growth, then the DNA encoding that provokes it may well have been designed by Sathanael for our ancestors.

The influence of Kundabuffer takes place to us personally while we are embryos. When Gurdjieff refers to "our ancestors" in *The Tales*, it is almost certainly allegorical—he means ourselves at a much younger age. We inherit what we are from our history. We are our own ancestors.

Kundabuffer impacts the moving/instinctive center directly, causing us to adopt an attitude of self-calming from a very early age. It doesn't appear to impact the emotional center until a baby's birth, when the octave of breath and impressions commence in earnest. For the early part of a baby's life, the moving-instinctive center dominates. Instinctive activity aimed at harmonizing feeding with the growth of the child dominates. After that, moving center imitation dominates a child's life. The longer a psychic habit is established, the more difficult it is to break.

Feeding in a Different Way

When Gurdjieff suggests that a particular class of life, say *Plants* (H192), is food for another class of life, *Vertebrates* (H48), he is not referring to specific species. He is referring to whole classes of life. Neither is he referring to the substances that comprise any given species but to the whole class of creatures. The Step Diagram makes no logical sense otherwise. He is talking about one class of cosmos feeding on another class of cosmos.

The nine lower squares of the Step Diagram divide into the two following "feeding chains":

1. **The Center-of-Gravity Series:** *Angels* feed on *Vertebrates* which feed on *Plants* which feed on *Metals*.

Gurdjieff's Hydrogens: The Ray of Creation

2. **The Transition Series**: *Archangels* feed on *Man*, which feed on *Invertebrates* which feed on *Minerals* which feed on *Kernel*.

A cosmos is a living entity with a specific lifetime and lives within its environment by feeding in three distinct ways. It ingests substances that go to form its body; it breathes in substances that energize its body and establish its being, and ingests substances in order to know. It eats, it breathes, it senses.

We can think of creatures being (what they are) and becoming (their evolution or at least their evolutionary possibilities). The center of gravity squares in the Step Diagram make up the center of gravity of each of the domains of Earth, Water, Air, and Fire. They are musical notes of being, the essence of the domain to which they belong.

- The essence of Earth is *Metals*, residing under 96 orders of laws and corresponding to the Moon in the Ray of Creation. *Metals* are the essence of the state of solidity. Their being is H768.

- The essence of Water is *Plants*, existing under 48 orders of laws and corresponding to our planet in the Ray of Creation. *Plants* are the essence of the state of being liquid. Their being is H192.

- The essence of Air is *Vertebrates*, constrained by 24 orders of laws and corresponding to Planets (the family of planets) in the Ray of Creation. Animals are the essence of the gaseous state. Their being is H48.

- The essence of Fire is *Angels*, limited by just 12 orders of laws and corresponding to the Sun in the Ray of Creation. *Angels* are pure—the essence of plasma in its invisible state. Their being is H12.

When Gurdjieff states that a specific class of life, *Metals*, for example, is food for *Plants*, we take it to mean that *Plants* sense and consume the knowledge of *Metals*. It is feeding at a psychic level, where evolution occurs by increasing both knowledge and being.

The Transition squares in the Step Diagram make up the transformation points between each of the domains. They are the points of evolution, expressed as classes of life that straddle two domains.

- *Kernel* marks the evolution from the Holy Firm to Earth. It is born into 96 orders of laws that correspond to the Moon in the Ray of Creation. *Kernel* is the first step, the foundation of evolution. It is H1536.

- *Minerals* bridge the evolution from Earth to Water, from the solid to the liquid state, from 96 to 48 orders of laws, from the Moon to Earth in the Ray of Creation. It is H384.

- *Invertebrates* span the evolution from WATER to AIR, from liquid to gas, from 48 to 24 orders of laws, from the Earth to the family of Planets in the Ray of Creation. It is H96.
- *Man* traverses the evolution from AIR to FIRE, from gas to the invisible form of plasma, from 24 to 12 orders of laws, from the Planets to the Sun in the Ray of Creation. It is H24.
- *Archangels* cross the line between FIRE and the Eternal domain, from the invisible form of plasma to its glowing and its incandescent forms, from 12 orders of laws to 6, from the Sun to the *Eternal Unchanging*. It is H6.

The Step Diagram's evolutionary squares form the path of evolution from the lowest point to the highest, to the *Sun Absolute* itself. Thus, in general, the Center-of-Gravity squares represent cosmoses without the inner triangle, without their own Law of Three. As such, those life-forms are not vehicles for evolution.

The Center-of-Gravity Squares

We regard the clearest example of a *Metals* life-form to be a volcano. We know too little about its internal workings to know whether it is a collective being, like an anthill, or an individual one, like a tree, but we suspect it is individual. It is a silicon being whose knowledge concerns how to gather various metals from the "foundry" at the center of the Earth, particularly iron and magnesium, and transport them to the surface, either into the domain of WATER or the domain of AIR.

Plants can be thought of as similar, although their roots are in *Minerals*, and their transport activity moves metal salts, not metal ores. They are carbon life-forms, and their transport activity leverages the power of liquids rather than the brute force of pressure and heat. *Plants* have transformed the knowledge of *Metals* to be appropriate to the domain of WATER, making it possible to interact positively with the domain of AIR.

The metals that are so prevalent in the magma that pours out onto the Earth's surface, magnesium and iron, are key to photosynthesis. Iron is instrumental in its creation, and magnesium sits right at the heart of the chlorophyll molecule. The being of *Plants* is H192 (Air). They manage the harmony of the composition of the atmosphere.

Gurdjieff said* in respect of Man's four bodies that it was usual to take the average *Hydrogen* (the *Hydrogen* of the middle story) to represent the being of

---

* *In Search of the Miraculous* by P D Ouspensky, p319

	Carnal Body	Kesdjan Body	Mental Body	Divine Body
Reason	H48	H24	H12	H6
Emotion	H96	H48	H24	H12
Mechanism	H192	H96	H48	H24

Table 12. Hydrogens of the Four Bodies

that particular body. In *Table 12,* we show the three stories of each of the bodies with their *Hydrogens.* Thus the being of man's Carnal body is H96, The being of the body Kesdjan H48 and so on.

The highest level corresponds to Reason, the middle to Feeling, and the lower to mechanism (activity). For *Invertebrates,* the three stories are the carnal body; for *Vertebrates,* the body Kesdjan and for *Man,* the mental body.

If we now consider *Vertebrates,* they can be regarded physically as the perfection of both the moving center in all its aspects and the lower emotional center. Gurdjieff once commented that "animals are specialized aspects of man," and indeed, you can look at it that way. The dog spotlights the sense of smell and the emotion of loyalty; the horse is speed and emotional power, the cat is bodily awareness, the hawk is eyesight, the elephant is strength.

The genius of *Plants* is to exploit and stretch the domain of WATER to its limit. Weighed down by 48 orders of laws, they still rise as sequoias, hundreds of feet into the sky, survive as cactus in the driest desert and hold fast as pine trees on steep mountainsides. In the tropics, they are the architects of the rain forests, and beneath the seas, they collaborate with coral to design coral reefs. Aesthetically, *Plants* are astonishing, which perhaps might lead us to consider that the Reason of *Plants* (H96) fully understands aesthetics and cannot help but exhibit beauty.

It is these impressions on which *Vertebrates* feed and which they are also capable of exhibiting. If you've swum above or through coral reefs or just seen footage of those extraordinary underwater gardens, you may be bewildered by their beauty. If so, you may also be bemused by the fact that these beautiful worlds existed for millions of years without human beings setting eyes on them. Perhaps they were not designed for our eyes, but for the eyes of the brightly colored fish that meander through the vivid seaweed and the brilliantly painted coral.

The same may be true of the rain forests with their wild orchids, their colorful reptiles, parrots and hummingbirds, and their leisurely sloths. Why, when the rain falls in the desert, does it bloom with such extraordinary beauty? Why do

The Domain of Fire

	Metals	*Plants*	*Vertebrates*	*Angels*
Reason	H384	H96	H24	H6
Emotion	H768	H192	H48	H12
Mechanism	H1536	H384	H96	H24

Table 13. *The Psychic Hydrogens of the Center of Gravity Squares*

the birds of the early morning choose to sing in unison to greet the rising Sun? Have they learned from the plants that the Sun is the creator of our world?

The *Vertebrates* learn from *Plants* to explore the bounds of the domain of AIR. Mammals and birds have penetrated as far North and as far South as any life-forms. Polar bears haunt the Arctic ice flows and penguins huddle together in Antarctica's snowy wastes. Plants gave seeds to the birds, and the birds dropped them on every island or landmass they encountered.

The being of *Vertebrates* is H48, impressions. Their reason is H24, affirmation and denial unhindered by intellectual discourse. Their essence expresses the idea of presence.

If the knowledge of *Metals* is food for *Plants* and the knowledge that *Plants* have acquired is food for *Vertebrates*, then it follows that *Vertebrate* knowledge will feed *Angels*. Curiously, all of the nine angelic categories of Christian Angelology are described as choirs. Perhaps like the birds of the dawn chorus, they sing for joy, not in praise of the rising Sun but to the glory of the *Sun Absolute*.

And while we choose the word "sing" for the angelic choir, perhaps the word is a metaphor for emanating a specific vibration that harmonizes with a yet higher vibration. In the domain of FIRE, we expect the *Angels* life-forms, constrained by only 12 orders of laws, to explore the whole of that domain, not just the domain of EARTH, but the whole of the solar system.

The Transition Squares

With no data, we can only speculate about the life-forms of the *Kernel*. We deduce that their role is to form and transmute elements. These creatures live under 96 orders of laws and experience only the domain of EARTH.

Mineral life-forms feed on the knowledge of *Kernel* life-forms. We can characterize their existence as wind, waves, and rock, in league with temperature and pressure. Their destiny on Earth's surface may be to become soil or sedi-

Gurdjieff's Hydrogens: The Ray of Creation

	Kernel	**Minerals**	**Inverte-brates**	**Man**	**Arch-angels**
Reason	H768	H192	H48	H12	H3
Emotion	H1536	H384	H96	H24	H6
Mechanism	H3072	H768	H192	H48	H12

Table 14. The Psychic Hydrogens of the Transition Squares

ment to feed plants and provide a medium of existence for some *Invertebrates*. They can be under 96 laws or 48 laws.

Fractures in their structure permit the entry of water which fragments rock through cycles of freezing and thawing. Waves and rivers wear them down; the wind throws dust in their faces, and the rain dissolves them. Their destiny may be to slide back into the Earth and be consumed by *Metals*. They have the possibility of evolution or involution, as do all of the life-forms in the transition squares.

Their being is H384 (Water), their mechanical activity is H768 (in the domain of Earth), and their Reason is H192 in the realm of Air. They understand the transition from solid to liquid. They know about structural strength; they experience organic material and compress it into coal and oil. They know of coral as limestone and of sand both as sediment and sandstone. Their knowledge is of considerable use to *Invertebrates*.

Table 14 illustrates the pattern of feeding between transition squares in the Step Diagram. What is achieved by Reason for the lower being becomes mechanical (habitual) understanding for the next being in the series.

Thus we see the *Invertebrates* implementing what *Minerals* life-forms know. Mineral life, with its tectonic plates, creates an exoskeleton for the domain of Earth. Volcanoes have individual exoskeletons. *Invertebrates* apply such understanding to create their own exoskeletons, and in doing so, they make physical movement possible. In some *Invertebrates*, the exoskeleton is forged from silicon; in others (crabs and insects), it is constructed from hard carbohydrate compounds. Irrespective of its substance, it is external.

The *Invertebrates* straddle the borderline between the domain of Water and the domain of Air. They can be under 48 orders of laws or just 24. Their primary innovation is that they can move. They have muscles that can contract and expand. Rather than create oxygen like their neighbors, the *Plants*, they consume it to generate energy. They have a fully functional moving center and a fully functional endocrine system that can respond to influences from the

The Domain of Fire

Planets. Like insects, they are capable of flight. They have mastered movement through the air.

Their Reason as H48 is of plasma, vibrating at the same speed as light. They are capable of perceiving the world directly, the moving center capable of creating images. They have eyes, organs of perception, and they can respond to external activity through motion.

Man straddles the divide between the domain of AIR and the domain of FIRE. He is under 24 orders of laws but is capable of putting himself under 12 orders of laws. He is under the influences of the Planets, the influence of fate, but may even escape that constraint. Like *Vertebrates* and *Invertebrates* he is food for the Moon, but he can, if he has the being and he so chooses, even alter his relationship with the Moon.

He is a creature of remarkable invention, able to mimic the achievements of lesser creatures. He builds cities into the sky, mimicking termite mounds and anthills. His machines dominate the land and the water and the air. While personally limited, he moves faster than the Cheetah, flies higher and sees further than the Eagle, and dives as deep beneath the waves as any fish.

He has mastered fire, not just in his ability to generate heat, but in his ability to harness it in engines and in his ability to generate electricity, FIRE itself. He harnesses electricity to communicate over greater distances than any other of Earth's creatures, to communicate even beyond the limits of his own planet. And he even builds machines that can think at the level of H48.

His Reason is or can be at the level of H12. It can be impartial, and it can be self-aware. His capacity for knowledge far exceeds that of *Invertebrates* or *Vertebrates*. Yet, while he can raise himself so much higher even than the domain of AIR, he can also fall lower to the domain of EARTH. The way up is also the way down. The Being of *Man*, H24, can be affirming or denying. It can ascend towards an individuality that subdues the ego or it can establish one that glorifies it.

Archangels represent the borderline between FIRE and the Higher Realm beyond the Sun, from the invisible form of plasma to its glowing and its incandescent forms, from 12 orders of laws to 6, from the Sun to the *Eternal Unchanging*. The Being of *Archangels* is H6. They are higher intelligence—immortal in the lifetime of the Sun itself.

The "Man of Light" does not seek dominance but to become submissive to a master that he knows only by his longing for unity with it. It is H6, the Being of *Archangels*, the Being of the higher intellectual center. And suppose he achieves the crystallization of the body Kesdjan. In that case, he will discover a fully formed conscience identical to that possessed by the *Archangels*. It will be a far higher individuality than is possible for "man incarnate" as its Reason is not confined even by the Reason of *Archangels* but can rise higher still. And yet, it is not an individuality separated from the higher realm, but one that rejoices in its participation with that world—at a level of perfection that we can only imagine.

The *Sun Absolute* and the *Eternal Unchanging*

The highest two squares in the Step Diagram represent the *Sun Absolute* and the *Eternal Unchanging*. The nine squares below these two fully encapsulate the lateral octave from the Sun, from *do* to *do*. These two higher squares represent the notes *si* and *la* in the Ray of Creation. The highest note in the Ray, *do*, the Absolute, appears in the Step Diagram only as a notation, as Holy God at the highest level and Holy Firm at its lowest level. The Step Diagram thus represents the space between. The top square represents the *Sun Absolute*, or All Worlds, and the lower square, *Eternal Unchanging*, the Milky Way in our Ray of Creation.

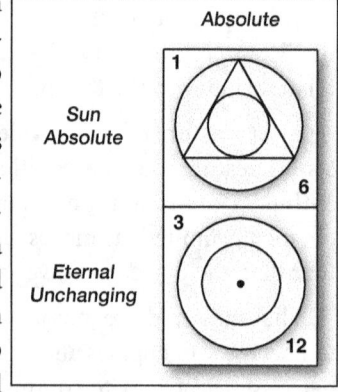

Figure 68. The Higher Realm

The Eternal Unchanging

The current estimate of the number of stars in the Milky Way is 400 billion. Historically this estimate was once much lower, but it increased as the power of our telescopes improved. Our galaxy is a barred spiral galaxy—the word "barred" meaning that there's a bar-shaped grouping of stars at the galactic center, as there is in about half of the observed galaxies.

It exhibits a central bulge surrounded by four large spiral arms, two of which are significantly bigger than the other two. Our Sun lies on the Orion Spur, a minor arm that lies between two major arms of the galaxy, the Perseus arm, and the Sagittarius arm. We lie about halfway between the galactic center and its edge. The galaxy is ellipsoid in shape and measures about 100,000 light-years across at its widest.

If light were the only means of passing information from the edge of our galaxy to the center, it would take 50,000 years for the center of the galaxy to know what was happening at its periphery. It is a time to distance ratio of roughly 1:300,000. In man, nerve impulses travel at a rate of about 100m/s, a time to distance ratio of 1:100. So it takes roughly 0.02 seconds for information to travel from the furthest part of the body (say 2 meters) to the brain.

The mathematics suggests that the speed of light is too slow to be the basis of a communication mechanism by which a galaxy senses its own body—just as the speed of man's nervous system is too slow to serve the essence and the mental body. We suspect, therefore, that another mechanism, a faster-than-light mechanism, serves that purpose.

Astronomical research* suggests that all galaxies, no matter how small or large, revolve at the same speed, with the galaxy's outer edge completing an orbit in about a billion years. Our solar system orbits the Milky Way at a speed of 515,000 mph (828,000 km/h). That's five times the speed of Mercury's orbit around the Sun—Mercury being the fastest moving planet. So our solar system takes 230 million years to complete a single orbit.

The symbol in the *Eternal Unchanging* square is of a point within a circle within another circle. We take the outer circle to represent the Absolute, within which a lesser circle represents the *Sun Absolute*, the domain of the Absolute, which he contains within himself, with the *Eternal Unchanging* represented by a single point.

The *Eternal Unchanging* is H3, a *Hydrogen* about which we assume we know nothing by experience. It feeds on H12, on *Angels*, on presence, on impartiality. The words *"Eternal Unchanging"* may seem to imply that this is a realm in which no change occurs. However, as it feeds on lower life-forms, change must indeed occur within it. Nevertheless, from the perspective of *Man*, it can be thought of as Eternal.

The *Sun Absolute* represents the universe in total. Current estimates suggest there are 2 trillion galaxies, an estimate that is a similar order of magnitude to the number of cells in a human body. Thus, the Absolute's relationship to our whole galaxy is roughly equivalent to our relationship to a single cell of our body. It may thus seem that we are irrelevant in the existence of the Absolute. However, remember that while our bodies comprise trillions of cells, we were no more than a single cell at our conception.

The *Sun Absolute* square shows a circle surrounding a triangle, which in turn surrounds another circle. We take the outer circle to represent the Absolute. We take the triangle to represent the Law of Three, and the separation of the single Law at the level of the Absolute himself into three laws. So while the

---

* *https://phys.org/news/2018-03-astronomers-galaxies-clockwork.html*

Gurdjieff's Hydrogens: The Ray of Creation

Eternal Unchanging is under six orders of laws, the *Sun Absolute* is only subject to three.

The *Sun Absolute* is H1, and it feeds on H6, on *Archangels*.

Fire in Overview

Some observers suggest that plasma accounts for 99% of the universe. Others suggest the number is more like 99.99%, and yet others put the percentage higher. It's likely to be the case since suns are plasma, and they are what you see no matter where you point your telescope. Regarding our solar system, the Sun makes up an estimated 99.8 to 99.9% of the whole, ignoring the solar wind and any planetary plasma.

It is awkward that we know so little about plasma, especially since all the *Hydrogens* above and including H96 are plasma. In this chapter, we have documented plasma's presence in many life processes and could easily have added more. The whole of life on Earth is dependent on photosynthesis, the transmission to Earth of energy from the Sun to create a supply of energy for all life on Earth.

Everything in the domain of Fire moves at speeds far greater than achieved in the domain of Air. Electricity moves at speeds from a third to a half the speed of light, depending on context. The body Kesdjan is said to move with a similar speed.

From objective science's perspective, the nine lower squares of the Step Diagram comprise all life on Earth. This idea aligns with the general direction of the Gaia hypothesis proposed in the 1970s by James Lovelock, but descends into far greater detail. It embraces all non-biological life-forms, from the *Kernel* life-forms at its lower end to the *Archangels* above.

The hypothesis is also distinct in that it explains why such a varied and complex system of life exists, linking it to both the Planets and the Moon. It also explains why we cannot detect a similar spectrum of life on any other planet in our system.

It offers a distinctly different explanation as to how this extensive ecosystem evolved, what man's position in that ecosystem is, and the origin of its intelligence, which it places higher than man. On its own, it does not answer all questions; indeed, you could say it poses more questions than it satisfies. In its view, the Earth is a living being, as are all planets and indeed all suns. This profound ecosystem is simply one very small and, in most ways, insignificant part of that.

Chapter 10

PLASMA COSMOLOGY

"Do not feel lonely; the entire universe is inside you."

~Rumi

A battle of ideas is underway in the sphere of the intellect. The last such conflict destroyed the Catholic Church's intellectual hegemony and appointed a distinctively different priesthood in its place. The newly-minted priesthood put its faith in a creed that had its genesis in empiricism rather than revered holy writings from centuries long gone. It was atheistic in spirit, not denying the existence of an absolute being, but regarding his existence as irrelevant. Its church was academia.

The creed that governs man's intellectual life is important because of the illusions it fosters. Mankind has a collective intellectual personality—an alliance of many "I"s. The intelligentsia naturally adopt its fashionable opinions and champion its "sainted" concepts. The priesthood of the day claims to profess a cast iron loyalty to the truth, hiding their hypocrisy as best they can when they betray it. Seekers of the truth inevitably see through the false facade of "knowledge" it erects, and turn away.

The priesthood that defends the Standard Lambda CDM Model of astrophysics is no different in spirit to the priesthood that forced Galileo to his knees. It is against their interests to give air time to any theory or intellectual movement that challenges their precious dogma. They will happily promote almost any absurdity—gravitational lenses, neutron stars, dark matter, any academic fantasy you please—to conceal their intellectual bankruptcy.

Nevertheless, their empire is crumbling and, if the past is prelude, a newly-minted priesthood will soon replace them. Objective science is unlikely to become the creed of this new intellectual empire—better that it doesn't. Nevertheless, it will serve man and objective science if a theory that approaches closer to reality replaces modern science's current dogma. A plasma cosmology is such a theory.

If you currently have little knowledge of what such a cosmology asserts, the following may serve as an introduction.

Birkeland Currents

Kristian Birkeland was a Norwegian scientist who studied the aurora, gathering magnetic field data in the far North. Identifying a global pattern of electric currents associated with the aurora, he concluded that the Sun was the cause. He theorized that the Earth's geomagnetic field guided energetic particles ejected by the Sun from sunspots towards Earth, to the poles, where they produced the visible aurora. He was correct.

Because of his pioneering work, such field-aligned currents are known today as Birkeland currents. The importance of these currents was ignored for decades—the general assumption was that nothing "electrical" happened in space. It wasn't until 1967, 50 years after Birkeland's death, that a US space probe proved that sunspot emissions indeed caused the aurora. Even then, this phenomenon provoked little interest.

The Sun-Earth interaction is just a small part of a very big picture; Birkeland currents pervade the whole universe. They are the highways and byways of plasma. Happily, such currents are easy to create and test in the laboratory at a much smaller scale, so their behavior is no secret. *Figure 69* illustrates the twisted spiral form they take. A spiral, like the one on the left side of the illustration, lines up with the direction of the Earth's magnetic field. The charged particles from the Sun create the aurora in alignment with the Earth's magnetic field at the North or South poles.

The mechanism is this. When a charged particle encounters a magnetic field at any angle other than 90^0, it moves in a spiral (a helical path) that follows the magnetic field's direction. This behavior tends to create not one, but two twisting plasma flows (i.e. electric currents) as illustrated on the left side of *Figure 69*. Any new charged particle which turns up falls into one of those two plasma flows.

Initially, these two plasma flows attract each other so that the radius of the spiral diminishes. The cross-section view from beneath, on the right side of *Figure 69*, represents this tightening of the two spirals. However, when a certain level of proximity occurs, a balancing force of repulsion is generated that holds the twin streams apart. The configuration is thus extremely stable,

Plasma Cosmology

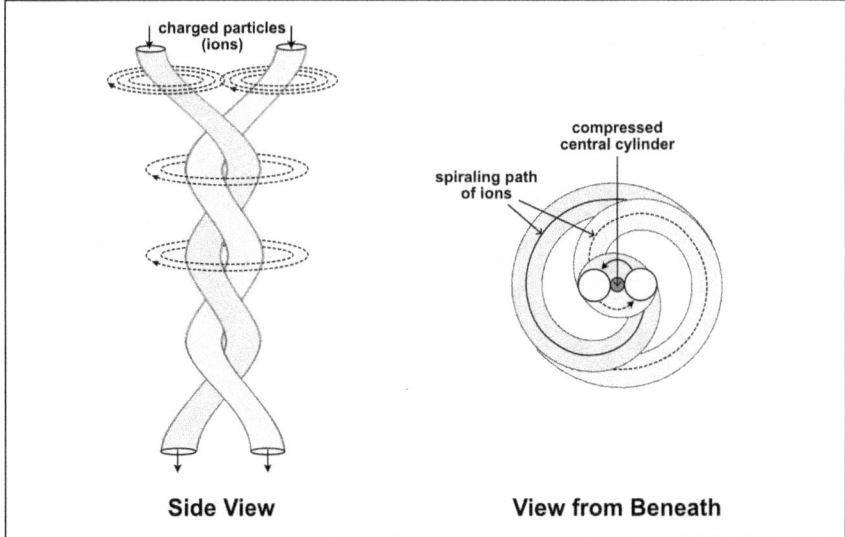

Figure 69. Birkeland Currents

with a centrifugal and centripetal force balancing each other. These tightly wound pairs that are called a Birkeland Current.

The Compressed Central Cylinder

Note that the view from beneath in *Figure 69* shows a dense (dark) circle between the two plasma flows. It represents a cylinder that naturally forms in the center of the spiral where compression occurs. The behavior of matter that resides inside this cylinder is significant. When several different chemical elements occur within such a compressed region, as often happens, they do not mix homogeneously. Instead, they distribute themselves radially according to their ionization potentials.

We illustrate this in *Figure 70* on the following page. The phenomenon, investigated by G T Marklund, is called Marklund convection. According to plasma scientist Anthony Peratt, the most abundant elements of cosmic plasma can be divided into categories of roughly equal ionization potentials (shown in parentheses) as follows:

– helium (24eV)

   hydrogen, oxygen, nitrogen (13eV)

– carbon, sulfur (11eV)

– iron, silicon, magnesium (8eV)

Gurdjieff's Hydrogens: The Ray of Creation

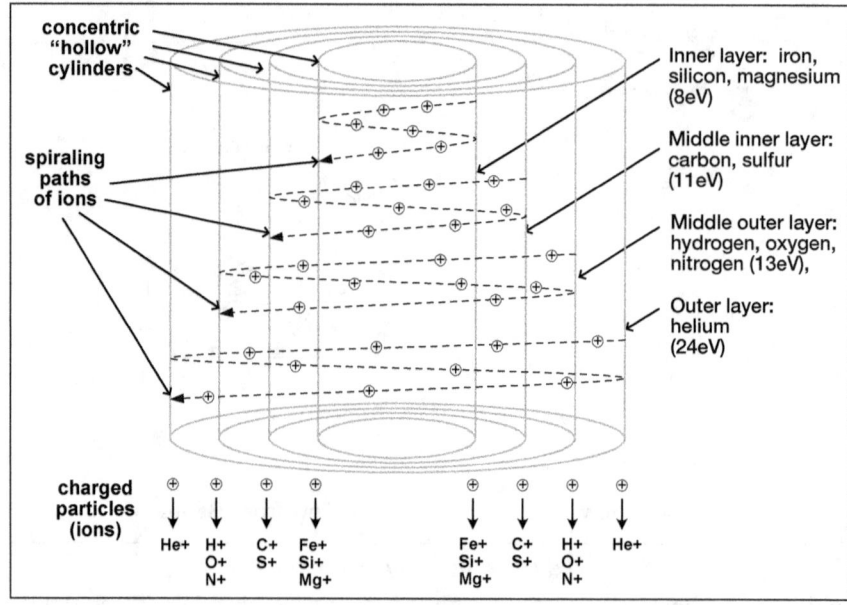

Figure 70. The Compressed Central Cylinder and its Ions

It is curious that, with the exception of helium, these are the elements of life. The only vital life element missing here is phosphorus, which is present in DNA. The elements form "hollow cylinders" whose radii are proportional to their ionization potential. Helium forms an outer layer; hydrogen, oxygen, and nitrogen an outer middle layer; carbon and sulfur an inner middle layer; and iron, silicon, and magnesium make up the inner layer. The layers are not perfectly homogeneous, as some mixing of elements occurs between them.

One of the beauties of this is that such behavior can be demonstrated and examined in a laboratory.

The Z-pinch

Figure 71 illustrates what is called a z-pinch. Electric currents flowing within a plasma tend to form twin spiraling Birkeland currents. They create a magnetic field that compresses the plasma between the twin currents into a central cylinder as shown. (It is called a z-pinch for the simple reason that if you mathematically represent the spiraling motion in respect of an x and y-axis, then the cylinder forms along the z-axis.)

You can think of this as an elaboration of the situation where two parallel wires carrying current in the same direction attract each other (as illustrated

Plasma Cosmology

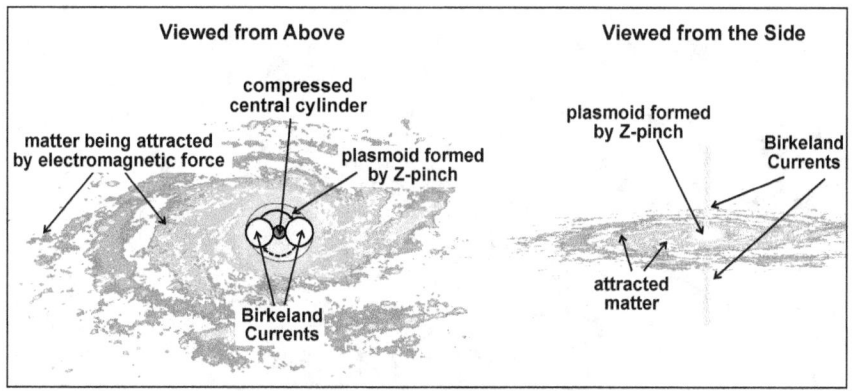

Figure 71. Z-pinch Plasmoid

in *Figure 41, p145*). Replace the wires with plasma, which can naturally carry a current, and you get spiraling Birkeland currents, which attract each other and compress the space between them.

In extremis, when the current density is very high in a Z-pinch, a specific kind of plasmoid (a plasma structure) will form. Technically, a plasmoid is any kind of coherent structure formed by plasma in conjunction with magnetic fields. In a Z-pinch, a spinning plasma torus forms, which produces a very strong attractive electromagnetic force in the plane perpendicular to the Birkeland currents. The Z-pinch attracts charged matter.

Looking down from above, it looks like a whirlpool of charged matter being attracted to the center by the electromagnetic force. The right side of the diagram shows a view of this from the side.

Galaxies

Birkeland currents occur at every scale, and as far as we know, function in the same way at every scale. They can be created in a laboratory at a small scale; they occur at planetary scale with the aurora borealis, they occur on the scale of the Sun. Astronomical observations demonstrate that they occur on the vast scale of galaxies. Plasma cosmology proposes that the Z-pinch within Birkeland currents is the mechanism by which galaxies form—solar systems too.

Circumstantial evidence for this is provided by *Figure 72* on the next page, which shows galaxies chained together by Birkeland currents. It is a fragment taken from a map assembled by scientists and published by the journal *Nature*. They created a map of over 8000 galaxies to provide a picture of our local part of the universe.

Gurdjieff's Hydrogens: The Ray of Creation

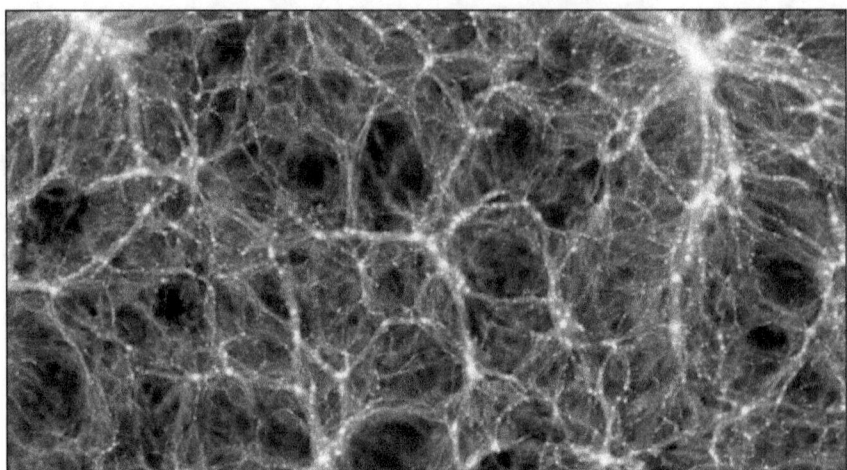

Figure 72. Birkeland Currents Connecting Galaxies

Our Milky Way lies in the borderlands of a galaxy supercluster that the scientists named Laniakea, from the Hawaiian words for "immense heaven." Galaxies fill the universe irregularly, clustering in some areas and avoiding others. (*Nature* also provides a Youtube video on this topic*).

If a universe formed by and kept active by plasma and electricity seems unlikely (at first blush), it becomes truly credible when you consider the nature of electromagnetic force. The EM force is massively more powerful than gravity, by a mind-boggling factor of about 10^{35}.

The implication of that numerical truth is not easy to conceptualize. In comparison, the force of gravity almost does not exist. Consider this. The gravitational attraction between the Sun and its nearest neighbor 4.3 light-years away is roughly equivalent to the gravitational attraction between two specks of dust that are 4 miles apart. Yes, 4 miles. Even dust particles an inch from each other show no indication of mutual attraction.

The weakness of the gravitational force is not a secret. Contemporary cosmology got itself into its current theoretical malaise because, in the early days, the only force anyone could imagine to account for the orbit of the planets was gravity. Before 1967 there was no evidence that space was permeated with plasma and that electromagnetic forces played any part in the solar system or beyond. Astrophysicists assumed that space was filled with widely dispersed uncharged atoms and molecules. Thus gravity was believed to be the only force responsible for the motion of galaxies, suns and so on.

---

* *https://phys.org/news/2018-03-astronomers-galaxies-clockwork.html*
The title of this Nature video is Laniakea: Our Home Supercluster

Problems quickly arose when astrophysicists tried to explain the shape and motion of galaxies using the force of gravity. Gravity was far too weak a force to prevent all of those orbiting suns from wandering off into empty space. It became necessary to invent invisible (dark) matter to provide the missing gravitational pull that would keep errant solar systems in line—and that sent astrophysicists off on a hopeless quest to detect the non-existent dark matter.

Although plasmoids and the Z-pinch effect do not on their own fully explain galaxy formation and evolution, it's almost certain they are part of the picture. Incidentally, the widely reported "image of a black hole" that gained attention in 2019 is probably an image of the plasmoid at the center of galaxy M87.

The Solar System

The fabric of the universe is almost all plasma. It's not just the shiny suns and galaxies. Space probes seem to find plasma wherever they go. Our solar system is teeming with it. Everywhere we point telescopes, we find evidence of plasma and little else. Matter in the form of AIR, WATER, or EARTH makes up only a minuscule proportion of the whole.

The scale of the universe is immense. Estimates suggest that there are one or two trillion (observable) galaxies. The estimate for the number of stars in the average galaxy is currently 100 billion—a vast number. In all likelihood, each of those stars supports a family of planets, moons, and comets too. All of these bodies exist within apparently endless space that appears to be overflowing with plasma.

To add to the complexity of the situation, none of this is static. The galaxies move relative to each other; the suns move within their galaxies in ways not yet well understood. The planets revolve around suns and the moons around their planets. Our solar system is the only one we can examine in detail. A plasma-oriented illustration of it is shown in *Figure 73* on the next page. Just as all suns within a galaxy gather in a single plane perpendicular to the Birkeland currents, so do all the planets, moons, and other condensed matter in a solar system. It is called the plane of the ecliptic. The heliosphere, a vast atmosphere that surrounds the Sun and determines the extent of its local influence, is almost spherical—in fact it is "apple-shaped."

The region referred to as the Kuiper belt is a kind of asteroid belt which lies beyond Neptune. It may contain as yet undiscovered "planets," and comets are believed to originate there. Some astronomers suspect that comets can also originate beyond the heliopause. They suggest a theoretical region called the Oort cloud as their origin. We show it in the diagram in the plane of the

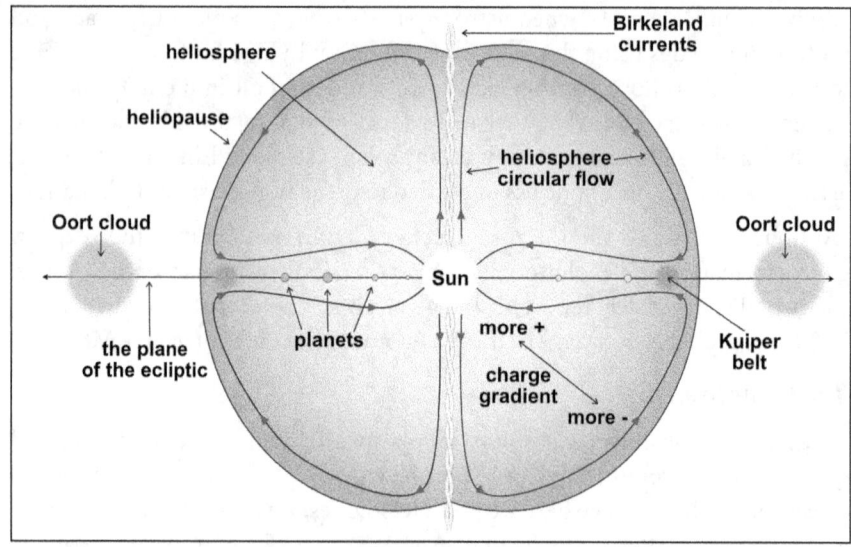

Figure 73. The Solar System

ecliptic, but it could as easily be a spherical shell that surrounds the whole heliosphere, and it may not exist at all.

The solar system is maintained by two electric circuits: an interstellar current and a local current. The interstellar Birkeland currents enter the Sun at its north and south poles and stimulate its activity. Protons and other positively charged particles ejected from the Sun create the local circuit. They travel through the heliosphere to its edge, the heliopause. Electrons attracted from everywhere within the heliopause back towards the Sun complete the circuit. This circuit occupies a whole sphere but is more concentrated in the plane of the ecliptic.

The Sun is an anode and the heliopause a diffuse cathode. Hence there is a charge gradient between the Sun and its surrounding space that reaches its peak at the heliopause. If it were possible to put one lead of a voltmeter on the Sun and another on the Earth, it would show the Earth as negatively charged in relation to the Sun. The same is true of the other planets.

The negative charge associated with each planet will vary according to its interaction with the local circuit (the solar wind). However, each planet also has a magnetosphere, which insulates it from this circuit. It deflects charged particles to its poles, where the planet receives some of them via the Birkeland currents. The extent to which the electric nature of the planets affect or determine their orbit is not known. However, there is likely an effect.

We calculate the masses of each planet and the Sun on the assumption that gravity is the only force that keeps the planet in orbit. If that assumption is incorrect, then so are those calculations.

The Electric Sun

In *Figure 74*, we depict the inner electric circuit of the Sun created by the interstellar Birkeland currents that arrive at its north and south pole. The Sun's surface is a sea of incandescent plasma powered by these interstellar currents, which ejects positive ions (protons mainly) in all directions.

In the illustration, we envisage a situation where the primary current entering the Sun's poles is increasing. When currents increase, the magnetic fields created by those currents, as illustrated in the diagram, also increase. The change in these magnetic fields will naturally induce a secondary current below the surface of the Sun. This circuit will flow from the pole to the Sun's equator and back to the pole. Scientists using the joint European Space Agency/NASA Solar and Heliospheric Observatory spacecraft detected "rivers" of hot plasma flowing beneath the Sun's surface. They could be such a current.

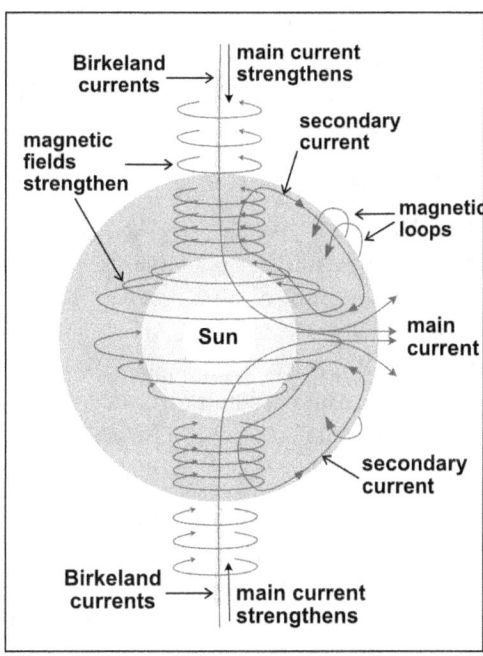

Figure 74. The Electric Sun

A river of hot plasma is, of course, a circuit of some kind. And such a circuit could explain the coronal loops—the huge plasma arches observed on the surface of the Sun that extend out of the photosphere into the corona. The loops, which can be 200,000 km long, are bipolar fields connecting opposite poles.

It works in the following way. If a secondary current "tube" flows southwards from near to the Sun's north pole and it is on or just beneath the Sun's surface, a looping magnetic field will emerge to the east of that current. It will create a north magnetic pole at the point where it emerges. The loop will move out above the Sun's surface and back down into the surface to the west of the "tube."

In the Sun's southern hemisphere, the opposite occurs because the magnetic fields run in the opposite direction. These magnetic loops reverse their orientation with the 11-year sunspot cycle. The explanation is that if the primary current begins to decrease rather than increase, the induced circuit will reverse its direction, and the magnetic loops will also reverse. The Sun appears to breathe in and out every 11 years.

The strength of the Sun's magnetic field has doubled during the Twentieth Century. There is no good theory as to why. Neither do we know how the constant bombardment of the Sun by cosmic rays affects it. Incoming cosmic ray protons are individual plasma currents that impinge on the Sun's surface, supplying it with energy.

The HR Diagram

Our Sun is an average star. According to plasma cosmology, all stars are electric suns emitting light either by plasma in glow mode or arc mode. If we know the distance of a star, we can compare its luminosity to the Sun. The measurement of a star's absolute magnitude records how bright it would look if it were 10 parsecs (32.6 light-years) away. This distance of 10 parsecs is a standard measure. A luminosity value of 1 indicates the luminosity of our Sun, which incidentally is more luminous than roughly 85% of the stars in the Milky Way.*

In 1911 Danish chemist-astronomer Ejnar Hertzsprung constructed a plot of the absolute brightness against temperature (spectral class) of the stars whose distances he could measure with reasonable accuracy. Princeton University astronomer Henry Norris Russell repeated the exercise in 1913. Because of their efforts, the graph they created, which is frequently updated, is called the Hertzsprung-Russell (HR) diagram. The HR diagram is a plot of actual observations, not something deduced from theory. Any theory about the nature of and life cycle of stars needs to conform with this.

In the HR graph (*Figure 75*), every dot represents data about a star. The vertical axis has two scales. The left side scale shows the absolute magnitude of a star's luminosity (at the standard distance of 10 parsecs) on an inverse logarithmic scale. The lower down on the scale, the dimmer it is. The right side scale is also logarithmic and shows the luminosity of the star compared to the Sun. Our Sun fits somewhere in the middle between very dim and very bright.

* https://earthsky.org/astronomy-essentials/stellar-luminosity-the-true-brightness-of-stars/

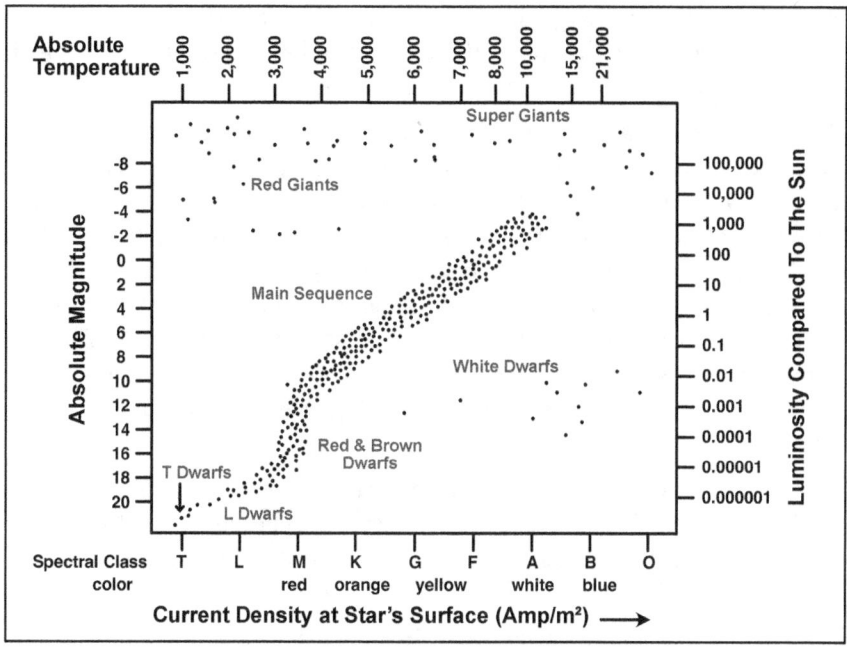

Figure 75. Hertzsprung-Russell (HR) Graph

The horizontal axis also is marked out with several scales. At the top is the Absolute Temperature of the star's surface, showing a scale that descends from 21,000°K to 3000°K. Again our Sun sits roughly in the middle, with a surface temperature of about 5800°K. The other horizontal scales are equivalent scales expressing different perspectives. In terms of color, our Sun is, of course, yellow. Spectral Class provides a more exact color classification using the Morgan-Keenan system, which runs from type O to type T.

The label at the bottom of the diagram highlights the electrical dimension of luminosity, suggesting a scale that represents the current (Amps/m2) on a star's surface. This idea, drawn from the work of retired professor Dr. Donald Scott,* adds the plasma cosmology perspective. A star's luminescence relates directly to the current passing through it. The strength of the current density on the star's surface is a measure of that.

The Electric Sun Perspective

For the moment, we'll set to one side the white dwarfs and red giants and discuss the main sequence of stars that rise from the lower left side of the HR graph. The diagram shows that the more luminous a star is (i.e., the greater its

---

* See *The Electric Sky* by Dr Donald Scott

electricity supply), the larger its mass. Conversely, the smaller stars are lesser lights, trailing off into the area of the T, L, red and brown dwarfs, all of which are very dim and quite small compared to our sun.

Fifty of the sixty nearest stars to the Sun are red dwarfs, although none of them are visible to the naked eye. Estimates suggest that red dwarfs make up about three-quarters of the Milky Way's stars, and the same is probably true for other galaxies. Red dwarfs have low surface temperatures (2000-3000°K) and are about a tenth the size of our sun or less. Brown dwarfs are smaller still. The L dwarfs and T dwarfs are usually regarded as brown dwarfs. They fit in between massive gas giant planets and the least massive red dwarfs. Think in terms of a body somewhere in the range 10 to 80 times the size of Jupiter. Incidentally, "brown" is a misnomer. Color-wise, they range from deep red to magenta.

Objective science considers the brown dwarfs to be evolving planets or involving stars. The theory is that as planets evolve to become Suns, the electric current passing through them increases, they acquire more mass, and their luminosity increases.

Jupiter and Saturn have intrinsic luminosity—a small amount in the infrared range. Both have a fairly substantial family of moons and are probably evolving towards becoming suns (although they could equally be involving). Interestingly, the T-type dwarfs can have temperatures below 1000°K, less than twice the temperature of the surface of Venus.

As we follow the main sequence of stars from the bottom left to the top right, we move from brown dwarfs, where the plasma on the surface of the star has not yet achieved the glow state of plasma, to where the surface of the star begins to exhibit the arc mode of plasma. It happens on the HR graph above spectral class M (temperature around 3000°K). For stars in this area, a small increase in electric current leads to a large increase in luminosity. Arc mode plasma is much brighter than glow mode plasma. There's an obvious difference between looking into a fire and looking directly at the Sun.

The stars at the upper right end of the main sequence are massive. Their surface temperature is very high (it can be 35,000°K or more), and their color is blue-white. Their absolute luminosities approach 100,000 times that of the Sun. Such stars are under intense electrical stress.

However, as we move to the left side of the HR graph, we find stars that are just as large but have relatively low surface temperatures. These are the red giants, not particularly luminous but visible because they are so large. Betelgeuse in the Orion constellation is a red giant. It is a variable star, which means that it pulsates, with its diameter varying from 480 to 800 million miles. Its diameter is thus somewhere between 550 and 920 times that of our Sun. However, its surface temperature is merely 1300°K, less than a quarter of our Sun's. Such stars are not particularly energetic.

To the lower right corner of the HR graph, we find the so-called white dwarfs, which as the color axis at the bottom of the chart indicates, can also be blue. They are relatively uncommon stars that have high current densities.

The Birth of Suns and Planets

Stars under extreme electrical stress—the white dwarfs and supergiants found to the right of the HR diagram—can and sometimes do explode in nova or supernova events. During such supernova events, a massive amount of light, UV light, and gamma rays are emitted, along with multitudes of neutrinos. It typically lasts around 100 seconds, but for that time it will emit more light than the whole galaxy to which the star belongs. Nova events are similar but less explosive. They increase a star's luminosity by anywhere between 1000 and 100,000 times. Some stars experience nova events at fairly regular intervals.

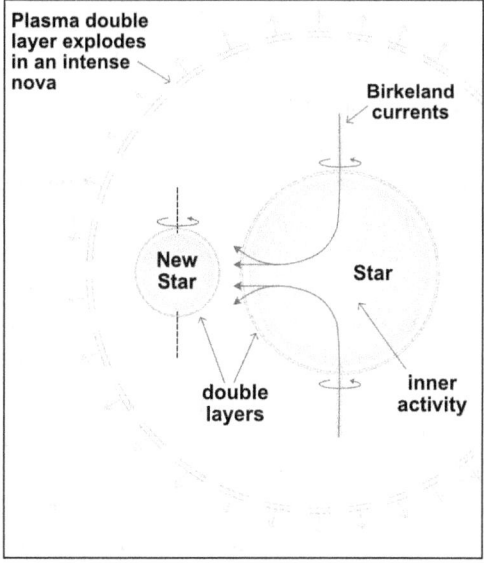

Figure 76. The Birth of Stars

Nova and supernova events can create new stars or planets, as illustrated in Figure 76. The theory* is that a star forms a series of double layers that act as membranes. One membrane surrounds the star while one descends into the interior and divides it into two. The exterior membrane becomes increasingly unstable and explodes. There is a burst of gamma-rays, then an explosion of visible light. It leaves a cloud (or nebula) and two stars rather than a single star in its wake. The two stars are wrapped in double layers similar to our Sun's heliopause. They have different surface current density, temperature, luminosity, and spectral type to their parent star, indicating a loss of energy.

---

* See The Electric Sky by Dr Donald Scott, Chapter 14

When a sphere splits into two smaller ones, their combined surface area is greater than that of the original sphere. By splitting a star in this way, the current density of the total surface reduces. If the resulting two spheres are of equal size, the increase in surface area is 26%.

The Crab Nebula provides an excellent picture of the aftermath of a super-nova—one that Chinese astronomers in 1054 AD observed. At the center of the nebula is a pulsar (CM Tauri) that pulses 30 times per second. It has a small companion star that is 1500 AU distant from it.

Lesser novas likely create planets—gas giants or even smaller bodies. The birth of the Moon may have involved this kind of electric fissioning process. It seems tangentially related that when a human or animal sperm fertilizes an ovum, there is a small flash of light.*

Galaxies and Quasars

Just as stars give birth, so do galaxies. The newborn galaxy is, it seems, a quasar. A quasar (a quasi-stellar object) is an extremely luminous body surrounded by a gaseous accretion disk. Extremely luminous is perhaps an understatement. All quasars are brighter than galaxies, and there are quasars with luminosities much greater than the whole Milky Way—thousands of times greater than the brightest of supernovas. And unlike supernovas, which are brief though brilliant events, quasars just keep on shining brightly.

Images from the Hubble telescope demonstrate that quasars commonly occur near the center of galaxies, and all are closely associated with a parent galaxy. Some host galaxies appear to be merging galaxies. (It's worth noting that we would need to observe galactic behavior for perhaps millions of years to know for sure that two galaxies were merging. All evidence of merging galaxies is circumstantial.)

Astronomers have not yet identified any images that can be interpreted as the birth of a quasar. Currently, about 750,000 quasars have been identified from trillions of galaxies. Therefore we suspect that the birth of a quasar is a very rare event.

Pulsars

Pulsars are stars that emit extremely rapid regular bursts of EMR, from light to radio waves. Electrically, the behavior resembles that of a simple relaxation oscillator that can be used, for example, to make an electric light blink. Most likely, the pulsing involves electric arc interactions between two closely spaced

---

* https://www.sciencealert.com/scientists-just-captured-the-actual-flash-of-light-that-sparks-when-sperm-meets-an-egg

binary stars (for example, in the Crab Nebula). In effect, the two stars act as capacitors and the space between them, permeated with plasma, acts as a resistor. One star (call it capacitor A) gradually charges up. When the voltage reaches a trigger value, it discharges to the other star (capacitor B). This behavior could oscillate periodically from one star to the other if the electrical conditions were right.

We could devote further time to examining the characteristics of various types of stars. Suffice it to say that plasma cosmologists offer credible electrical explanations* as to what is taking place with a broad variety of different types of stars without resorting to scientific fantasies like stars composed entirely of of neutrons. The plasma cosmology theories for such objects can be tested in the lab.

The SAFIRE Project: An Electric Sun

The idea of a plasma-driven electric universe gradually grew in importance through the work of Kristian Birkeland, Irving Langmuir, Nobel prize winner Hannes Alfvén, Charles Bruce, Ralph Jeurgens, and Halton Arp. More recently, Anthony Peratt, David Talbott, and Wal Thornhill have carried the baton. Until recently, their work and ideas were ignored and even derided by contemporary science. The scientific establishment invested tens of billions of dollars every year into mainstream astronomy and astrophysics. The study of the electric universe received no funding at all. It was seen as an unimportant "fringe" idea.

In 2012 there were just two scientific hypotheses concerning the nature of the universe.

- One asserted the universe was gravity-driven and the stars were thermonuclear phenomena. This Standard Model of the universe was taught in schools as if it were true. Yet, despite a vast amount of funding, it had never been proven.

- The alternative theory asserted that the universe is electrically driven at every level from galaxies down to planets. In particular, it maintained that stars were electrical phenomena. It had never been proven.

Until that time, nobody had formulated an experiment that could reliably test either hypothesis. That changed in 2012. Montgomery Childs formulated a means of testing a physical model of an electric sun. This gave birth to the

---

* Online resources include https://www.thunderbolts.info/wp/home/ and https://www.everythingselectric.com/

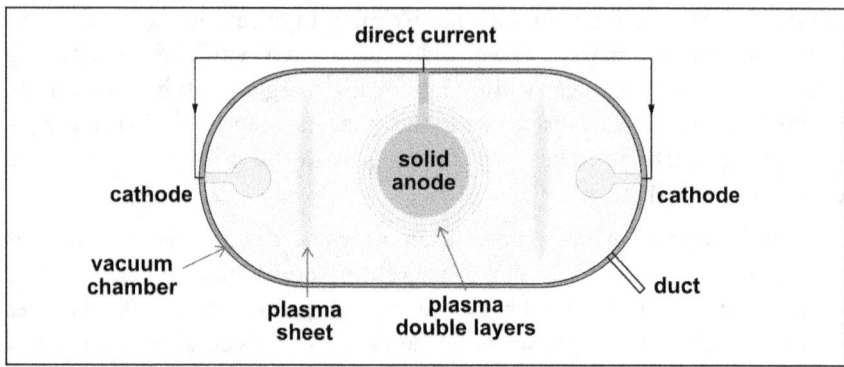

Figure 77. The SAFIRE Reactor Chamber

SAFIRE Project, which attracted private funding from the Mainwaring Archive Foundation.

The SAFIRE Project has been in progress since then. It has confirmed the Electric Sun Model, having mimicked an electric sun in laboratory conditions. It found no disparities. The project continues because it has made many surprising discoveries.

Figure 77 provides a simple diagram of the experimental reactor chamber used in the SAFIRE project. Towards the center of the chamber is a solid spherical metal anode, and towards the edges of the chamber are two cathodes. The anode and cathodes are supplied with a direct current, as shown. The vacuum chamber can be emptied almost completely using a vacuum pump attached to the duct. If desired, small amounts of gas can be introduced into the chamber. So it is possible to vary the gas mixture, the pressure, and the voltage and current. This facilitates a variety of plasma discharges.

The diagram illustrates a situation where several plasma double layers form around the anode, and plasma sheets form in the chamber between the anode and the two cathodes. We provide a link below* to a 20-minute video that offers much greater detail.

The project has established the Electric Sun and Electric Universe as a credible, soon to become a dominant, theory. However, it has yielded other important and surprising results.

- The design of a new, clean and efficient power generation capability to build power plants based on plasma interactions. This has very promising commercial application.

---

* https://www.youtube.com/watch?v=7GFFfmBGb5U (Safire Project YouTube video)

- The discovery of a means to neutralize nuclear waste. This is achieved by exposing the radioactive waste to the nuclei of hydrogen isotopes within the vacuum chamber. The process generates energy and could be the basis for a power generating device with nuclear waste as its fuel. It has obvious commercial application.

- A means of transmuting elements. From a commercial perspective, the most important result here is that the SAFIRE team has found a means of creating the rare earths lanthanum and cerium. The cost of doing so is far lower than the cost of mining these elements.

The most important result from the SAFIRE Project is its validation of the Electric Sun model. Mainstream science might want to ignore the science, but they cannot ignore these commercial applications. Neither will they be able to ignore the transmutation of elements. A revolution in mainstream astrophysics will thus occur in time.

The Transmutation of Elements

The discovery of elemental transmutation in the vacuum chamber by the SAFIRE team was accidental and unexpected. Of the 11 most abundant inorganic elements found in the interstellar medium, nine (including sodium, magnesium, aluminum, silicon, phosphorus, chlorine, potassium, and titanium) were detected in the SAFIRE reactor chamber after an experiment. This is dramatic evidence for the Electric Sun theory.

According to NASA,* while the solar wind is primarily a mixture of protons and some helium nuclei (He2+) and electrons, a small proportion of it (0.1%) is composed of other elements, including almost all of the known elements. These elements are probably being manufactured by the Sun as part of its surface activity.

It may be relevant here to draw attention to a passage** from *The Tales*. Because the passage may be difficult to read for some readers, we paraphrase parts of it here.

> *The scientist Gornahoor Harharkh demonstrates the process of Djartklom, which separates the ever-present substance Okidanokh into three active parts, and then demonstrates how these parts can reblend. This reblending causes the arising of mineraloids, gases, metalloids and metals of varying densities from minerals within the planet. He demonstrates how the vibrations flowing from these transformations of substances, one into*

---

* https://genesismission.jpl.nasa.gov/science/module4_solarmax/SolarWind.html#
** Chapter XVIII, The Arch-preposterous, p169-176

another, constitute the totality of vibrations which gives the planets themselves stability in the solar system's harmonious movement.

Gurdjieff asserts that there is a process by which minerals are transformed within our planet and that these transformations provide vibrations that are necessary for the harmonious movement of the solar system. *The Tales* proceeds from there to describe GH's experiment, which demonstrates the transformation of copper.

> *Gornahoor Harharkh takes some copper and places it on a plate within his experimental apparatus. He says he will demonstrate an artificial and accelerated transformation of this copper, both to a denser or less dense form. He does so by introducing appropriate amounts of the active parts of Okidanokh that have an affinity of vibration with copper and which blend with the copper. Its nature changes. The copper becomes either denser or less dense according to the proportions of the active elements of Okidanokh, which Gornahoor Harharkh introduces. As a final act Gornahoor Harharkh transforms the copper into gold.*

It is clear from previous pages of *The Tales* that Okidanokh relates directly to electricity. In some way, Gurdjieff appears to be describing a means by which elements can be transformed from one to another, using electricity.

We have little doubt that the transmutation of elements occurs in the core of our planet and in the core of suns, and perhaps even in their atmospheres.

Chapter 11

Planets, Earth and Moon

"It is impossible for a man to learn what he thinks he already knows."

~ Epictetus

The modern scientific narrative about the solar system assures us that the Earth is about 4.54 billion years old and has occupied its current orbit since it was born. Should we have any confidence in these suggestions?

The 4.54 billion estimates come from radiometric dating—meteorite samples, Moon rocks, and ancient Earth rock. There are too many anomalies in radiometric dating to have confidence in that specific figure, but the Earth is undoubtedly old. If you look at layers of sedimentary rock—there are 40 visible in the Grand Canyon walls, for example—it is hard to doubt that each layer took millions of years to form.

Consider the mountains. Surprisingly, given that it is the tallest mountain, many marine fossils have been found in the upper regions of Everest. It is not igneous rock; it is a sedimentary formation composed of limestone, marble, shale, and pelite. Geologists suspect that it is about 400 million years old. Incidentally, many tall mountain ranges show evidence of similar marine fossils. Whale bone fossils were found at 5000ft in the Andes, and whales only appeared 50 million years ago. Sea beds have risen; it is hard to imagine that happening quickly.

However, when it comes to the Earth's orbit, there are reasons to doubt that we have always swung around the Sun at the current distance and that other planets have been similarly well-behaved. Indeed there is good evidence that this has not been the case. In this chapter, our primary focus is on the Earth and the Moon, but to better understand Earth and Moon, we also need to consider other planets.

The Planets

The first exoplanet (a planet not in our solar system) was discovered in 1995. Now many have been identified. As of October 2020, there were 4,354 confirmed, with 712 solar systems identified with more than one planet. They range in size from small Mercury-sized planets to giants larger than Jupiter. Some are very cold, some orbit in days, some orbit twin suns, and some gas giants have been detected closely orbiting their star, with surface temperatures as high as red stars.

So far, very little can be deduced from this data other than that few solar systems are like ours. The prevailing model for our solar system considers planets to be of three types: gas giants (Jupiter and Saturn), ice giants (Uranus and Neptune), and terrestrial planets (Mars, Earth, Venus, Mercury). Theories of planet formation (distinctly different from the planets-caused-by-novas theory discussed in the last chapter) involve the accretion of dust by each planet roughly in its current orbit.

The exoplanet data offers little support to this accretion theory. That was always a stretch anyway. There is no known mechanism whereby dust and gravel-sized particles could 'stick together' to begin an accretion process—gravity just isn't a powerful enough force to explain that. It's hard to imagine how the collision of small stones (assuming they somehow managed to accrete from gas clouds) would cause any kind of clumping. More likely, each collision would just send the two stones off in different directions.

The planetary formation theories focus on how small terrestrial-type planets form close to the Sun and how giants form at greater distances. Sadly this is not a pattern that emerges from the exoplanet data. So the assertion that planets, once formed in some way, remain in their orbits with little variation is suspect. Evidence from our solar system suggests a completely different picture:

- The planets have distinctly different axial tilts. They are $0.1°$ (Mercury), $3°$ (Venus and Jupiter), $23°$ (Earth), $25°$ (Mars), $27°$ (Saturn), $30°$ (Neptune), and $98°$ (Uranus). Why?
- Planets spin on their axes at distinctly different rates. Jupiter 10 hrs, Saturn 11 hrs, Neptune 16 hrs, Uranus 17 hrs, Earth 24 hrs, Mars 25 hrs, Mercury 58 days, and Venus 243 days. Why?
- Venus spins in the opposite direction to all the other planets. Why?
- Neither Mars nor Venus has intrinsic magnetic fields, but the other planets do. Why?

- It appears that a planet disintegrated to create the asteroid belt. Given how little material there is in the asteroid belt (about 3-4% of the Moon's mass), what exactly happened there and how?

We have sent many space probes to distant parts of our solar system. We have flown past and gathered data from every planet, mapping a good deal of most planets' surfaces—those where the atmosphere did not prevent it. We have taken a close look at some of the moons, aside from our own. Astronauts have even visited our Moon and brought back samples of rock. We have sent robot vehicles to Mars to provide close-up pictures of the terrain and analyze samples. Nevertheless, we are not close to answering any of the questions posed above.

The Velikovsky Heresy

Since we intend to consider whether the planets, particularly the Earth, were previously in different orbits and have interacted directly with each other, we need to discuss Immanuel Velikovsky's work. Velikovsky shook the scientific world when he published *Worlds in Collision** in 1950. The historical evidence Velikovsky presented challenged many academic fields, particularly the field of astronomy.

The scientific establishment made every attempt to discredit his theories and blacken his name into the bargain. The flack he attracted was intense, perhaps record-setting. To be fair, there was not the least inkling in anyone's mind that plasma flows drove the universe. Gravity managed everything and what Velikovsky suggested was not credible behavior for gravitationally obedient planets.

In *Worlds in Collision*, Velikovsky asserts that in the 15th century BC, Jupiter gave birth to the planet Venus which initially appeared as a massive comet or comet-like planet. Subsequently, Venus came close to and interacted with the Earth, changing Earth's orbit and axis and causing global catastrophes. Velikovsky's primary evidence for this surprising assertion are worldwide historical records found in myths and holy writings. They appear to describe such an event, often in detail.

However, that was not his only evidence. He ruffled geologists' feathers by insisting that there was no ice age but that the Earth's axis had shifted. That's why the ice cap stretched south over North America and Europe, but there was no evidence of a Siberian icecap (which remains unexplained by geologists). He noted that Siberia's flora and fauna changed dramatically as the temperature

---

* *Worlds in Collision* by Immanual Velikovsky

fell, destroying mammoths in less than a day. Frozen mammoths were found with food still undigested in their stomachs. He asserted that catastrophic tsunamis swept trees into piles and buried them to form coal seams. It provides an explanation for coal seams 80 feet thick, which are difficult to explain otherwise.

In other books* he made assertions about various planetary events, including Noah's flood—a myth found in many traditions. He asserted that the flood related to interactions with Saturn. He disputed Egyptologists' and archaeologists' and geographers' theories, and he did so in an excellent academic fashion, quoting sources for pretty much every piece of data to which he referred. The scientific establishment was horrified.

His suggested that the Sun and the Planets are electrically charged. Also that electromagnetic and electrostatic forces are responsible for interplanetary behavior, cushioning possible collisions, altering rotational motions, tilting axes, and damping orbital eccentricities over relatively short periods. Incidentally, this is partly why the scientific establishment currently shuns plasma cosmology. It has the taint of Velikovsky on it.

Negatively Charged Planets

Since its discovery over 200 years ago, efforts to explain the cause of Earth's atmospheric electric field have only produced bad theories. It is why geologists usually assert that Earth's core is mainly iron without an atom of evidence. The Earth carries a negative charge. Professor Erman of Berlin demonstrated that in an experiment in 1803. In a similar experiment in 1836, French physicist Jean Peltier came to the same conclusion and suggested that this negative charge causes the Earth's atmospheric electric field.

Nikola Tesla's early research led him to the same result; the Earth harbors many free electrons. One of his goals was to transmit electric waves through the ground, which he thought possible because of the Earth's electricity and it did indeed prove possible. Interestingly, NASA accidentally achieved the transmission of waves through the Moon's surface. Winfield Salisbury and Darrell Fernald of the Smithsonian Astrophysical Observatory reported receiving signals from the command module of Apollo 15 when it was behind the Moon.** The signals were carried around the curvature of the supposedly radio-opaque Moon by electric waves in the Moon's surface layers. Incidentally, this suggests that the Moon carries a negative charge when it is outside the magnetotail of the Earth.

---

* *Ages in Chaos, Earth in Upheaval*
** *Nature* magazine, November 12, 1971

The planets Mars and Venus, like the Moon, do not have an intrinsic magnetic field and magnetosphere. However, the solar wind confers one on them. It seems to work as follows. The solar wind carries with it magnetic field lines from the Sun. These field lines do not pass through electrically conductive objects (like a planet or a moon) but instead organize themselves around them, forming a different kind of magnetosphere, which serves to deflect the bulk of the solar wind. We will revisit this when we discuss the Moon in greater depth.

The current harmony among the Planets is undoubtedly assisted by the fact that none of the magnetospheres contact each other. Pressured by the solar wind, all planetary magnetospheres point away from the Sun. The closest approach occurs between the magnetospheres of Jupiter and Saturn. However, Jupiter's magnetosphere still misses the perihelion of Saturn's by a comfortable 546 million km.

You can imagine what would occur if two planetary magnetospheres touched. Sparks would fly; interplanetary thunderbolts would flash between the bodies, continuing to do so until they equalized the two electric potentials. The electromagnetic forces in play would be very large, and it is not beyond the bounds of possibility that the axial tilt of both planets would change. The length of the day might change, and so might the orbits of the planets. And, of course, Velikovsky pointed at evidence that this happened to the Earth.

Planetary Interactions: Canyons

There is other evidence. Consider the Grand Canyon. It is over 6000 feet (1857 meters) deep, and it varies in width from 4 to 18 miles. Geologists have long puzzled over how it could have formed. It cannot have been water erosion. There are no alluvial deposits to be found of the rock that once occupied the 230-mile-long, 18-mile-wide, and one-mile-deep canyon. If you filled it up to the top with water, you'd have a lake, not a fast-flowing river.

Look at the canyon from above. It does not look anything like the typical path of a river. The canyon has a jagged shape that has nothing to do with water erosion. To add to the mystery, the Colorado River would have had to flow uphill to cut the canyon into the Kaibab plateau. To cap it all, dating suggests that the canyon is a mere five to six million years old. If a river can erode a canyon in that way, then all the great rivers of the world would flow through similarly deep or even deeper canyons. The rock of the Grand Canyon appears to have melted away like snow.

The Grand Canyon is most likely a large rocky scar carved out by interplanetary thunderbolts. It is difficult to think of any other credible explanation. When a geological feature cannot be explained by local planetary phenomena, an extra-planetary explanation is indicated.

And the theory seems even more plausible when you turn your attention to the canyon on Mars named Valles Marineris. It is the largest known canyon in the solar system. Running roughly west to east, it cuts a wide channel, over 3000 km in length, across the face of Mars. That's about one-seventh of the circumference of the planet. It is also 600 km (373 miles) wide, and at points, it is 8 km (5 miles) deep. There is no possibility that this was created by river erosion, especially not on a planet where water is hard to find.

The electric discharge that carved the Grand Canyon out of the rock cannot have been a single thunderbolt. For it to carve out such a zigzag path, the current must-have flowed while the Earth was turning. Maybe the event took many minutes or even hours. It is almost impossible to know what speed the Earth was turning when it occurred, but clearly it had to be turning at some speed to carve such a long and deep scar into the plateau. As for the carving of the Valles Marineras, that must have taken far longer and been more spectacular.

Planetary Interactions: Craters

Such planetary discharges cannot be common. The evidence of individual thunderbolts is a little more common. The scar that they leave is not a canyon but a crater. The Moon images returned by early space probes showed a surface far more pocked with craters than expected—and riddled with long-sinuous channels. Geologists and astronomers assumed that meteor impacts caused these features. They don't quite resemble meteor impacts, but that was brushed aside because "what else could have caused them?"

Almost all solid bodies in the solar system exhibit craters—planets, moons, and even comets. Close observation reveals that many of these craters have features not found with volcanic craters or impact craters. In the 1960s, Brian J. Ford, an amateur astronomer who conducted laboratory experiments of electrical discharges and their effect on rock, suggested they were mainly electrical phenomena. His work was published over 50 years ago in a British journal* and generally ignored.

Ford had reproduced some of the puzzling lunar crater features in miniature, including craters with central peaks, small craters perched on the high rims of

---

* *Spaceflight 7, January, 1965*

larger craters, and craters strung out in long chains. There has been further such experimentation which we can summarize.

The following craters and surface features have been reproduced electrically in the lab: craters with central bumps, linear chains of craters, hexagonal craters, craters with central peaks, with "twin peaks," "bulls-eye" craters, rampart craters, "pedestal" craters, domed craters, and aligned craters.

All of these have been produced in the lab by plasma physicist C J Ransom. He also investigated Martian "blueberries," tiny spherules embedded in vast numbers on the Martian surface, and managed to reproduce replicas of them by blasting hematite with an electric arc.

Few of these features can be reproduced by meteor impacts.

Plasma Cosmology as a Predictive Discipline

Scientifically, a theory is normally judged by its ability to predict or model new situations that arise. Astronomers and astrophysicists have been astonishingly unsuccessful in this respect. That is not the case with plasma cosmologists.

Most sources of information about Jupiter's moon Io will insist that it is volcanic, in fact the most volcanic body in the solar system. That opinion severely stretches the definition of volcano. There are indeed eruptions from the surface of Io, most notably from its largest "volcano," Loki Patera. But there's no sign of volcanic ash or smoke, no glowing red rivers of lava.

The volcanic plumes take the form of filaments rather than clouds, and the plume glows blue. Io's "volcanoes" exhibit another interesting and unprecedented phenomenon—they move around the surface. For example, the Prometheus plume moved 75 to 95 km west in 20 years.

Based on the electric universe model, Wal Thornhill predicted much of what was later observed on Io by space probes. His projections were accurate, and his work was ignored. Io's "volcanoes" are electrical phenomena caused by Io's orbit within Jupiter's magnetosphere, a region of highly energized plasma that surrounds Jupiter.

The plumes are electrical discharges similar to discharges created in laboratory conditions at a smaller scale. Large plasma discharges are carving up Io's surface.

Io's surface shows craters in the proximity of these plumes. There are reasons to believe that they are also electrical phenomena.

Objective Science and the Moon

In describing the Ray of Creation, Gurdjieff likens it to a tree with many branches and twigs.* If you think of the Absolute as the tree's trunk, each lower note in the Ray represents a branching. At the bottom of the Ray are the Earth and Moon. The Moon is the growing end of a twig that sprouted from the Earth.

The energy that the Moon needs to grow comes from the Earth. It is created on the Earth by the joint action of the Sun, the other Planets, and the Earth itself. This energy is collected and stored in a large accumulator—organic life on Earth. In that respect, all biological life serves to feed the Moon. There is a mutual dependency in this. Gurdjieff said that without organic life, the Moon would die, and conversely, if there were no Moon, organic life would cease.

There is some circumstantial evidence for this. Space exploration has found no evidence of organic life on any other planet in the solar system (beyond bacterial life, which we cannot discount). Nevertheless, there may once have been organic life on Mars. There is a great deal of evidence of sedimentary layers in the rock on Mars, which only occur where there is an abundance of flowing water. It is difficult to imagine the creation of significant amounts of sedimentary rock without assistance from organic life. Mars does not have a substantial moon. Its two satellites, Phobos and Deimos, are respectively 13.8 miles (22.2 km) and 7.8 miles (12.6 km) in diameter. They are more akin to orbiting asteroids than moons. It may be, then, that Mars once nursed a moon, but it failed. Without a robust magnetosphere to shield it from the solar winds, Mars appears to be losing its atmosphere.

We can think of the Moon as an embryo that is gradually growing. It exhibits no movement of its own—it does not spin on its axis, and the moment when it does may correspond to the fetal quickening. If so, then calculations suggest that the Moon will need to move twice as far from the Earth as it currently is for that to occur.

Gurdjieff explained that the Moon is growing and that it might in the future become a planet on the same level as the Earth. The Earth might then become a Sun, and the Moon would become one of its planets. The Moon itself might give birth to another moon and be host to organic life of some form.

The Origin of the Moon

It is natural to ponder the origin of the Moon. In *The Tales*, Gurdjieff describes it as the result of a comet's impact on the Earth.* In our view, his account is better read as an allegory than literal fact. Nevertheless, contemporary

---
* In Search of the Miraculous by P D Ouspensky, Ch 7, p134

science provides credible evidence that the Moon is indeed a child of the Earth, in the sense that it has "similar DNA."

There are multiple theories as to how the Moon came to be. One suggests that it was captured as it flew by. Another suggests that the Earth and Moon formed together from the primordial gas cloud circling the Sun as a double system. The currently favored theory is that Earth initially had no moon, but a collision with an early "protoplanet" caused the ejection of a great deal of material from Earth, with some of it forming the Moon and the rest drifting off into space.

In Greek myth, Theia was the mother of Selene, the Moon goddess, and thus astronomers named the theoretical protoplanet after her. The theory fails to explain what happened to the protoplanet Theia after the collision. There is no evidence of it to be found within the solar system, so the presumption must be that it either drifted away beyond the heliopause or it disintegrated and gradually fell onto the Sun and other planets.

There are different collision opinions, too; one suggests that Theia gave the Earth a glancing blow and another that it hit the Earth head-on. The theory has been modeled using computer simulations. However, as we have no data about what happens when planets collide (if they ever do), the simulations are, at best, educated guesses.

Regarding the DNA of moon rock, NASA's astronauts brought back enough to enable extensive chemical analyses to be carried out. It provides the best and most credible evidence of the Moon's origin. The isotopic signatures of lunar rocks are almost identical to Earth's igneous rocks. They differ significantly from what is known (from meteorites) of rocks from other parts of the solar system.

A study published in 2013 indicated that water in lunar magma is almost the same in isotopic composition as water found in Earth magma. Other published analyses reach the same conclusion: that the Moon comes from material that originated on Earth. The only observed difference is that lunar rocks seem to have more aluminum and titanium, which may be due to local effects occurring over millions of years either to the Earth or the Moon.

Given this information, the protoplanet collision theory is hard to defend. If there ever was a Theia, lunar rocks ought to show some evidence of material from Theia. The likelihood of Theia having an identical isotopic signature to the Earth is very small. The only get-out is the possibility that rocks from other sites on the Moon might reveal a different picture.

The Moon's Internal Structure

Evidence suggests that the Moon's internal structure is similar to Earth; it has a geochemically distinct crust, mantle, and core. Thus it is presumed to have once been a sphere of magma shortly after its formation that gradually cooled like the Earth to create its current structure. *Figure 78* shows a simple comparison of the Earth's supposed internal structure with the Moon, based on the current scientific models. If the "sphere of magma" presumption is true, then the protoplanet theory is even less credible.

There are significant differences between the two models of internal structure. The Moon shows no evidence of the tectonic plates that comprise the surface of the Earth. Its crust is entirely igneous rock, so the whole idea of what a crust is has to be rethought in respect of the Moon. Just as there are earthquakes, there are moonquakes.

We have detected moonquakes occurring 1,000 km below the surface. They occur at a monthly frequency and are possibly caused gravitationally by the Moon's eccentric orbit around the Earth. Shallow moonquakes about 100 km below the surface have also been detected. The cause of these is uncertain. As there is no evidence of tectonic plates, it's hard to argue that they are tectonic plate movements. Earthquakes caused by tectonic plate movements have epicenters near the surface, although they can be as deep as 700 km.

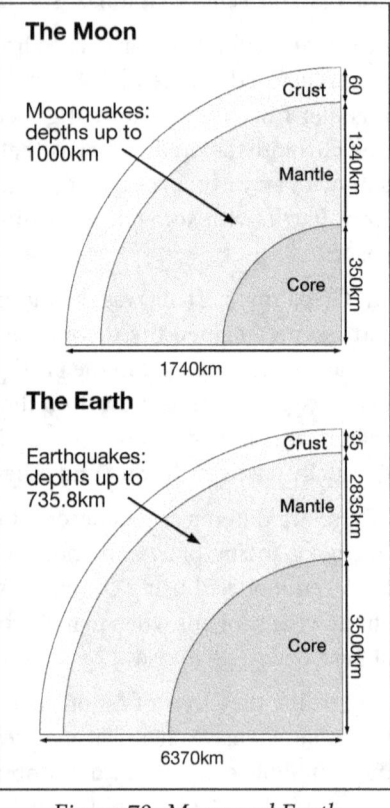

Figure 78. Moon and Earth

Analyses suggest that the Moon's core is partly molten, and it is likely to be eccentric, pulled by gravity towards the Earth as the Moon orbits. The lunar core is believed to be much smaller than Earth's, its radius being about 20% of the surface's radius. The Earth's core's radius is supposed to be 55% of the radius of the surface. Other planets are believed to have a core to surface ratios similar to the Earth.

The Moon's Atmosphere

Evidence suggests that the Moon has a very thin atmosphere. At sea level on Earth, there are about 10^{19} molecules per cc, whereas, on the Moon's surface, the lunar atmosphere has less than 10^6 molecules per cc. That very low density is similar to the Earth's atmosphere 300-400 km above sea level on the atmosphere's outermost fringes. However, it is significantly denser than the solar wind that pervades the space between planets. It has just a few protons per cc.

Three noble gases, argon, helium, and neon, dominate the Moon's atmosphere. There are also traces of hydrogen, ammonia, methane, carbon dioxide, carbon monoxide, nitrogen, sodium, and potassium. Lunar soil contains water in very small quantities. Scientists suspect that there is ice in craters at the Moon's poles. There may be an extremely weak water cycle that transports water to the poles. It would be very weak indeed; sunlight quickly decomposes evaporated water on the Moon.

The supposed causes for the Moon's atmosphere are:

- high energy photons and solar wind particles ejecting atoms from the lunar surface
- chemical reactions between the solar wind and lunar surface material
- evaporation of surface material
- material thrown out by comet and meteoroid impact
- gases released from chemical interactions below the Moon's surface
- electrostatically raised dust.

Thin atmospheres exist elsewhere in the solar system: on Mercury, the larger moons of other planets, and large asteroids. Such atmospheres don't retain lighter gases. Hydrogen and helium tend to escape into space.

Planetary Atmosphere

There is a distinct difference between the atmosphere of the Moon and that of a typical planet. Most planets have a self-created magnetosphere. As already mentioned, the Moon has a magnetosphere provided for it by the Sun but only when it lies outside the Earth's magnetosphere. *Figure 79* on the next page illustrates Earth's magnetosphere. It shields the Earth's atmosphere and biological life from the impact of the solar wind, solar radiation, and cosmic rays.

The boundary and the outermost layer of the magnetosphere meet the solar wind head-on, deflecting it around the Earth and slowing it down. The region between the bow shock and the magnetopause is called the magnetosheath. It

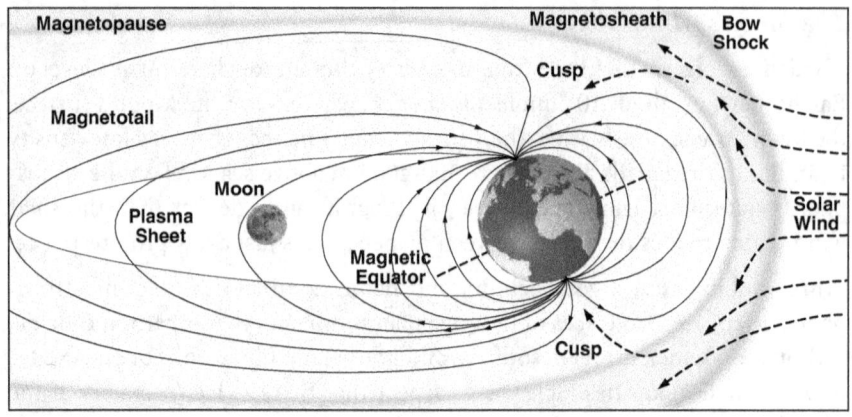

Figure 79. The Moon, Earth and its Magnetotail

consists primarily of solar wind intermixed with small amounts of magnetosphere plasma (charged particles). In this region, the direction and magnitude of the magnetic field vary erratically.

The magnetopause is the magnetosphere's boundary where the repellant pressure from the Earth's magnetic field balances the solar wind pressure. It changes size and shape with the fluctuation in the solar wind pressure. Some solar wind penetrates the magnetopause at the points labeled "cusp," descending into the Earth's atmosphere near the poles. The release of energy causes the phenomena known as the northern lights and the southern lights.

The magnetosphere is not spherical; rather, it is a very elongated malformed spheroid extending in the opposite direction to the Sun. Earth's Sun-facing bow shock is a mere 56,000 miles from Earth and about 11 miles thick. The magnetopause behind it is just a few hundred miles above Earth's surface. However, on Earth's nightside, the magnetotail extends 3,900,000 miles from Earth, a distance that is roughly 15 times as great as the orbit of the Moon. Within it, extending in the same direction, is a plasma sheet, a denser and hotter plasma region than is present in the other regions of the magnetotail. The orbit of the Moon passes through this plasma sheet.

The local magnetic field of the Moon is extremely weak. The detected magnetization is almost entirely crustal in origin. Large impact events may generate local magnetic fields. The largest crustal magnetizations are near the antipodes of the moon's "giant impact basins." As we shall discuss later, they are not necessarily impact basins. Nevertheless, their formation may be the cause of those magnetic fields.

Feeding The Moon

It is clear that if organic life feeds the Moon, it must do so, physically, by sending it plasma—substances with the density of H96, H48, and H24. Such substances could conceivably find their way to the Moon by electromagnetic attraction.

We can think in the following way. When a man or other animal dies, its corpse begins to behave differently at the microbiological level. The immune system, which had previously governed the microbiological world, has vanished. The internal temperature of the body is no longer controlled and bacteria—agents of deconstruction—multiply. The skin ruptures allowing the entry of air and more bacteria and even insects. The corpse becomes food and gradually releases various useful substances into the environment that can be nutrients for other life-forms.

However, while alive, the being also used and maintained some quantity of higher *Hydrogens*. H96 is the energy of the immune system, and more vivifying *Hydrogens* (H48, H24 and H12) are the substances of the psyche.

In respect of man, we can think of personality, which includes H48 and H24, as food for the Moon. The personality forms through imitation, education, movies, radio, and books, all the hypnotic influences of life. They create the mechanical backbone of the personality, furnishing different behaviors for different situations, evoked by everyday impressions and frequently repeated.

The personality's role is to shield essence as a kind of skin. Its mechanics are like the dead skin covering the body that acts as a barrier to external pathogens. As the epidermis cells (the outer layer of skin) mature, they produce keratin, a fibrous, waterproof compound. The body also uses keratin to create nails and hair and (in other life-forms) claws, horns, hooves, scales, shells, and beaks. By the time epithelial cells reach the skin's surface, they are dead, scale-like structures that form an organic barrier to the outer world. When ruptures of the skin occur, epithelial scar tissue grows over the wound.

Similarly, the personality acts as an automatic barrier that prevents external influences from reaching essence. We tend to identify with the personality despite its defects (vanity, lies, deceit, imagination, negative emotions, etc.), which we rarely observe. We believe that the mechanical (dead) personality is "I." Just like the surface of the skin, our usual behavior is to continually renew our personality, which we do through repetition. In those who work on themselves, essence may learn from personality, but in others, it rarely happens.

Gurdjieff's Hydrogens: The Ray of Creation

In *In Search of the Miraculous*,* Gurdjieff connects the Moon's feeding to the death of organic life. Everything organic being provides some of the energy that animated it in life to the Moon. He refers to this energy as the "souls" of living things, attracted to the Moon as if by an electromagnet. In respect of men, he suggests that even a certain amount of consciousness and memory goes to the Moon, where it finds itself under ninety-six orders of laws, in the conditions of mineral life, from which there is no escape except by the evolution of the Moon itself. In some way, personality and some of the structures that support it go to the Moon.

In *The Tales*, Beelzebub refers to a conversation with Archangel Looisos in respect of feeding the Moon, saying:**

> "And further, His Highness also explained that this cosmic substance, the Sacred Askokin, exists in general in the Universe chiefly blended with the sacred substances 'Abrustdonis' and 'Helkdonis,' and hence that this sacred substance Askokin in order to become vivifying for such a maintenance must first be freed from the said sacred substances Abrustdonis and Helkdonis.
>
> "To tell the truth, my boy, I did not at once clearly understand all that he then said, and it was only later that I came to understand it all clearly, when, during my studies of the fundamental cosmic Laws, I learned that these sacred substances Abrustdonis and Helkdonis are just those substances by which the higher being-bodies of three-brained beings, namely, the body Kesdjan and the body of the Soul, are in general formed and perfected; and when I learned that the separation of the sacred Askokin from the said sacred substances proceeds in general when the beings on whatever planet it might be transubstantiate the sacred substances Abrustdonis and Helkdonis in themselves for the forming and perfecting of their higher bodies, by means of conscious labors and intentional sufferings.

Put simply, the Moon feeds on Askokin, which comes from the substances Abrustdonis and Helkdonis, which Men use to perfect their higher bodies. There is a distinction between the substances that man can provide and that the rest of nature can provide. Men who are trying to perfect themselves generate Askokin directly and thus feed the Moon while alive. Man is the only type of being on Earth that can feed the Moon in this way. All others feed it through the process of Rascooarno, the breaking up of their body when they die.

---

* In Search of the Miraculous by P D Ouspensky, Ch 5, p85
** The Tales, Ch XLIII, Beelzebub's Opinion of War, p1106

The Electrical Nature of the Earth

The limit of the Sun's influence, the heliosphere, encompasses the whole solar system and is permeated by the solar wind, an ionic stream ejected by the Sun, which consists primarily of protons and free electrons. As already discussed, the Earth lives within that stream but is protected from it to some degree by the magnetosphere.

The outermost layer of the Earth's atmosphere, the exosphere, consists entirely of ions loosely held in place by gravity. Below is the thermosphere, a hot layer of ionized gas where atmospheric gases encounter solar radiation (X-rays and ultraviolet rays) and become ionized. Part of the thermosphere is fully ionized and is called the ionosphere. It is the lower boundary of the magnetosphere.

Because of the solar wind, the ionosphere is neither positively nor negatively charged. It can fluctuate between slightly positive or negative but soon reverts to a net neutral charge under the solar wind's influence. If the Earth becomes positive, it repels protons and attracts electrons; if it becomes negative, it does the opposite.

Below that layer, the atmosphere carries a positive charge surrounding the negatively charged Earth with a net negative charge of about half a million coulombs. There is an electrical gradient from the surface into the air, which averages out at about 100 volts per meter. The total potential difference from the Earth's surface to the top of the atmosphere is about 300,000 to 400,000 volts.

The lowest layer of air is mildly conductive, and thus there is a perpetual electric current from the atmosphere to the surface of roughly 1800 amperes, which reduces the voltage. Air is conductive because of its ionic content, with its conductivity rising at higher altitudes because of increased interactions with gamma rays, X-rays, and ultraviolet rays. When we reach the ionosphere, the air is rarefied and fully ionic, so the ions flow freely. It has become perfectly conductive, and there is no lateral variation at all in voltage.

The downward current would fully discharge the Earth's surface in about half an hour were it not for lightning. Lightning has the opposite effect of increasing the negative charge of the surface, and it balances the effect of the downward current. There are about 40,000 thunderstorms per day.

Thunderstorms work in the following way. Temperature, moisture, airflow, and the air's natural electric gradient cause positive charge to rise to the top of the thunder cloud. Negative charge congregates at the bottom. A lightning strike occurs as soon as the negative charge finds an ionized path to the surface. When lightning strikes, an opposite flow of charge occurs from the top of the cloud into the ionosphere.

Magnetotail and The Moon

If there is a feeding process between the Earth and the Moon, there will be a connecting structure roughly corresponding to a placenta through which nutrients pass. The nutrients coming from organic life must comprise substances that correspond to *Hydrogens* H96 *or higher*. *Hydrogens* denser than that would stay close to the surface by gravity alone.

It is likely then that the "placenta" between Earth and Moon is the Earth's Magnetotail. Personality and essence are constructed from very rarefied materials, and they disconnect from the bodies of men or animals at death. Such materials, less dense than the air, will naturally rise upwards. They carry a negative charge, as does the Earth's surface. Retaining their negative charge, such substances may rise to the top of the ionosphere and enter the magnetosphere.

At maximum, Earth's exosphere stretches approximately 190,000 kilometers (120,000 miles) towards the Moon and makes no contact. However, the Earth's magnetotail extends far beyond the orbit of the Moon. The Moon passes through it, spending about six days in every lunar month inside it.

The electrical state of the Moon varies according to its orbital position. When it is outside the Earth's magnetotail, the weak magnetosphere bestowed upon it by the solar wind insulates it. We suspect that the protection this affords the Moon is critical to it retaining its negative charge. If it were directly exposed to the solar wind, it would become electrically neutral from the influence of positive ions.

When the Moon enters the magnetotail, the solar wind deflects away, and the plasma sheet within the Earth's magnetotail takes over. The plasma sheet is in a constant state of motion and hotter (i.e., more energetic) than the solar wind. During the six days as the Moon passes through the magnetotail, the magnetosphere's plasma sheet sweeps across it many times, with encounters lasting anywhere from minutes to hours or even days. Electrons pepper the Moon's surface, increasing the Moon's negative charge. On the Moon's Sun-facing side, the sunlight counteracts this. Photons of ultraviolet light displace electrons from the surface. Thus the nightside of the Moon is negatively charged compared to the dayside.

NASA's Lunar Prospector spacecraft, which orbited the Moon in 1998-99, gathered the best data we currently have about magnetotail crossings. During some crossings, it detected significant changes in the lunar nightside voltage, typically rising from -200 V to -1000 V. In 2017* Japanese researchers, analyzing data from Japan's Moon-orbiting Kaguya spacecraft, reported that oxygen ions from the Earth's atmosphere made their way to the surface of the Moon during the Moon's passage through the magnetotail.

The passage of negatively charged plasma from the Earth to the Moon is thus well established. Gurdjieff said that the Moon was a large electromagnet for organic life.** It is during the Moon's passage through the magnetotail that it behaves as an electromagnet. Outside the magnetotail, under the solar wind's direct influence, it will not attract anything from the Earth.

The Moon's Evolution

The elliptical orbit of the Moon varies in radius between 238,900-225,700 miles (384,400-363,300 km). Laser-based measurements indicate that it is slowly receding at a rate of 1.5 inches (3.8 centimeters) per year. Models of the mutual gravitational influence suggest that the Moon's rate of recession will decrease with time. The theory is that while the Moon recedes, the Earth's spin gradually reduces, and the day gets longer. So projections suggest that the Earth and Moon will eventually find a stable orbit with the Moon and Earth locked into permanently facing each other. In theory, the Moon's orbit will then take approximately 47 (24 hour) days, and Earth's rotation will have slowed down to a single rotation taking 47 (24 hour) days. These projections are almost certainly wrong.

However, it is not at all obvious exactly what will happen. In *Table 15*, on the next page, we list some of the characteristics of the eight largest moons in the solar system. It is clear from simply perusing the table that our Moon is distinctly different from all the other satellites. All the others are part of moon families. Jupiter has 53 or more moons, Saturn 53 or more, Uranus 27, and Neptune 14. It might indicate that these planets are on their way to becoming solar systems.

Earth has but one large moon and its orbital period is 27.3 days. This may suggest (it seems logical) that our Moon is more mature in its development than any of these other moons, even though the Earth looks less mature in respect of becoming a sun with its own solar system. That seems odd, but it's not the only odd detail about these 8 moons.

---

* https://phys.org/news/2017-01-moon-periodically-showered-oxygen-ions.html
** *In Search of the Miraculous* by P D Ouspensky, p85

Gurdjieff's Hydrogens: The Ray of Creation

Satellite	Planet	Diam. (km)	Orbit (days)	Temp. (°K)	Atmos. Gas	Ice	Craters/ Grooved
Ganymede	Jupiter	5,262	7.15	110	Oxygen	Yes	Yes, Yes
Titan	Saturn	5,150	15.93	93.7	Nitrogen	Yes	Yes, Yes
Callisto	Jupiter	4,821	16.69	134	Carbon Dioxide	Some	Yes, No
Io	Jupiter	3,643	1.77	110	Sulfur Dioxide	No	No, No
Moon	Earth	3,474	27.30	250	-	No	Yes, No
Europa	Jupiter	3,120	3.55	102	Oxygen	Yes	No, Yes
Triton	Neptune	2,710	5.88	38	Nitrogen	Yes	Yes, Yes
Titania	Uranus	1,578	8.71	70	-	Yes	Yes, Yes

Table 15. The Solar System's Large Moons

Our Moon is by far the warmest. Its mean temperature, 250 °K at its equator, varies from 100 °K, far colder than it ever gets on Earth, to 390 °K, significantly hotter than it ever gets on Earth. Of course, that is due to its proximity to the Sun and its extremely thin atmosphere. It gets hot, but cannot retain the heat.

This high temperature explains the Moon's lack of ice. All the other moons, except Io, have substantial amounts of water ice on their surface. If they were hotter, the ice would become water and then water vapor and, for the moons with thin atmospheres, the water would be split by radiation into hydrogen and oxygen, and the hydrogen would escape from the atmosphere.

You may have noticed in the table that the atmospheres of Ganymede and Europa are dominated by oxygen. This is almost certainly because of the action of shortwave radiation on surface ice, and not organic life. It is on the icy moons where scientists have their greatest hopes of finding organic life. Ganymede has a salty ocean beneath areas of its icy surface, and both Europa and Enceladus, one of Saturn's moons, appear to have oceans beneath the ice. Such oceans are assumed to be created by the inner heat from the moon's core melting the surface ice from below. Ganymede's ocean is estimated to contain more water than the Earth's oceans.

All of these moons show evidence of craters except Io and Europa, and most have grooves which could be canyons caused by electrical interactions. However, there are other credible geological theories for the cause of some of these grooved features. The Moon has no canyons. Most of the moons have thin atmospheres with the notable exception of Titan, which, surprisingly, has a denser atmosphere even than Earth.

Only one of the moons listed, Ganymede, has its own magnetosphere (although Europa may have a very weak one), which seems surprising, because like all these other moons Ganymede does not spin on its axis. Clearly, a moon can be tidally locked and still have a magnetosphere. The other distinctly unusual moon is Io, with its roaming "plasma volcanoes" and lack of craters or grooves. We assume Io has these features because of its rapid orbit and its proximity to Jupiter.

We have too little data to suggest what path the evolution of a moon follows. It remains an unanswered question, as indeed does the evolution of a planet. Objective science asserts that should the Earth and Moon's evolution proceed positively, the Moon will eventually attain the status of a planet and the Earth that of a sun. Consequently, it should be the case that the Earth and the Moon are growing.

Is there evidence for this?

The Growing Earth

There is abundant evidence in the fossil record that the Earth is growing. Even a cursory examination of the fossil record makes it obvious that creatures were much bigger in the past. About 300 million years ago, there were dragonflies with 2.5 feet wingspans, and millipedes six feet long. Their biological structure was not particularly different from that of the same life-forms in our era. They were simply much larger.

Take a long step forward, about a hundred and fifty million years, and you land in the middle of the Mesozoic Era, the time of the dinosaurs. There you meet with the sauropods, the largest dinosaurs that ever roamed the Earth. With long necks and tails, they could measure as much as 80 feet from end to end. They stood on elephantine legs and were said to have weighed 50 tonnes, about the same as six full-grown African bull elephants.

Later, dinosaurs were smaller. The carnivorous Tyrannosaurus Rex was half that size—about 40-43 feet long and 15 to 20 feet tall, and may have weighed as little as 8 or 9 tonnes. They strutted their stuff about 66 million years ago. And when that era ended and mammals dominated our planet, they were large mammals. In the Eocene period, about 38 million years ago, there were giant pigs 9 ft long and large dog-like animals 16 ft from head to toe.

All of this and much more is well explained and documented by Stephen Hurrell in one of his books.* In a nutshell, the explanation for all of this gigantism is that the Earth was smaller, gravity was less, and thus creatures of those sizes were possible in their time. If you could, by some act of time travel, bring any of these creatures to the modern era, they would not survive. Their biological structure (bones in mammals, exoskeletons in insects) would be completely inadequate.

Hurrell's arguments to support this hypothesis are particularly convincing because he is an engineer who knows how to scale structures— bridges and buildings. You may not be familiar with the problem of scale, but it is quite simple to explain. A perfect cube with a side measuring 1 ft has a volume of 1 cubic foot. A cube with a side of 2ft has a volume of 8 cubic feet. The length of the cube doubles, but the volume and weight of the cube multiply by 8. The same applies to any three-dimensional object; if you scale it up by a factor of 2, its volume and its mass will multiply by 8. In building anything large, it is necessary to understand scale.

The heaviest land animal of our era is the African elephant, weighing about 8 tons and standing 12 feet in height. Double the height, and it will weigh 64 tons. Its bones would break. The same rules apply to the Patagotitan mayorum (also called the Titanosaur). It is thought to have been 37.2 meters long and weighed 70 metric tons—far too big for our Earth.

Sand Dunes and the Growing Earth

Professors Mann and Kanagy studied sandstone blocks that had once been ancient dunes, analyzing blocks where the dune's shape could be determined and measured. They published their work in 1990.** In various samples, they measured the "angle of repose." That is the angle between the horizontal plane and the surface of the sand that forms the dune. Dunes are created from the action of the wind. At a steep angle, the sand slides down the dune until eventually a point of stability is reached. At that point, the dune has acquired an "angle of repose."

The professors noted that the older the specimen, the steeper was the angle of repose. Their opinion was that: "... steeper angles may have been recorded in ancient sediment because Earth's acceleration of gravity was less than now." They did not conclude that the Earth was growing in size, perhaps for fear of the reaction such a conclusion might provoke; they simply suggested that the

---

* *Dinosaurs and the Expanding Earth: Solving the Mystery of the Dinosaurs' Gigantic Size* by Stephen Hurrell.
** *Geology*, magazine of Geological Society of America, Mann and Kanagy, 1990

force of gravity was different in prior times without explaining how that could be so.

Continental Drift and the Growing Earth

Evidence suggests the Earth's landmasses were once a supercontinent, a single landmass surrounded by an ocean. Geologists have named that supercontinent Pangaea and the super-ocean that supposedly surrounded it Panthalassa. Pangaea is believed to date back 336 million years and have had its center on the Equator. The theory is that Pangaea fragmented and various parts drifted slowly apart, eventually producing our current globe with all its continents.

In 1958 Sam Warren Carey demonstrated, using a large globe and continents made from perspex that could slide over the globe, that South America and the African continent fit together remarkably well. Including the continental shelves in the shapes of the continents achieves an even better fit. Refining the model, he later demonstrated that the continents only fit together really well on a smaller diameter Earth. He thus concluded that continental drift had been caused by the Earth expanding in size.

The expanding Earth hypothesis was treated as a genuine possibility for a few years. However, with the advent of plate tectonics in the 1960s, it fell out of favor. The effect of parts of the Earth's crust sliding below different but adjacent areas of the crust during earthquakes is detected and observed. So plate tectonic enthusiasts carved up the Earth into distinct plates and concluded they could explain continental drift without any need to consider the possibility of the Earth growing in size.

However, although you can create theoretical continental drift models based on plate tectonics, there is no easy way to explain why the continents fit together so poorly if the Earth has always been the same size. It is also difficult to explain how some mountain ranges came to be with the plate tectonics model.

In an expanding Earth model, mountain ranges are wrinkles in the crust formed due to the expansion of the Earth's core; with plate tectonics, the idea is that mountains form at the border between two plates, with one plate gradually lifting the other into the air.

Earth's Speed of Growth

In his book about the growing Earth and the scale of prior life-forms, Stephen Hurrell presents the graph shown in *Figure 80*, representing his estimate of the

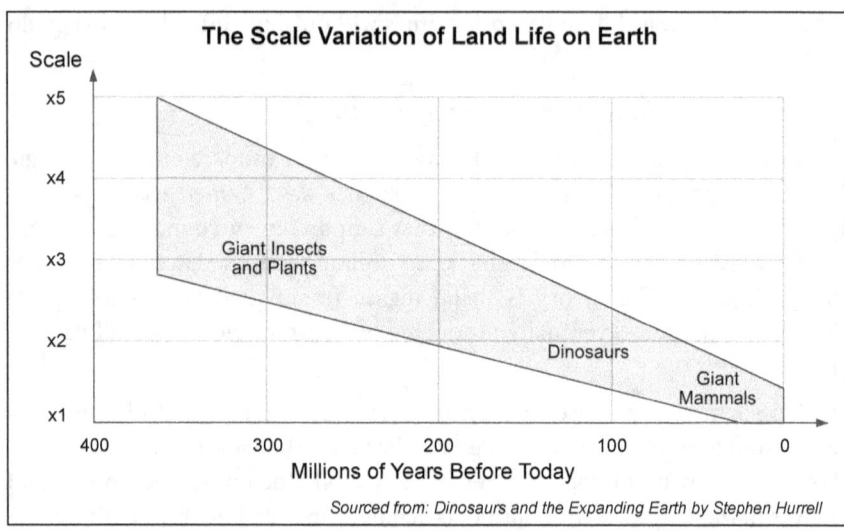

Figure 80. Earth's Growth based on the Size of Land Life

increase in the Earth's gravitational force based on the size of land-based life-forms. Due to the fossil record limitations, it presents a range of values (the shaded area) rather than a simple line. The graph's left side indicates that 360 million years ago, life-forms were between 3 and 5 times larger than today.

Over millions of years of geological time since the Carboniferous period, life-forms have diminished in scale. First came giant insects and plant life. Much later came the massive sauropods and then lesser dinosaurs. After the dinosaurs became extinct about 66 million years ago, several large mammals grew to sizes similar to the smaller dinosaurs. Millions of years later came super-sized versions of today's animals. These gave way to their smaller present-day versions.

That graph provides a view of the growth in the Earth's gravitational force, whereas *Figure 81* on the next page shows an estimate in the change of the Earth's radius. It is based on computer reconstructions of the ancient continental and ocean crust that imply an increasing radius. It is important to note that a relatively small change in radius leads to a much greater volume change. According to the graph, 200 million years ago, the Earth had an estimated radius of 4200 km. The calculated volume with that radius (rounded) is: 310,000,000,000 cubic km.

Taking the current radius of the Earth as 6367 km, the calculated volume now works out at 1,081,000,000,000. It suggests a 3.6x increase in volume. It

PLANETS, EARTH AND MOON

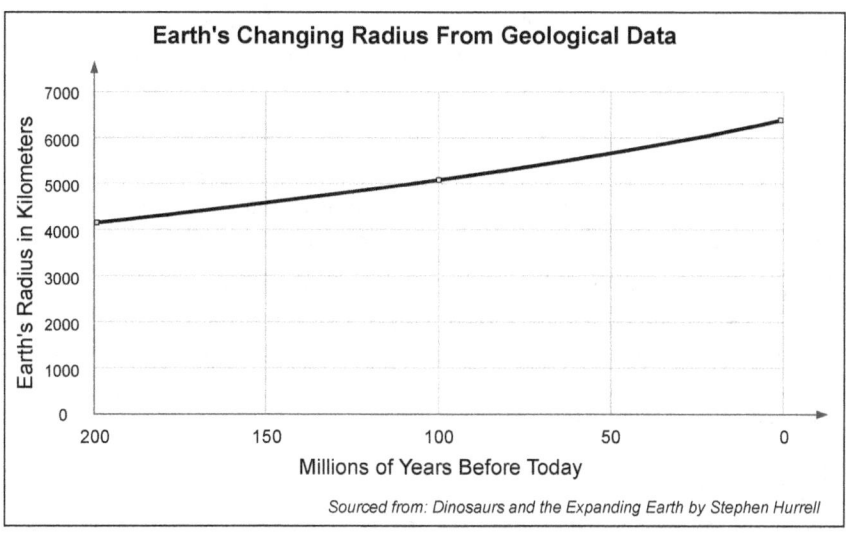

Figure 81. Earth's Growth based on Continental Drift

corresponds well with *Figure 80*, suggesting a figure slightly above the upper range of gravity increase suggested for 200 million years ago, shown in that graph.

Thus we have two different projections of the Earth's growth that conform reasonably well with each other. They also suggest that the Earth's density (mass per volume) is slightly less now than it was 200 million years ago.

If you wonder whether growth in the Earth's mass would cause its orbit to change mathematically, the answer is: no.

Growth by Dust and Meteorite?

Opposition to the theory of a growing Earth centers on questions about the mechanism for growth. The Earth's accumulation of material from space dust and meteorites, while regular, appears to be insignificant. There are varying estimates and no meaningful consensus. Estimates are as low as 37,000 tons annually and as high as 1000 times that figure. More than anything else, this indicates that there is no easy way to measure it.

Naturally, estimates for the Moon's possible accumulation are much lower. However, compared to the Earth's mass (5.9722×10^{24} kg) and the Moon's mass (7.35×10^{22} kg), these amounts are so small you can ignore them.

Either body could experience large impacts that would add much more mass in a single event, but there is no evidence of frequent impacts of this kind. They would leave geological evidence. They devastate the biosphere and thus leave the evidence in the fossil record. And while it's clear that there have been such events, there have not been many.

Estimates of the amount of asteroid and comet material in the solar system are very low. By way of illustration, if you took all the known asteroids in the asteroid belt and dropped them on the Moon, it would add only about 4% to the Moon's mass. Add them to the Earth, and you add just 1%.

It is possible that at times the whole solar system drifts through areas of space that are dense with dust, and thus the level of mass accumulation increases in that way. However, if that happened it would be visible in the geological record, and there is no obvious evidence of it. It leads to the conclusion that if the Earth is growing, it is growing from within.

Growth by Flood?

As mentioned earlier, there is very strong evidence of a great flood in the mythical and historical record occurring in fairly recent times. The flood myth occurs in the traditions of Assyria/Babylon, Persia, Syria, Asia Minor, Greece, Egypt, Rome, Russia, India, China, Australia (Aborigine), Fiji, Hawaii, Canada (the Cree), the US (Cherokee, Apache) Mexico, Peru—it's in almost every tradition. These myths were either a collective synchronous act of self-delusion by tribes in every part of the globe, or they refer to a real and dramatically important event. The latter seems far more likely.

If we accept that there was a massive inundation of the Earth, it clearly cannot have been caused by local weather. The water must have come from elsewhere—which can only mean that it came from space, raining down from the sky, day after day after day. Most of the flood myths record small groups of people surviving in boats and eventually finding land.

One of the more interesting oral records comes from the Australian Aborigines because it is quite precise. It reports that the sea level rose about 120 meters in the period estimated to be between 18,000 and 7000 years ago.* The record includes the Bundaba Flood Story, which speaks of rain, thunder, and a terrible hurricane-level wind that uprooted trees. The flood rose higher and higher until the land was covered except for the tops of two or three mountains. The floodwaters eventually receded.

The 120 meters rise in sea level can be attributed to ice melting from the poles over a long period. But this great flood, present in so many mythological

---

* *https://listverse.com/2016/10/01/10-fascinating-discoveries-from-ancient-australia/*

records dating from approximately 7000 years ago, cannot be explained in that way. It is a mystery that poses two distinct questions:

– Where did the water come from?

– Where did it go?

Note that the Moon and probably other planets would necessarily also be affected by such an inundation. In the case of the Moon, the water would soon evaporate into space.

Because of this event's mythical record, we are obliged to entertain the possibility that the growth of the Earth itself could include a certain amount of accretion from space, but how it occurs is not clear.

Proof and Opposition

Some ideas provoke a level of abhorrence from the scientific establishment that is difficult to comprehend. The growing Earth hypothesis is currently such an idea. If we assume that the Earth is still growing, it is possible to test the hypothesis by measurement.

In his book,* Stephen Hurrell reports that the data used in Global Positioning System (GPS) measurements ought to help resolve the argument. By 1998 it was clear that preliminary results taken from different parts of the Earth's surface indicated expansion.

Sadly, "scientists" arbitrarily began to modify the raw data, removing anything that indicated Earth expansion, saying that those particular measurements were "considered unreliable." The modified results (showing no Earth expansion) were then published in various scientific journals supporting the Constant Diameter Earth model.

Some people cried foul. But as the results were what the establishment expected to see, objections were ignored. The Australian geologist James Maxlow reported** that scientists had measured an average upward (radial) motion of 18 mm per year using measurements from Very Long Baseline Interferometry (VLBI) stations and chose to ignore those results.

There are, of course, complications in determining whether radial growth is occurring if measurements are limited to short periods. There is the reality that earthquakes can make significant differences in measurements, increasing and decreasing the land height in various places. There is the impact of glacial

---

* *Dinosaurs and the Expanding Earth,* by Stephen Hurrell
** *Terra Non Firma Earth* by James Maxlow

melting, which has been occurring for over 100 years, removing weight from the land and causing small local upward movement.

In his book, Maxlow explains that eliminating any small annual increase in radius has resulted in relatively large periodic adjustments in charts of the published data. A chart of a location near Canberra in Australia provides an extreme example, embodying an arbitrary adjustment of 71 mm from 1993 to 1994. Only an earthquake could explain such an adjustment, and there was none.

The Nature of Planetary Growth

It is important to avoid the temptation to extrapolate when we contemplate the growth of the Earth, or any moon, planet, or sun, for that matter. The evidence is very strong that the Earth's radius has grown by over 2000 km in the last 200 million years. But as we go further back, the picture is less clear, and we currently have no data for the period before 400 million years ago.

In objective science, we conceive of moons, planets, and suns as living beings. When we survey Nature, we notice that growth is rarely constant from a single seed to a completed living thing. Plants and trees have growing seasons. Butterflies grow as grubs then transform surprisingly in a chrysalis to become flying creatures. Mammals grow up to a point and then cease growing. To assume that moons, planets, or suns grow constantly may be completely wrong.

We have very little idea about how a moon becomes a planet or how a planet becomes a sun. To even hazard a guess, we need a foundation from which to begin. Modern astrophysics does not provide us with such a basis, but happily, plasma physics does.

The evidence suggests that the evolution of a planet and a moon's birth is not a foregone conclusion. Although we cannot assert that planetary disharmony caused all the great extinction events found in the fossil record, it seems likely. Earthquakes and volcanic eruptions are not violent enough and are anyway under the control of the Earth. But large rogue asteroids or even direct planetary interactions may be the cause of such extinctions.

If we set that to one side, we also face Gurdjieff's assertion that peopling the planet to provide the Moon with the substances it requires for its growth is also fraught with hazard, since three-brained beings have to be hoodwinked into providing those substances. A consequence of the method chosen to achieve this has been that man has degraded to the point where he is no longer able to operate in a manner that consciously provides those substances (the Fulasnitamnian principle). Nature adapted, and now Man operates under the Itoklanoz principle, providing the necessary substances only at death.

Gurdjieff began teaching the Ray of Creation and its attendant concepts to Ouspensky and others in his two Russian groups in 1914. The idea that other galaxies of stars, aside from the Milky Way, existed was merely a hypothesis among astronomers at that time. It was proved true about ten years later, in 1924.

None of the assertions Gurdjieff made in his articulations of objective science has, to our knowledge, been proven incorrect. There is a contentious one, which we already discussed,* the existence of an aether. Contemporary science insists it does not exist but has yet to answer the obvious question about EMR: if it travels in waves, what is the medium?

---

* Chapter 1, p22

Chapter 12

A New Model of The Universe

—⚶—

"And those who were seen dancing were thought to be insane by those who could not hear the music."

~ Nietzsche

This final chapter is a summary dressed up as a narrative. The narrative describes the universe in a way that reflects Gurdjieff's cosmological teachings. It is an objective science perspective supported by the information furnished in this book. It offers a new model of the universe.

Modern science's Lambda CDM narrative with its absurd 'Big Bang' origin story and its immense baggage of dark matter and dark energy is finally losing its grip on the mind of man. It isn't because new astronomical data from space probes and telescopes have failed to support this collection of tired theories. That has been the case for decades. Neither is it because some bright new theory has suddenly emerged to offer a better alternative. Plasma cosmology has provided a plausible option for decades.

No. It is because of the SAFIRE Project, which among other important results has experimentally verified the "Electric Sun" theory. On its own that would not be enough to cast the Lambda CDM theory into the trash can. But the Safire project is destined for commercial success, and in this world, that's impossible to ignore. To find out more, watch the video.*

It may take a few years for a plasma-based Standard Model of the universe to unseat the modern scientific creed, but it will happen. Such a model will most likely be atheistic, irrespective of whether it is framed by atheists. An atheistic stance has long become de rigueur in academia, so it will probably continue by intellectual momentum. Here we offer a model of a different kind.

---

* https://www.youtube.com/watch?v=7GFFfmBGb5U (Safire Project YouTube video)

Gurdjieff's Hydrogens: The Ray of Creation

The Narrative

Laws govern the universe, and everything within it can be weighed and measured, including the Absolute (God himself). There are laws, which form a hierarchy according to the level or world in which they act.

For example, a crystalline substance like a stone belongs to the level of the Moon, World 96. A stone cannot flow like a liquid. A Law at its level prevents it from acting as a fluid. For that to change, the stone would have to raise its level one higher to World 48. Thus levels are important in respect of the laws of the universe. The primary levels of the universe are notes in the Ray of Creation. The Ray of Creation is an octave that descends as illustrated in *Figure 82*: *do-si-la-sol-fa-mi-re-do*.

The labels of the various notes indicate what each level is in relation to the universe itself. So the Absolute, the higher *do*, is, in theory, the unified intelligence that created the universe. The level of *si* is All Worlds, which can also be called the Sun Absolute. It is "the abode" of the Absolute. Whatever exists at that level is in direct contact with the Absolute.

Below that, the note *la* is the level of All Suns, of galaxies. There are multitudes of galaxies in the universe, and each has Rays of Creation passing through. We do not really know what a galaxy is beyond the observation that it is a collection of billions of suns that seem to orbit together in a spiral manner. We cannot tell whether it is alive and, if so, what kind of life it leads.

The next level, *sol*, is the level of suns and, by extension, solar systems. In our Ray of Creation, this note is our Sun. There are countless numbers of such rays. The family of planets of our solar system is the note *fa* and forms the next level. Below that, the note *mi* is our planet, the Earth. At this point, at the level of Earth, we have narrowed the Rays of Creation down to our specific ray. The final note and level, *re*, is the Moon, our planet's satellite.

The octave completes with the final *do*, which is also the Absolute—the serpent eats its tail. While appearing to be an opposite, the lower Absolute is another aspect of the higher Absolute. It acts as the single foundation of stability for every Ray of Creation in the universe.

Laws and Atoms

As illustrated in *Figure 82*, a specific number of laws exist for each note or level: 1 for the first level, 3 for the second, 6 for the third, and so on. Absolute is subject to a single Law expressed by the words "I am"—he exists. From this one Law, he creates three laws that govern his "place of existence." They also apply to every subsequent level in the context of that level. Levels below the

A New Model of the Universe

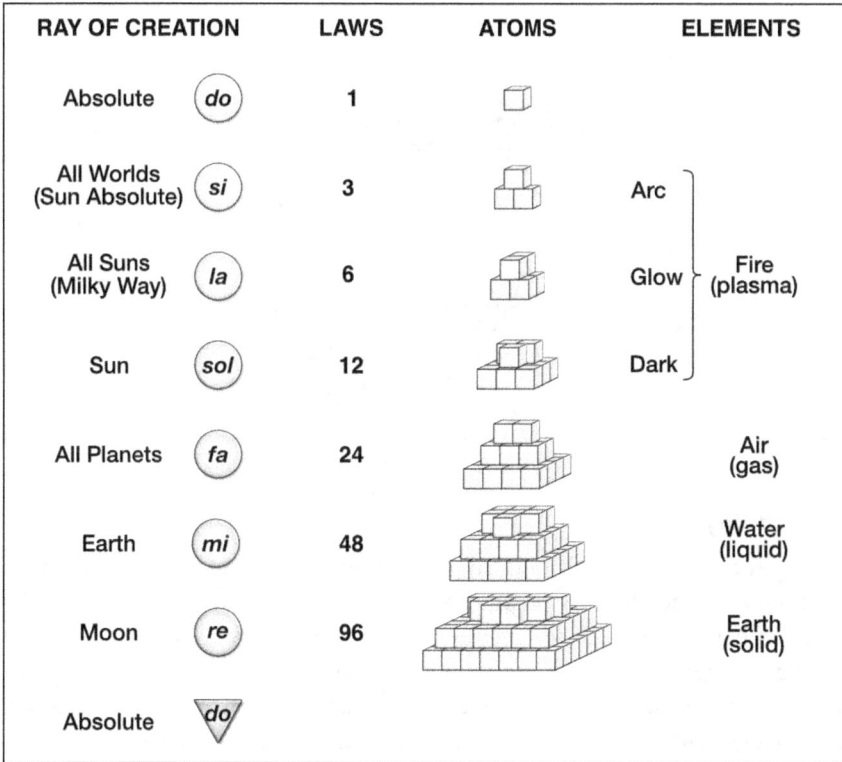

Figure 82. The Ray, Laws, Atoms and Elements

Sun Absolute create their own laws. Such laws apply to the levels below in their context. For this reason the laws double up mathematically for each lower level. All the laws are immutable forces—identical in concept to scientific laws.

Figure 82 also illustrates atoms associated with each of these notes or levels. The only genuinely indivisible atom is an atom of the Absolute. As we descend from one level to the next, the materiality of atoms increases. The density of atoms at each level is parallel to the number of laws. Thus an atom of the Sun is twelve times as dense as an atom of the Absolute. An atom of planets is 24 times, of the Earth 48 times, and so on.

On the right of *Figure 82*, we designate atoms of various densities in accord with the four elements of the ancient Greeks. These designations are as relevant today as they were in the past. However, the element of FIRE, which we nowadays call plasma, has three different physical manifestations: dark (or invisible), glowing, and arc (or incandescent).

Gurdjieff's Hydrogens: The Ray of Creation

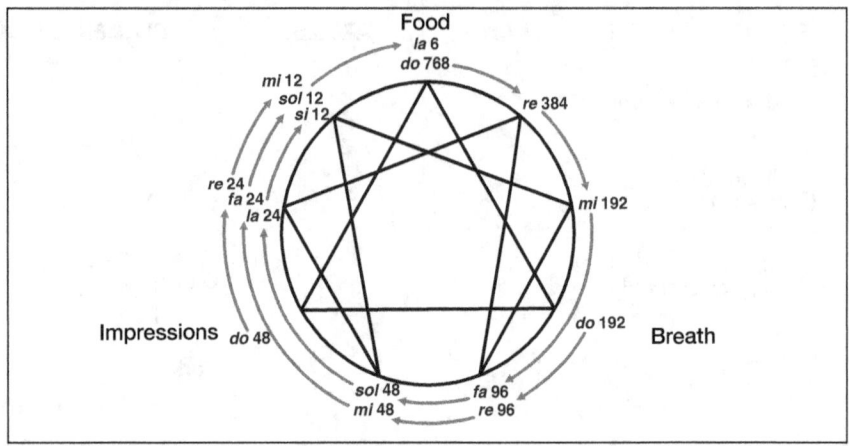

Figure 83. The Enneagram

Materiality

Everything is material, from the densest atom of matter to the most energetic photon with the shortest possible wavelength. There is no need for a distinction between matter and energy. All is material; energy is simply a force acting through matter of some kind. Many things that we rarely think of as material are nevertheless material. For example, information is material, knowledge is material, a thought is material, an emotion is material.

All events in the universe happen according to the Law of Three, which is as follows: Events occur only when three different substances with three specific forces (active, passive, and neutralizing) acting through them mix. It applies to everything at every level and scale. The Law of Three is one of two primordial laws that the Absolute determined.

The other is the Law of Seven. It states that all series of events unfold in octaves that are either creative or receptive. Creative octaves, such as the Ray of Creation itself, descend from high to low: *do-si-la-sol-fa-mi-re-do*. An example is building a house, which begins with an intention and results in a physical object. Receptive octaves ascend from low to high: *do-re-mi-fa-sol-la-si-do*. An example of a receptive octave is a man's education. If the octave completes, the man is transformed from ignorance to become skilled and perhaps even wise.

All living things embody these two laws. Consequently, they can be represented by an enneagram that combines the Law of Three and the Law of Seven, as illustrated in *Figure 83* above, which shows the octaves of food in Man.

A New Model of the Universe

The Living Universe and the Trogoautoegocrat

At every level, the universe is composed of cosmoses, living beings—cosmic units that eat, breathe, and perceive. Each such creature, large or small, has a definite lifespan: a birth and eventually a death. Only the Absolute and the Sun Absolute are eternal and immortal. All other levels, galaxies, suns, planets, and moons, experience death.

The universe is Trogoautoegocratic, which means that it is governed by eating. From microbes to galaxies, every living thing consumes food, air, and impressions, and reciprocal feeding regulates the whole. Cosmic units (creatures) consume the substances of other creatures, sometimes destroying them to do so. Nature may be "red in tooth and claw," but so are cosmic units at every level. Thus, all the universe's substances are either part of a being or matter discarded by a being that some other being will eventually consume.

The Moon is the growing tip of our Ray of Creation. In time it may become a planet, and similarly, a planet, like the Earth or Jupiter, may become a star. Beings at all levels can evolve to higher levels or descend to lower levels. The way up is also the way down. Thus a star could lose its luminosity and become a planet, and a planet could lose its atmosphere to become as lifeless as a moon.

Sexual reproduction is a facet of organic life. At other levels, reproduction is asexual. Galaxies can give birth to other galaxies in a process analogous to cell division but at a much larger scale. When that occurs, the newborn galaxy is called a quasar by astronomers. It is an extremely energetic mass residing in the center of a broad disk of energetic matter. Such infant galaxies shine more brightly than their parent galaxy.

Similarly, stars can give birth to new stars, with the new star tending to be a white dwarf that outshines its parent. Suns can give birth to planets, and planets can give birth to other planets and moons. When planets or moons die, they become food for other planets or suns.

The suns, planets, and moons are points of stability, which tend to attract substances that enter their sphere of influence. At the center of these bodies is the lower Absolute. The universe thus proceeds as a flow between two aspects of the Absolute. One aspect is the Absolute as an energetic but diffuse being whose body saturates the universe. The other aspect, referred to as Holy the Firm, is scattered among the vast number of suns, planets, and moons, occupying a point of stability at the center of each.

Plasma and Plasma Structures

Estimates suggest that 99.99% of the material of the universe is plasma. The age-old idea of the universe composed of four "elements," EARTH, WATER, AIR, and FIRE, equates precisely to the modern concepts of solid, liquid, gas, and plasma.

However, plasma can have three states: invisible or dark (e.g., ions drifting through the air), glowing (e.g., a neon light), and incandescent (e.g., arc welding). Because of this, we can think in terms of six distinct types of substance.

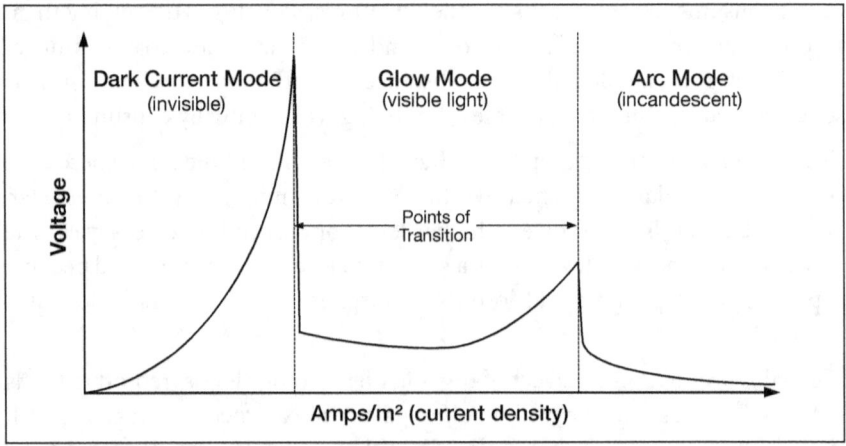

Figure 84. The Three States of Plasma

The difference between the three states of plasma is as distinct as the difference between ice and water. The graph in *Figure 84* above illustrates it. The transformation from a less energetic state to a higher one requires a specific amount of additional energy before a state change occurs.

Because less energetic plasma is invisible to the eye, we fail to notice many of its effects. Astronomers ignored it for many years. Nowadays, space probes test for it, and some, like the EAS's XMM-Newton Observatory, are built to monitor plasma activity. In 2012, a team led by Philipp Kronberg of the Los Alamos National Laboratory discovered that a gigantic plasma cloud surrounded our galaxy with a width 80 times greater than the galaxy itself. The cloud, which appears to be a galactic "heliosphere," contains as much matter as the galaxy itself.* Recent readings suggest that the cloud has a temperature of 10,000,000°K.**

---

* https://www.space.com/17734-milky-way-galaxy-giant-gas-halo.html
**http://www.sci-news.com/astronomy/hot-milky-ways-halo-08492.html

A New Model of the Universe

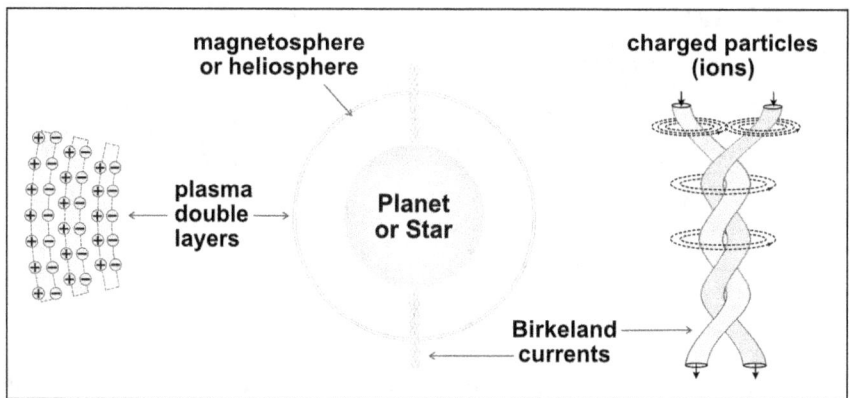

Figure 85. Plasma Structures

That cloud of plasma may be an equivalent structure to the Sun's heliosphere and the Earth's magnetosphere but at a much larger scale. Note that while gases rarely form coherent structures, plasma does. *Figure 85* illustrates two important plasma structures: double layers and Birkeland currents. Double layers, usually multiple ones, form around planets and solar systems, and galaxies too, in a way that preserves their electrical state.

The Earth is electrically negative with respect to the Sun. Its magnetosphere, surrounded by double layers, preserves its electrical state. It deflects the positively charged solar wind. Evidence suggests that plasma double layers surround the heliosphere. In 2019 when Voyager 2 entered the heliopause, it reported temperatures of nearly $50,000°K$, hotter than the Sun's surface.

Birkeland currents are of primary importance. Illustrated in *Figure 85*, they resemble a twisted pair of wires that carry charged particles in the same direction. The "sleeves" of the Birkeland currents are double layers that isolate the currents from the surrounding environment. Birkeland currents provide a continuous stream of positively charged particles to the Earth's poles. The current's strength varies according to the activity of the Sun. Similar Birkeland currents, but much more substantial, flow into stars and also galaxies, delivering an interstellar and intergalactic electricity supply.

The Z-pinch

The final plasma structure we need to explain here is the z-pinch. Z-pinch formations occur in Birkeland currents at points where the current flows exceptionally strongly. They give rise to a plasma structure called a plasmoid—a twisted ball of plasma. Z-pinch plasmoids are instrumental in the creation of galaxies and solar systems. The plasmoid attracts plasma at a particle level in a

Gurdjieff's Hydrogens: The Ray of Creation

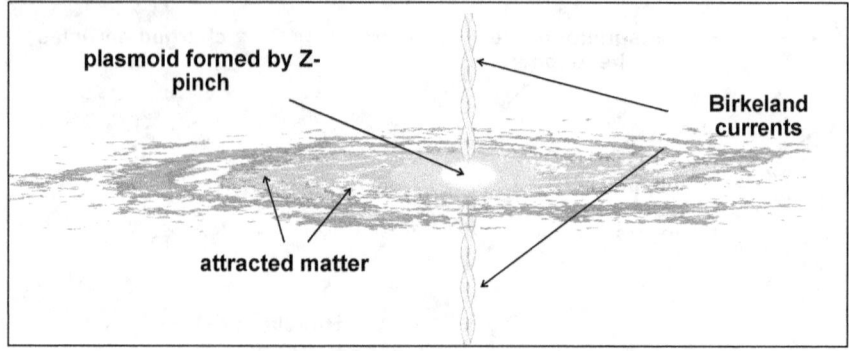

Figure 86. The Z-pinch at a Galactic Center

plane perpendicular to the Birkeland currents, thus creating a disk structure around its axis, as shown in *Figure 86*. In a galaxy, this evolves into the galactic disk with its multitude of stars. In a solar system, it becomes the plane of the ecliptic where planets orbit.

Conveniently for plasma cosmologists, the behavior of plasma structures scales up without significant change. What can be observed in laboratory conditions seems to occur in precisely the same way at the level of a planet or a solar system, or a galaxy. As below, so above.

The Creation of Elements

The Earth's interior is a crucible for creating new heavy and light elements that can bubble up from its core—within which the Absolute acts as a point of stability. When we connect this with evidence that the Earth is growing in size, it becomes even clearer that the Earth does not have a sedate stable existence.

The evidence of the Earth increasing in size comes from multiple sources.

– Fossils going back 400 million years record that species, in general, were much larger in the past and have scaled down in recent years, indicating that gravity has increased. Dinosaurs could not survive on the current Earth.

– Geological evidence of the angles of repose of ancient sand dunes tells the same story of a growing Earth.

– In continental drift modeling, the continents fit together far more accurately if you assume continental drift began on a smaller Earth.

– Current satellite measurements suggest that the Earth is gradually expanding.

If the Earth is growing, then, most likely, the other planets are too. And if planets grow, then their magnetospheres probably also increase in size beyond the variance that the solar wind causes.

Planetary Behavior

Most of the planets have intrinsic magnetic fields and corresponding magnetospheres. Only Mars and Venus lack them, but the solar wind generates a magnetosphere for them, as it does for the Moon. The cause of the Earth's magnetic field is unknown. Internal electrical activity is responsible in some way, perhaps related to Birkeland currents from the Sun.

The magnetic field's behavior is mysterious. It reverses regularly, on average, every 200,000 to 300,000 years. It hasn't flipped in the last 780,000 years, so some people suspect it may soon do so. No-one has a good theory for why this happens.

Planets do not have immutable orbits. Their orbits can and occasionally do change. The Earth's magnetic field anomalies are just one of several facts that suggest this. The fossil record indicates that periodically, every 25 million years or so, a mass extinction destroys enormous numbers of species. The planets' fundamental characteristics: their varying axial tilts, their different spin rates, and the existence of the asteroid belt all suggest that interplanetary interactions of some kind occurred in the past.

Planets can become suns. Jupiter may be in the process of doing so. It exhibits a low level of luminescence, and it already possesses a mini solar system consisting of four large moons and a plethora of small ones. It would have to move out beyond the heliopause and become far more luminescent to evolve into a solar system. Similarly, for the Moon to become a planet, it will need to move much further away from the Earth.

What would happen if the magnetospheres of two planets touched?

An immediate and mighty plasma discharge would most likely occur, and given the attractive and repulsive powers of electromagnetic forces, the orbits of the planets would most likely be disturbed.

There is strong evidence that this may have happened in the past. The proof of this on Earth is the Grand Canyon and several similar geological features, which water erosion does not explain. Most likely, interplanetary electrical discharges carved out the Grand Canyon. Valles Marineris is an example of a much larger canyon on Mars, which is surely the result of a much more powerful series of lightning bolts.

Gurdjieff's Hydrogens: The Ray of Creation

Mars presents an interesting but perhaps sad picture. It has extensive amounts of sedimentary rock created by water action over many millions of years and even shows evidence of rivers and river deltas. Most likely, it once had life. However, now, there is little evidence of water, and it appears to be losing its atmosphere relatively rapidly. It may have had life, but if it did, there's nothing left now except, perhaps, fossils and bacteria. It's a speculative idea, but it may be the case that cataclysmic interactions with other planets brought Mars to its current state.

In this universe, nothing stays in the same place. Whether it's a sun or a planet or a human being, it either ascends or descends.

The Unimportance of Humanity

Advances in human technology have proven invaluable in extending our knowledge of the universe. A hundred years ago, we had no idea of its extent —we never even knew whether other galaxies existed beyond the Milky Way. Now with space probes and orbiting telescopes, we can see further than we ever imagined possible.

The universe, estimated to be at least 93 billion light-years in diameter, comprises at least 2 trillion galaxies. A human body is estimated to have 30-40 trillion cells. So our Milky Way in relation to the universe is similar in scale to one of your cells in relation to your body. You are not aware of the cells of your body. Somewhere between 50 and 70 billion of them die every day, and you don't even notice.

The Milky Way has about 400 billion stars; Andromeda about a trillion. The estimate of the number of atoms in a human cell is 100 trillion. So an atom in relation to one of our cells is similar in scale to a sun in relation to a galaxy. The estimated number of suns in the universe is only a thousand times less than the number of atoms in a man's body.

Our star is unimportant. It is an average star on a minor arm of the Milky Way—closer to the galaxy's edge than the center. The Milky Way is itself an unimportant galaxy, part of a group of 54 galaxies that are in turn part of the Virgo supercluster. There are estimated to be 10 million superclusters in the observable universe.

Organic Life on Earth

Homo sapiens emerged an estimated 200,000 years ago, hundreds of millions of years after organic life first developed on land (an estimated 485 million years ago). The basis for organic life, eukaryotic cells, date back 1.5 - 2 billion years, with the first traces of multi-cellular life dated to 600 million years ago. While these dates may not be precise, it's clear that organic life has been a long-term project for planet Earth and humanity is a very recent innovation.

Organic life serves two specific roles:

1. It serves as a mechanism for other planets in the solar system to communicate with the Earth.
2. Its higher substances serve as food for the Moon

The theory is that the planets sound the note *fa* in the Ray of Creation and the Earth sounds the note *mi*. There is an interval between these two notes, which needs to be bridged by an outside force. The external force is provided by the notes *la*, *sol*, and *fa* of a lateral octave originating in the Sun, as illustrated in *Figure 87*.

The notes *la*, *sol*, and *fa* are organic life on Earth. Specifically, the Earth fulfills the role of a digestive system and a womb in the context of the solar system. The child of its womb is the Moon. With the assistance of organic life, the Earth absorbs substances from other planets. We can think of organic life on Earth as analogous in some ways to gut bacteria in man, which assist the digestion of food. Just as there is a lateral octave in the Ray of Creation, there is also a lateral octave in man's internal Ray of Creation. As above, so below.

Other Rays of Creation for planets such as Venus or Mars do not involve organic life on those planets. The data we have on other planets suggests that, in our solar system, only the planet Earth has a lateral octave of this kind. We presume that Nature arose (in its current form) just after the genesis of the Moon.

Logically this has to be so. With the birth of the Moon, the Earth had to prepare to sustain the Moon. We don't know what the composition of the Earth's

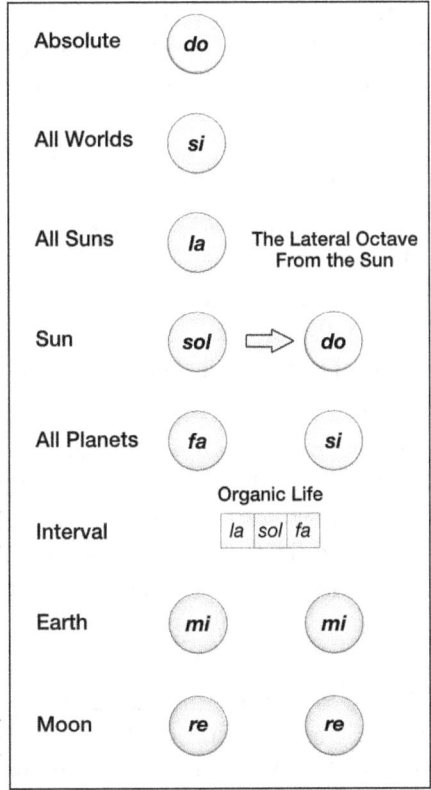

Figure 87. The Lateral Octave

atmosphere was then. Still, it is unlikely it was anywhere near its current nitrogen and oxygen balance—the atmosphere needed to support carbon-based life. No other planet currently has an atmosphere that could support life-forms like those on Earth.

The Genesis of Nature

The level of intelligence (or reason) of a cosmos relates directly to its range of vibration. The Absolute is the pinnacle of reason, with an intelligence far beyond the level of man. With only half the Absolute's level of vibration, the Sun also significantly exceeds man's level of reason. *Man*, **la** in the lateral octave that commences in the Sun, has a relatively low level of reason.

The biology of *Man* provides a perspective. The instinctive brain, which manages the body in real-time, is far in advance of any man-made system. It is blindingly intelligent and extraordinarily adaptive. It can deal with bodily health holistically and at a cellular level—capable of managing everything from an infection to a broken bone. It is engineered to digest a wide variety of food, collaborating intelligently with the large population of symbiotic bacteria that inhabit the digestive tract.

It is extraordinary in its ability to power the body through breath and food, distributing oxygen throughout the body along with a brilliantly managed supply chain of glucose. It presides over the body's astounding sensory systems, ensuring its supply of information and helping to digest it. It is responsible for and manages the continuation of the species through the miracle of sexual reproduction. And in tandem, it renews every part of the body to keep it working effectively for decades of life.

Such sophisticated system design is far beyond the capabilities of man. And yet this kind of system and others like it that govern the other life-forms on Earth is only a part of Nature's skill. Nature, as an aspect of a planet's life, manifests in the form we know when the planet gives birth to a Moon. Nature serves both the Earth and the solar system, enabling the Sun and the planets to participate in the Moon's growth.

We know from Moon rock analysis that the Moon is indeed a child of the Earth, although we have no idea of its birth date. Bacteria probably initiated the evolution of organic life. They are found almost everywhere on Earth: in the upper atmosphere, deep down within the Earth's crust, including places where the temperature is very low or very high. They are adaptable, and adaptability is the hallmark of intelligence. They likely existed on Earth before the

Moon's birth, and they probably exist in many places throughout the solar system.

The fossil record indicates that the first evolutionary step that led to the growth of organic life was the evolution of eukaryotes. These single-celled creatures contained either mitochondria or chloroplasts. A microbiological theory called endosymbiosis suggests that a fundamental evolutionary event occurred about two billion years ago. A large cell (archaea or bacteria) engulfed some smaller, quite different cells. Rather than choosing to feed on them, they decided to let them persist symbiotically as organelles (i.e., organs within the cell).

That is the current and only story of how mitochondria and chloroplasts came to be, and it rides on the fact that both mitochondria and chloroplasts have their own distinct DNA. There are some important facts to note here.

- If endosymbiosis did indeed occur as hypothesized, it is curious that it no longer happens—there is no evidence of it having occurred in over a billion years since the genesis of eukaryotes.
- Almost all multicellular life came, by way of evolution, from eukaryotic cells.
- Without mitochondria, cells would be incapable of combining glucose with oxygen to power animal life.
- Without chloroplasts, there would be no multicellular life capable of photosynthesis.
- Eukaryotic cells are capable of sexual reproduction and introduced sexual reproduction into multicellular life.

A Tale of Four Elements

Figure 88 on the next page provides a detailed view of the lateral octave, identifying each note as a class of being. We tend to think of Nature as the biological mass that constitutes life on Earth. An alternative view is that Nature constitutes the whole lateral octave from the Sun. In which case, there are life-forms higher than *Man* (referred to as *Angels* and *Archangels* in the diagram) and lower non-biological life-forms (*Minerals*, *Metals*, and *Kernel*).

We have labeled this diagram with the four elements, EARTH, WATER, AIR, and FIRE, according to the nature of each square's life-forms. Before we describe those, let us note that each of the four elements manifests as a circulation. Thus the realm of EARTH involves the circulation of substances from

Gurdjieff's Hydrogens: The Ray of Creation

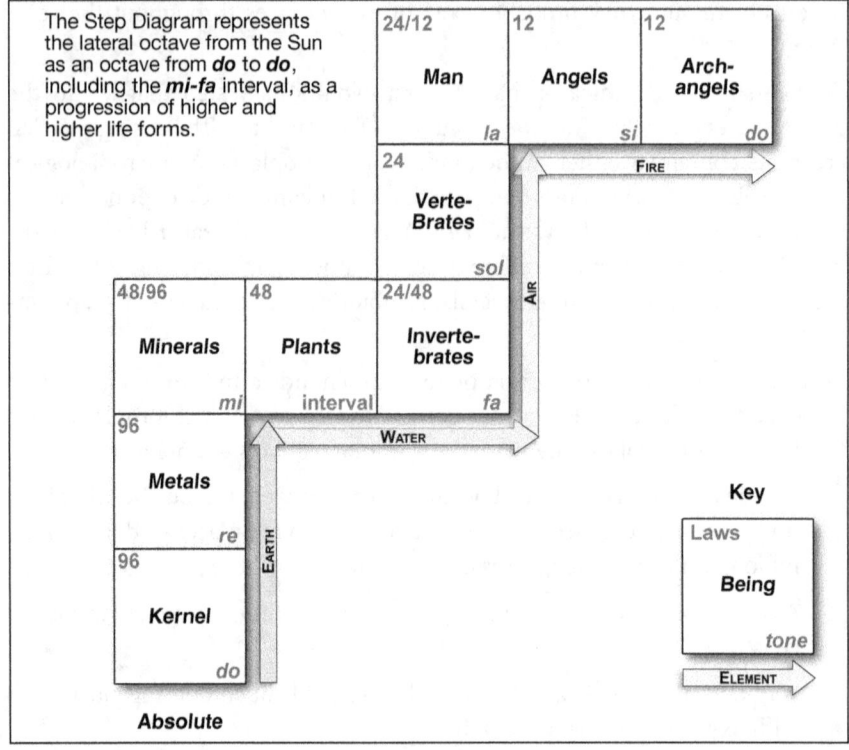

Figure 88. The Step Diagram

the core to the Earth's crust. WATER's realm is governed by the circulation of water from the seas through the atmosphere to the land where it forms lakes and rivers and aquifers. The realm of AIR circulates nitrogen, oxygen, and carbon through the air, the seas, the soil, and biological life. FIRE's realm manages the flow of plasma through the planet, through life, the sea, and the air, and into the magnetosphere.

When we consider the squares of *Figure 88*, we note that *Kernel* and *Metals* are fundamentally solid, and *Minerals* straddle the border between EARTH and WATER (solid and liquid). *Invertebrates* lie on the border between WATER and AIR (liquid and gaseous), and *Man* stands on the border between AIR and FIRE (gaseous and plasma).

In the top left corner of each square in *Figure 88* is a number that indicates the number of laws that act on that square's creatures. Thus *Metals* are under 96 laws, *Plants* under 48 laws, *Vertebrates* under 24 laws, and *Angels* under 12 laws. You can broadly think of such laws like this. *Metals* are crystalline. They live a more constrained existence (are under more laws) than *Plants* which live

Class	Realm	Description
Archangels	Magnetosphere	*Archangels* have the highest level of Reason of creatures on Earth. They manifest as ideals and feed on *Man*'s purer aspirations.
Angels	Ionosphere	*Angels* manifest emotional purity. Their Reason is at the level of *Man*'s perfected body Kesdjan.
Man	Anthroposphere	*Man* constructs his own realm, the anthroposphere, using knowledge he has acquired and powers it with plasma. He has the possibility of personal evolution.
Vertebrates	Biosphere	The biosphere is the realm of vertebrates, whose highest manifestations are emotions. *Vertebrates* dominate both the land and the oceans, and as warm-blooded creatures they can adapt to almost every environment.
Invertebrates	Hydrosphere	The hydrosphere sits above the pedosphere and is the environment inhabited by many *Invertebrate* life-forms, mollusks, worms, crustaceans and so on. Other invertebrates, particularly insects, span the border between the hydrosphere and the biosphere.
Plants	Pedosphere	The pedosphere, the soil on land and sediment underwater, is the natural medium for the roots of plants. *Plants* receive the light of the Sun and convert the atmosphere so that higher life-forms can breathe.
Minerals	Lithosphere	*Minerals* life-forms provide the Earth with its rocky crust. They are eroded by wind and water to become sediment that can serve plants or, in time, return below the surface.
Metals	Asthenosphere	These creatures are best thought of as having blood of molten magma. Volcanoes are probably *Metals* creatures. Their primary role is to transfer required elements (salts) to the surface to serve organic life.
Kernel	Earth's core	*Kernel* life-forms forge heavy and light elements through fusion and fission processes, some of which are delivered to the Earth's surface to sustain other life-forms.

Table 16. The Step Diagram and Relationships of Life-Forms

through water circulation. *Plants*, unable to move location, are far more constrained than *Vertebrates*, some of whom can even fly through the air. *Vertebrates* are, however, more constrained than *Angels*.

Each class is less adaptable and less intelligent than those in the square above them and correspondingly superior to the classes below. *Table 16* provides an annotated outline of each class of life.

A curious aspect of this staircase is its dependence on two specific developments that occurred untold millions of years ago. We have already discussed one, the evolution of eukaryotes. The other is the constitution of magma. Naturally, magma composition varies according to its source. Nevertheless, aside from oxygen and silicon, which dominate magma composition (47% and 28%, respectively), the two most common elements it contains are iron and magnesium. These are precisely the elements required for life.

Magnesium is required for the chlorophyll that makes photosynthesis possible. Iron is needed to transport oxygen in the blood of animals. The other common magma constituents: aluminum, calcium, sodium, potassium, and phosphorus, are all biologically important except for aluminum. Magmas also include hydrogen, carbon, and sulfur that become gases (steam, carbon dioxide, and hydrogen sulfide) when magma cools.

It seems then that the first three squares of the Step Diagram create the chemical foundation for biological life and deliver appropriate amounts of the required minerals and salts to the surface. Organic life then takes over organizing the domains of WATER and AIR, which give way to the realm of plasma, FIRE.

The Force of Evolution

The fossil record chronicles the biological progression from single-cell eukaryotes to multicellular life. Initially, photosynthesis-capable life-forms took the lead. Life thrived in the realm of WATER with the evolution of *Plants*, *Invertebrates*, and *Vertebrates* in that realm before they emerged into the realm of AIR. That evolutionary step was about 400 million years ago. The emergence of an animal with intellectual capabilities is a very late addition—the period at the end of the sentence. Nature may have attempted the evolution of such a creature earlier, but there is no evidence.

Lack of evidence is a fundamental problem for all discussions of evolution. Sadly the fossil record is extremely thin. Fossils usually preserve just bones, teeth, and shells. Cataclysmic events (asteroid impacts, volcanic explosions, earthquakes, landslides, and tsunamis) are most frequently the cause of fossil creation. Such events gather a random snapshot of biological life in a relatively small area. Currently, there are less than 600 fossil sites worldwide.

The fossil record suggests that few creatures from the distant past survived up to our era. This select group comprises a few worms and various sea creatures: sponges, jellyfish, horseshoe crabs, elephant and frilled sharks, coelacanths, and a few other fish. We do not know whether the fossils match the modern species precisely, because there is no DNA, only the impressions of bone, shells, and teeth. Such sea creatures may have evolved, for example, to tolerate a more or less salty ocean, but we cannot tell. Neither, incidentally, do we know for sure whether the seas were more or less salty hundreds of millions of years ago.

These survivors offer evidence that the force of gravity was less in the past because the select group contains no land-based animals except for a few small worms. A gradual increase in gravity that creates no problems for most sea species would doom most land creatures.

Contemporary science continues to have faith in the survival-of-the-fittest myth, which has not changed much since Darwin defined it. The discovery of DNA provoked a minor adjustment—it became "the survival of the fittest genes"—but the idea still persists by momentum, unassisted by anything so radical as evidence.

The fundamental misconception in Darwin's theory is that a cosmos, whether it is the biological system of an animal or the even more remarkable system of Nature itself, can be improved by randomly adjusting a feature here or there. It cannot. These are finely tuned systems where all the parts interact in multiple ways and do so harmoniously. Evolution is not a dice game—it unfolds in a coordinated manner.

Fragile Ecosystems

Managing a diverse ecosystem is currently beyond the capabilities of modern man. In times past, some tribes may have lived harmoniously with Nature, but modern man lacks such intelligence and sensitivity. Australia provides two telling examples of this: rabbits and cane toads.

The rabbit infestation of Australia began with the release of 24 wild rabbits in 1859. Within a decade, there was a population explosion that set records. Rabbits are now endemic, and their ecological impact has been disastrous. It has led to the destruction of plant species and the loss of other species from overgrazing.

Farmers introduced cane toads into sugarcane plantations in Puerto Rico, Hawaii, and the Philippines, where they happily feasted on beetles that damaged the crop. Thus, it seemed possible that it could do a similar job in the sugarcane fields of Queensland. One hundred and two toads were introduced "in controlled conditions," and after a year of observation, Australia's Commonwealth Department of Health decided to allow large-scale release. What could go wrong?

It proved to be a disaster of similar magnitude to the introduction of rabbits. The cane toad population increased exponentially and has spread right across Queensland and into other states. In areas where it has prospered, lizards, land snakes, and even crocodiles have declined. And ironically, it did little to reduce the population of grey-backed cane beetles it was supposed to control. If evolution happened by random mutation, it could and probably would devastate whole ecosystems. But it doesn't.

Ecosystem Evolution

When we gather and examine data about ecosystem evolution and behavior before *Man*'s appearance, we discover interesting patterns. Complete biological ecosystems form in areas isolated by the sea. A few hundred years ago, New Zealand, Australia, Madagascar, and the Galapagos Islands were independent ecosystems.

Consider New Zealand. Before human intervention, the only indigenous (non-marine) mammals were several species of bat. Geologists suggest that New Zealand became an island about 65 million years ago. Since then, its ecosystem has consisted of flora (at least 75% local), insects, reptiles, and local birds.

That balance never changed until humans intervened. The fossil record in New Zealand is very thin indeed. Fragments of fossilized dinosaur bones are found, but no fossils of mammals (except for whales and other sea mammals). There are, of course, fossils of extinct birds.

New Zealand demonstrates that ecosystems, rather than species, evolve. Nature did not feel compelled to introduce marsupials or any other kind of mammal in New Zealand. The mammals that swam by its shores didn't get the urge to evolve into land-based mammals. Even the sea snakes that occasionally visit its waters never felt the desire to come ashore to become land snakes. A few of its birds chose to migrate, but if they introduced anything new by their travels, it was a few non-indigenous plant species.

And yet, we know for sure that Earth's gravitational force has increased in the last 65 million years. Evolution must have happened in New Zealand, but it happened to the ecosystem as a whole, and Nature stuck with birds.

Collective Life Forms

Some life-forms, termites, ants, and bees are examples, live in such tightly-knit collective social groups that it is easier to think of the collective as the creature rather than individuals. Termites are the oldest of such eusocial species, dating back 200 million years—50 million years before ants and bees. Some are farmers—they farm fungus that they grow within their termite mounds. Their mounds are brilliantly engineered to maintain a specific inner temperature so they can efficiently cultivate the fungus they consume.

Individual termites know when to open and close vents to regulate the mound's temperature, but it's uncertain how they learn. They exhibit both individual and group intelligence that is hard to fathom.

A colony has a large queen who lays eggs all her life, producing a 1:1 ratio of males to females. Their society has four classes; the king and queen, who monopolize reproduction unless the queen develops a problem. Then female termites, a subsidiary reproductive class, are allowed to conceive. The largest class is the workers who farm, build and care for the young. The class of soldiers defends the colony.

Termites can change roles. The larvae molt to become workers and can molt again to become soldiers and yet again to become reproductive members. Soldiers have large jaws but cannot chew, so they are dependent on workers to feed them. Individual termites appear to communicate with each other and sometimes behave altruistically to other termites.

They usually have a positive impact on ecosystems. They populate deserts, savannas, tropical rain forests, and temperate woods. Their digging helps with moisture in the ground, and they enrich the soil. In dry areas, well-spaced termite mounds can form plant oases and even improve the lot of local animals that tend to prosper near termite mounds. Parts of Africa boast termite mounds 30 ft high and 80 ft in diameter. They are skyscrapers housing millions of termites.

Termite self-organization is reminiscent of large human communities. The primary difference is that there is no evidence of a government (a central organizing brain) in a termite community. The termite mound is remarkably intelligent, but we're not sure how it works.

Gurdjieff's Hydrogens: The Ray of Creation

The Human Collective

A major city like London, New York, or Tokyo, is somewhat analogous to a termite mound. Looking from above, you are observing a eusocial species. It is not clear how it is organized or how any individual knows what to do. Nevertheless, its residents manage to feed. They have an extensive transportation system. There's waste disposal, construction, and repair. A human country provides an even better example of a collective form of life.

It becomes evident when you examine a human economy. There is food production for the populace (farming and fishing). Raw materials from mining, quarrying, and forestry enable construction. Various transport systems circulate people and materials. The manufacture of vehicles and other machines gives rise to factories. Factories are required for textile manufacture, electronics manufacture. Industrial plants produce chemicals and pharmaceuticals and supply clean water.

The factories need power, as indeed do the houses in which people live. There is thus power generation and distribution. The financial sector collaborates with retail operations to distribute food and other goods. Education and entertainment collaborate to broadcast information to individuals, and there are the interactive capabilities of the phone and the Internet for more precise information passing. Then there's healthcare in all its variety, to which you can add clean-up activities: garbage collection, sewage systems, and recycling. Above all of this, you can add government activities, including policing and the armed forces.

That whole set of activities occurring in every nation, large or small, is characteristic of humanity. Humanity creates its extensive termite mounds and has done so as far back as archaeology can record. The only thing that varies is how it carries out those activities. Go back 10,000 years, and you find small tribal termite mounds with much less sophisticated organization than at present. But the people of those times are biologically identical to modern man. The cells of the termite mound are the same; only the mound has changed. The sophistication of our modern termite mounds comes from the knowledge humanity has accumulated.

Humanity exhibits a collective intelligence. It's unlikely that any individual, whatever their profession, thinks that they are contributing to the creation and maintenance of humanity's "termite mound." However, that is very likely to be the most consequential thing they are doing. Nowadays, humanity is building a global termite mound.

A New Model of the Universe

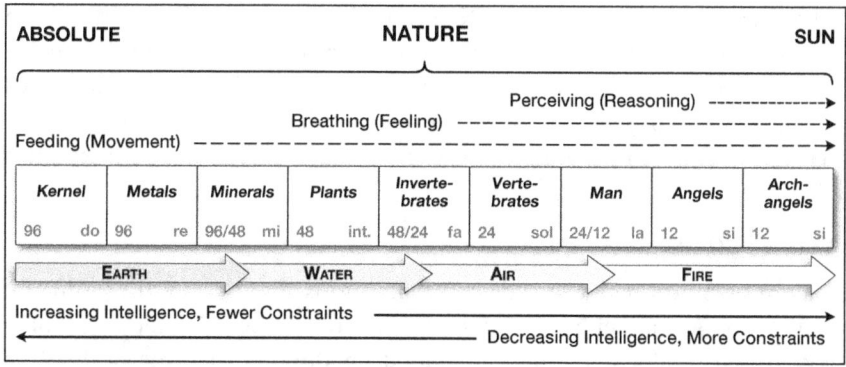

Figure 89. Nature as a Living Being

Mother Nature

The evolution of species is a part of the life of Nature as a whole. Nature's purpose is to enable planetary interactions with the Earth and provide a regular supply of particular plasma substances to the Moon. Like other cosmoses, Nature feeds, breathes, and perceives. It is illustrated in *Figure 89*, which shows Nature as comprised of nine different classes of creature.

Nature's "feeding" occurs in the realm of Earth. It consumes the layers of minerals that form on the crust, the detritus of biological life, its shells and bones, and other unconsumed organic matter. These disappear back into the Earth as food for Metals. It is also likely that Kernel life-forms provide a kind of foundry for heavy elements and, through transmutation, participate in preparing lighter elements for delivery to the surface.

Planet	Atmospheric Gases
Earth	N_2 - 78%, O_2 - 21%, Ar - 1%, Other - <1%
Mercury	O_2 - 42%, Na - 29%, H_2 - 22%, Other - 7%
Venus	CO_2 - 96%, N_2 - 3%, Other - 1%
Mars	CO_2 - 95%, N_2 - 3%, Ar - 1.5%, Other - 1%
Jupiter	H_2 - 90%, He - 10%, Other - <1%
Saturn	H_2 - 96%, He - 3%, Other - 1%
Uranus	H_2 - 83%, He - 15%, CH_4 - 2.5%
Neptune	H_2 - 80%, He - 19%, CH_4 - 1%

Table 17. Planetary Atmospheres

Contemporary science has no explanation for the composition of the Earth's atmosphere compared to other planets—as listed in *Table 17*. It is the percentage of nitrogen that is most anomalous. It seems as though this abundant gas must have come from within the Earth. But if it did, what was unique about Earth that led to its massive concentration of nitrogen?

Gurdjieff's Hydrogens: The Ray of Creation

The answer is almost certainly Nature itself. The theory is that the Earth produced nitrogen beneath its surface because Nature needed it to sustain life. Once created, organic life can recycle it to maintain the level. The same is true of atmospheric oxygen, although the Earth still produces oxygen. 50% of magma is oxygen (in the form of oxides), but it contains almost no nitrogen at all.

Nature's Breath

Nature breathes once every day. The life-forms that occupy the realms of WATER and AIR are either creating oxygen from carbon dioxide or burning oxygen and producing carbon dioxide, depending on the time of day. The oceans' surface presents a vast phytoplankton sheet to the Sun, and they spend the day transforming carbon dioxide into oxygen. These microorganisms are responsible for somewhere between 50 and 85 percent of oxygen production. Land-based plants and trees produce the rest. When the Sun goes down, the plant world sleeps, and while it does, it breathes oxygen in and exhales carbon dioxide. However, it consumes only a tiny proportion of the oxygen it created during the day, leaving the rest for animals. Nature maintains this balance and must do so whatever circumstances arise.

Planets' external influence on Nature's creatures operates through the endocrine systems in the bodies of *Invertebrates*, *Vertebrates*, and *Man*. Each endocrine gland is sensitive to a particular planet,* both to the radiations (EMR) and the magnetic influence of the planet, and also the magnetic influence of the planets in combination. It happens through the radiations we receive and the air we breathe.

The influence of a planet is greatest when it is closest and at its zenith. It is the mechanism by which the planets communicate with the Earth through Nature. Organic life on Earth acts as a receiving station collectively and individually.

The endocrine glands influence behaviors by secreting hormones, so the creatures in the realm of Air act as a kind of skin through which the Earth senses and receives influences from the planets. The creatures themselves have no understanding that this is taking place, just as the bacteria in our guts probably have no knowledge that they help us digest food.

Nature's perceptions and reasoning come from *Vertebrates*, *Man*, *Angels*, and *Archangels*. Together the *Vertebrates* and *Man* take in impressions from most parts of the planet. Warm-blooded animals can exist in even icy climates, and

---

* *The Theory of Celestial Influence* by Rodney Collin, p144 - p146

A New Model of the Universe

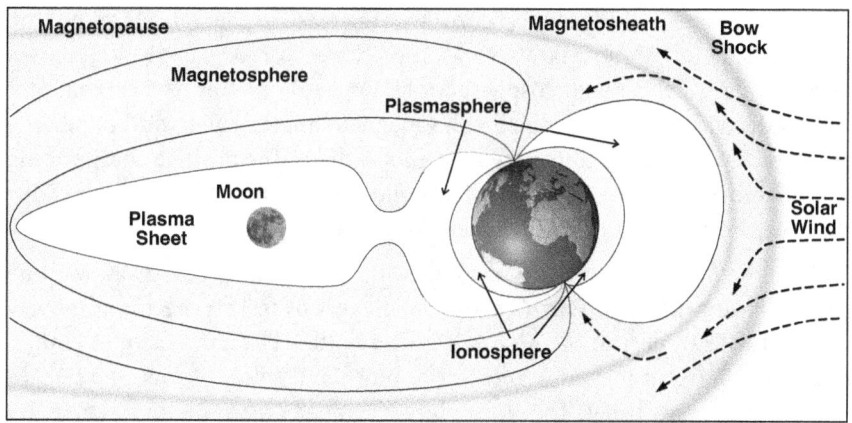

Figure 90. The Flow of Plasma to the Moon

Man has spread to almost every corner of the Earth. Collectively animal life constitutes a nervous system, which keeps Nature informed about almost all of its body. Nature, acting as the instinctive brain of the planet, thus knows about and can respond to changes in various parts of the globe.

Feeding the Moon

The invisible (or if you prefer theoretical) life-forms, *Angels* and *Archangels*, inhabit the most rarefied areas of plasma surrounding the Earth. Their depiction as winged life-forms suggests as much. Among the roles traditionally assigned to them in many religious traditions, aside from their inspirational roles as messengers, is to preside over the souls of the departed.

Objective science asserts that a process, termed Rascooarno, which means splitting apart, occurs to a man at death. His lifeless body, gradually consumed by lower life-forms, returns to the dust from whence it came. The instinctive brain's plasma structures that governed his body and his personality are magnetically attracted to the Moon. The first of the three octaves (food) that occupy the body of Man is capable of "immortality" only through reproduction. What does not descend back into dust begins to make its way to the Moon.

There is a feeding process between the Earth and the Moon. A connecting plasma structure supports it, fulfilling the function of an umbilical cord. *Figure 90* illustrates the situation. The outer sheath of the umbilical cord is the Earth's magnetotail. The life forces of animals and *Man* carry a negative charge. For that reason, they gradually rise to the top of the ionosphere and enter the plasmasphere.

The plasma sheet (the umbilical cord) that the plasmasphere forms within the magnetosphere extends far beyond the Moon's orbit. The Moon passes through the Earth's magnetosphere for about six days in every lunar month. When the Moon is outside the Earth's magnetotail, the solar wind bombards its surface with plasma, inducing a magnetosphere. The positive charge of the solar wind tends to reduce the Moon's negative charge giving it a positive charge in relation to the Earth's magnetosphere.

When the Moon enters the magnetotail, the solar wind veers away, and the plasma sheet within the Earth's magnetotail takes over. This sheet is negatively charged with respect to the Moon. It is also hotter (i.e., more energetic) than the solar wind and in a constant state of motion. It sweeps across the surface of the Moon many times. The encounters last anywhere from minutes to hours or even days.

Electrons pepper the Moon's surface, giving the Moon a negative charge. On the Moon's Sun-facing side, the negative charge is counteracted by sunlight. Photons of ultraviolet light displace electrons from the surface. Thus the Sun-facing side of the Moon attracts the plasma substances that the Earth is donating. This is how the Moon feeds.

The Possibility of Failure

According to the fossil record, the Earth has experienced five mass extinctions in the last 500 million years—cataclysmic events that destroyed 60% or more of life. There have been less severe extinctions too. Geologists estimate that a major extinction event occurs every 25 million years or so. It is an extraordinary hazard that somehow the Earth and Nature need to handle.

We can only guess at the cause of such extinctions: impacts from large asteroids, massive volcanic eruptions, nearby supernovas irradiating the Earth with gamma rays, a massive Coronal Mass Ejection (of plasma) from the Sun itself, a close approach by another planet causing earthquakes, interplanetary thunderbolts, and tsunamis. And there may be other possible causes that no-one has yet proposed.

Major ice ages are cataclysmic events of a different ilk. Geologists insist that there have been at least five. The first, called the Huronian ice age, is dated to around 2.4 to 2.1 billion years ago, just before the emergence of eukaryotic life. The subsequent one, 720 to 635 million years ago, named the Cryogenian period, appears to have been the most severe, with glacial ice sheets reaching the equator. The evolution of life began in earnest when it was over. The next ice age, called the late Paleozoic icehouse, produced ice sheets at intervals between 360 to 260 million years ago. The Quaternary ice age that began about 2.6 million years ago followed. Since then, there have been regular glacial periods with roughly 100,000-year frequency.

Geologists tend to ascribe the ice ages to atmospheric change. The first one was supposedly provoked by cyanobacteria oxygenating the atmosphere and eliminating a great deal of methane (a greenhouse gas). Variances in carbon dioxide may also cause ice ages. No scientists favor the possibility that the Earth's orbit changed, but it seems the most likely explanation. The ice ages decreased the abundance of life on Earth, but they do not appear to have caused vast extinctions, except for the second one, which froze the planet over.

Because the Earth is gradually increasing in size, the evolution of the Step Diagram creatures is a moving target that Nature must perpetually try to balance. And its work could be wrecked by some cataclysmic event.

It is not easy to understand what challenges this creates for the Earth and Nature. Viewing the solar system as a living being, we should perhaps think of this as a task for the Sun and planets too. There are more questions here than answers.

For example, it is clear from the Moon's craters that it has been scarred many times by thunderbolts. That's not unusual for a large moon—three of Jupiter's large moons have extensive cratering—but is that a natural and necessary process for a moon?

Are the ice ages and mass extinctions part of a natural growth process, or are these environmental perils whose causes could be fatal to the Moon or even the Earth?

We do not know.

Man

In recent centuries *Man* gradually constructed a sophisticated world peppered with all the systems and structures required for a global human economy. In doing so, *Man* appears to have subjugated the classes of being below him to his service: *Metals*, *Minerals*, *Plants*, *Invertebrates*, and *Vertebrates*.

The most populous animals are those in *Man*'s service: sheep, cows, pigs, goats, and donkeys and his favorite pets, dogs and cats. Chickens are the most populous bird by far. When it comes to fish, the most populous are species that Man does not hunt. The most common species of plant life are the grains and trees that he farms. He pillages the Earth's crust for minerals, metals, and fuel, with little care for the damage he may be doing.

Gurdjieff's Hydrogens: The Ray of Creation

Nevertheless, it's possible that Nature is pleased with this state of affairs and may even have encouraged it. Global warming may threaten *Man*'s future and that of whole ecosystems, but Nature has recovered from far worse perils and has the means to fix climate change quickly if she so chooses.

According to the data, volcanic activity caused nearly all the extreme cooling events in the northern hemisphere in the past 2,500 years. Volcanoes loaded the atmosphere with volcanic dust that obscured the Sun. It was the cause of crop failure, famine, and disease in 43 and 42 BC and again in 535 and 536 AD. It provoked the "Little Ice Age," which began in 1300 AD. If the Earth needs to reduce the temperature, a series of explosive volcanic eruptions will do the trick.

It's implausibly unlikely that modern man will meet the climate challenge, despite his command of reason and his treasury of technological knowledge. There is no global governance, just national governments with their agendas. Their inability to act in concert when facing a common threat was laid bare by the Covid-19 pandemic. Humanity is incapable of unity.

More pertinently, Mankind is utterly unaware that it is farmed in the same way that *Man* farms cows, pigs, and sheep, and termites farm fungus. A woeful decrease in his humility has accompanied his accumulation of knowledge. Atheism decrees that if there is nothing higher than Man, then Man is God, and atheism is his most popular creed. He dreams of conquering the universe when in reality, he has not conquered himself—nor does he know how to.

Automation, which began with the march of mechanized factories, stole from *Man* the joy of physical labor. In the early nineteenth century, the Luddites' protests were more than rage against the economic machine; they were a desperate cri de coeur. It is now just a footnote to a quaint bygone age—a century before the advent of electricity. Electricity is gradually destroying *Man*'s emotional life.

The radio, the phonograph, and the movie camera squashed the individual pursuit of art in favor of mass distribution. There were fewer musicians, fewer dancers, fewer singers, and fewer actors. Computers and the Internet accelerated all of that while assaulting the intellectual life of *Man*. The automation of process management has diminished the number of people engaged in actual thinking. A thousand skills have become redundant, and more are lost each day as *Man* becomes ever more dependent on electricity.

By this dependency, he has put his civilization in the greatest jeopardy. Humanity's web of electricity and electric devices is vulnerable to mass ejections of plasma from the Sun, which regularly occur. If one occurred with the strength of the one in 1859, it would destroy the world's power systems, create massive famine and send *Man* back to Napoleonic times. Humanity can wreak the same havoc himself if the mood takes him, with weapons of war designed for that purpose. He can reap the same whirlwind without any help from the Sun.

The Trogoautoegocrat and Plasma Beings

The implications of the Trogoegoautocratic universe are more profound than initially meets the eye. In our natural day-to-day activities, we emanate plasma in respect of thoughts, emotions, and imagination. These are substances we discard. So the natural question is, what beings feed on those substances that we so regularly discard?

Indeed we can pose the more general question: Assuming we accept the existence of *Angels* and *Archangels*, what other plasma beings are there?

It is an area of inquiry that is beyond the reach of modern scientific methods. It requires apparatuses able to examine plasma structures in fine detail, which do not currently exist and may never exist. For the purpose of inquiry, *Man* is the only apparatus capable of interacting with plasma life-forms. *Man* can be the crucible, but he will be unable to claim anything he learns as "objective" in the scientific sense.

In *Figure 91*, (next page) we employ the Ray of Creation to help us ponder the realm of plasma beings. The diagram illustrates a second lateral octave, formed by the note *sol* in the first lateral octave—the note that corresponds to *Vertebrates*. The *do* of the first lateral octave, at the level of the Sun, corresponds to *Archangels*. It is the level from which "messengers from above" descend to Earth to teach *Man* and leave their indelible traces.

In the second lateral octave, we assign Sathanael (Satan) to the note *do*. Sathanael is the Moon's representative on Earth, Lord of *Vertebrates*, and all other organic life. Typically he is represented as cloven-hoofed, sexually lascivious, with goat's horns and a tail. While the Catholic Church and most Protestant sects depict God and Satan as adversarial, it is clear from this diagram that Satan has a distinct role in respect of *Man*.

He farms *Man*—to the benefit of the Moon.

We suspect the other notes of the second lateral octave to be plasma beings of various orders. We can refer to these in general as elementals. Most likely, some of them feed on Man's psychic substances (dreams, hopes, fears, anger, etc.) Some will feed on the psychic excretions of beings at lower levels, from

Gurdjieff's Hydrogens: The Ray of Creation

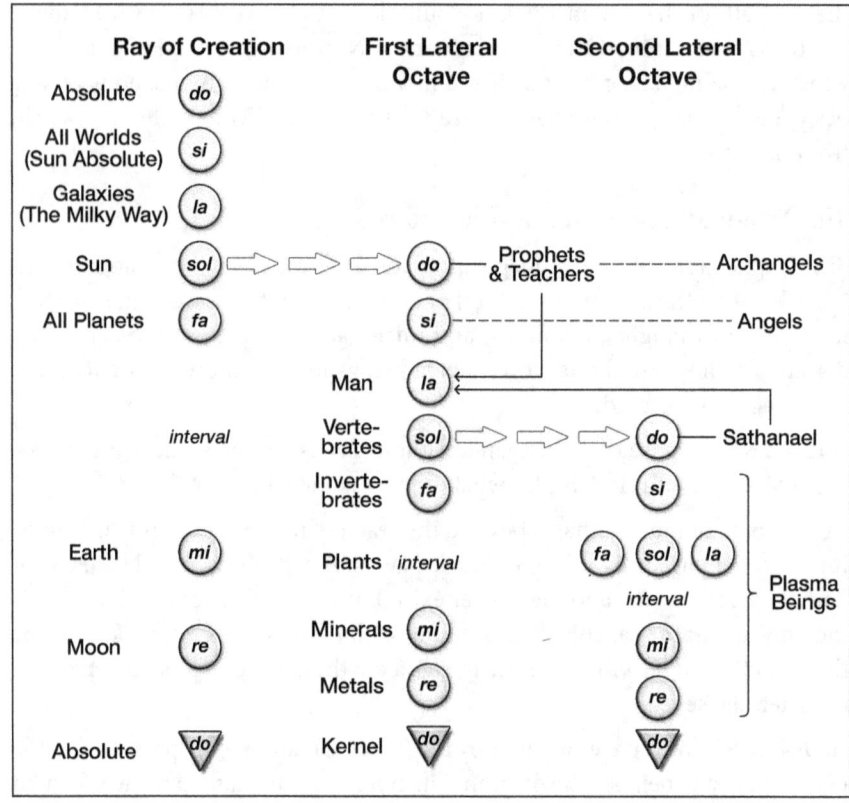

Figure 91. The Ray and the Two Lateral Octaves

Metals to *Vertebrates*. There are likely many such plasma creatures at every level, contributing to the Trogoautoegocrat in their own way. Folk traditions claim that there are 'deities' or 'spirits' associated with rocks, standing stones, volcanoes, mountains, streams and rivers, plants, trees, groves, and animals of all kinds. None of these 'Nature spirits' play a significant role in the life of Man.

Sathanael presides over *Man*'s animal nature. *Man* has a physical body subject to the same biological needs and tendencies as any vertebrate, including the need to procreate. On the other hand, *Man*'s aspirations, for the evolution of his body Kesdjan and mental body are the concern of *Archangels* and "messengers from above." In the struggle between those forces, a man's psychic efforts are the third force determining which direction he will choose. He either takes Sathanael's path or the path of the Truth Seeker.

The organ Kundabuffer serves Sathanael's interests. Gurdjieff discusses this in-depth in many places in *The Tales* and in lectures he gave before founding his famous Institute. This organ grows at the end of our spine for a few weeks while we are embryos and then disappears. It inculcates the habit of self-calming. It was added to our ancestors' DNA by Sathanael to ensure that *Man* could be farmed.

Man is thus pulled in two directions, embodying a ready-made arena for struggle. He has a choice; he can take up his cross or not.

Further Questions

This book is incomplete. It touches on some important questions only lightly, leaves several questions untouched and one or two unmentioned. According to the author's notes, the list of things still to be explored or more deeply investigated are:

- How did the creation occur?
- What is the relationship between the Absolute and the Aether?
- What is the nature of time?
- What are all the possibilities of the Law of Three?
- What can we learn from the Enneagram?
- What does the Food Diagram tell us?
- What is the Omnipresent Okidanokh?
- What do the Gornahoor Harharkh experiments in *The Tales* mean?
- What happens to *Man* at death?

All this subject matter belongs to the second volume of *Gurdjieff's Hydrogens*.

Bibliography

The author consulted all of these books:

- Beelzebub's Tales to His Grandson by G I Gurdjieff
- Views From The Real World by G I Gurdjieff
- Gurdjieff's Early Talks 1914 - 1931 by G I Gurdjieff
- In Search of the Miraculous by P D Ouspensky
- The Fourth Way by P D Ouspensky
- The Theory of Celestial Influence by Rodney Collin
- Perspectives on Beelzebub's Tales and Other of Gurdjieff's Writings by Keith A Buzzell
- Gurdjieff: Mysticism, Contemplation, and Exercises, Joseph Azize
- Gurdjieff and the Women of the Rope by Solita Solano and Kathryn Hulme
- Tao Te Ching by Lao Tsu
- The Tibetan Book of the Dead by Padmasambhava
- The Science of Mechanics: A Critical and Historical Exposition of Its Principles by Ernst Mach
- Weber's Electrodynamics by A.K. Assis
- I Contain Multitudes by Ed Yong
- Seeing Red: Redshifts, Cosmology and Academic Science by Halton Arp
- Spurious Correlations by Tyler Vigen
- The Big Bang Never Happened by Eric J Lerner
- The Static Universe by Hilton Ratcliffe
- Dinosaurs and the Expanding Earth, by Stephen Hurrell
- Terra Non Firma Earth by James Maxlow
- The Body Electric by Robert O Becker and Gary Selden
- A Beginners View of Our Electric Universe by Tom Findlay
- The Electric Sky by Dr Donald Scott
- The Electric Universe by David Talbott & Wallace Thornhill
- Thunderbolts of the Gods by David Talbott & Wallace Thornhill
- Worlds in Collision by Immanual Velikovsky

Gurdjieff's Hydrogens: The Ray of Creation

- Ages in Chaos by Immanual Velikovsky
- Earth in Upheaval by Immanual Velikovsky
- Ghost and Ghoul by T C Lethbridge
- Ghost and Divining Rod by T C Lethbridge

About the Author

Robin Bloor was born in 1951 in Liverpool, UK. He obtained a BSc in Mathematics at Nottingham University and took up a career in the computer industry, initially writing software. From 1989 onwards, he became a technology analyst and consultant. He has thus been a writer, author and blogger of a kind ever since. In 2002, he was awarded an honorary Ph.D. in Computer Science by Wolverhampton University in the UK for "services to the UK computer industry."

In 1988, after drifting through several work groups, Bloor met and became a pupil of Rina Hands. Rina was a one-time associate of J. G. Bennett, a student of Peter Ouspensky's, and later, a pupil of George Gurdjieff. Following Gurdjieff's death, she remained part of J. G. Bennett's group for a while. Subsequently, she formed groups both in London, where she lived, and in Bradford in the North of England – initially in conjunction with Madame Nott. She was both an accomplished movements teacher and an inspirational group leader.

Bloor's respect for Rina Hands is hard to overstate. She was a powerful teacher with few equals in the Gurdjieff line. It was because of her influence that he became a devoted student of Gurdjieff's writings and of objective science. Rina died in 1994 and is buried in a grave adjacent to that of Jane Heap in a cemetery in North London.

In 2002 Bloor emigrated to the United States. He currently resides in Austin, Texas from where he runs a Gurdjieff Group, the Austin Gurdjieff Society. He is also a member of a UK group, The Bradford Gurdjieff Society, which was, in times past, run by Rina Hands.

Acknowledgements

The author acknowledges the following individuals who participated in the creation of this book:

- Paula Schmidt edited this book meticulously and also provided a critical review of the content, and at times the style.
- Derek Sinko acted in a similar manner providing much needed edits and comments.
- Richard Webb, Steven Aronson and Stuart Goodnick (of The Mystical Positivist Radio Show) all provided important critical input.
- Bobbie Pennock, Arthur Zevura, Lenny Schwartz, David J and David Corcoran carried out first draft editing work.
- X J Bao provided a scientific edit.
- Attendees of the Seeker's Café (SeekersCafe.org) helped by discussing salient topics.
- Rina Hands

The intention is to update this book with new information as it appears. We have set up an objective science study resource on ToFathomTheGist.com. The URL is: https://tofathomthegist.com/osr-homepage/, which we hope will grow with time.

INDEX

Astronomy

Astronomy, 19, 20, 22, 29, 32, 33, 41, 43, 48, 58, 287, 289, 294, 316

Astrophysicist, 28, 29, 31, 32, 33, 35, 36, 38, 39, 41, 278, 279, 297

Big Bang, 20, 21, 28, 29, 35, 36, 37, 38, 39, 40, 41, 319, 349

Black Hole, 20, 29, 33, 34, 47, 279

Cepheid, 36

Cosmic Microwave Background Radiation, 21, 33

CMBR, 21, 34, 35, 39

Cosmology, 4, 14, 19, 20, 21, 26, 28, 29, 31, 32, 33, 37, 115, 273, 274, 275, 277, 278, 279, 281, 282, 283, 285, 287, 289, 294, 297, 349

dark energy, 21, 33, 34, 39, 40, 319

Dark matter, 20, 21, 31, 32, 33, 39, 40, 273, 279, 319

galaxy, 4, 21, 22, 23, 26, 27, 28, 29, 30, 32, 36, 37, 38, 39, 40, 41, 48, 56, 59, 60, 62, 63, 76, 77, 78, 79, 90, 91, 94, 95, 101, 102, 139, 150, 164, 236, 270, 271, 277, 278, 279, 284, 285, 286, 287, 317, 323, 324, 325, 328

Lambda CDM, 20, 23, 29, 31, 273, 319

planet, 4, 12, 20, 21, 23, 32, 35, 37, 56, 59, 60, 61, 62, 63, 66, 67, 70, 75, 76, 77, 78, 79, 80, 81, 86, 87, 88, 89, 90, 91, 92, 93, 94, 96, 99, 100, 111, 131, 146, 159, 164, 165, 166, 167, 169, 176, 177, 182, 185, 187, 194, 196, 200, 201, 206, 214, 220, 230, 231, 232, 233, 234, 236, 237, 238, 239, 240, 241, 242, 243, 245, 246, 248, 262, 264, 265, 269, 271, 272, 278, 279, 280, 281, 284, 285, 286, 287, 289, 290, 291, 292, 293, 294, 295, 296, 297, 298, 299, 300, 301, 303, 304, 305, 307, 308, 309, 311, 313, 315, 316, 317, 323, 325, 327, 329, 332, 339

Jupiter, 79, 166, 206, 220, 241, 284, 292, 293, 295, 297, 307, 308, 309, 323, 327 339

Mars, 90, 166, 185, 206, 220, 241, 292, 293, 295, 296, 298, 327, 339

Mercury, 91, 206, 241, 271, 292, 301, 339

Neptune, 79, 90, 91, 147, 206, 220, 241, 279, 292, 307, 308, 339

Pluto, 79, 90, 91, 220

Saturn, 90, 206, 220, 241, 284, 292, 294, 295, 307, 308, 339

Uranus, 91, 206, 220, 241, 292, 307, 308, 339

Venus, 91, 166, 206, 207, 220, 241, 284, 292, 293, 295, 327, 339

Proxima Centauri, 79

Ptolemaic, 32

quasars, 27, 28, 29, 30, 38, 40, 286

solar system, 4, 14, 15, 25, 26, 30, 32, 35, 36, 61, 62, 64, 76, 79, 81, 90, 94, 110, 111, 115, 163, 177, 185, 200, 206, 207, 219, 220, 233, 239, 243, 267, 271, 272, 277, 278, 279, 280, 290, 291, 292, 293, 296, 297, 298, 299, 301, 305, 307, 308, 314, 325, 327

space, 14, 15, 16, 20, 23, 24, 25, 26, 28, 30, 34, 35, 36, 39, 40, 51, 52, 53, 61, 62, 75, 77, 81, 87, 89, 90, 98, 120, 121, 125, 126, 127, 129, 142, 149, 150, 172, 176, 177, 187, 188, 189, 208, 236, 243, 270, 274, 277, 278, 279, 280, 281, 286, 287, 293, 296, 297, 298, 299, 301, 313, 314, 315, 319, 324, 328

sun, 32, 56, 60, 61, 62, 63, 79, 87, 89, 90, 122, 159, 176, 177, 214, 236, 237, 262, 272, 279, 282, 284, 288, 290, 292, 307, 309, 316, 323, 327

supernova, 20, 33, 37, 40, 139, 219, 236, 285, 286

Biology

algae, 155, 159, 195

anatomy of wood, 196

ATP, 192, 215, 245, 246, 247, 254, 257

bacteria, 61, 81, 109, 110, 156, 161, 164, 165, 166, 168, 174, 182, 185, 191, 192, 193, 194, 195, 204, 205, 208, 213, 214, 217,

Gurdjieff's Hydrogens: The Ray of Creation

218, 238, 239, 243, 244, 248, 253, 254, 303
botany, 19, 58
Breath, 40, 59, 96, 107, 154, 155, 156, 160, 161, 164, 167, 186, 191, 194, 200, 203, 207, 209, 210, 213, 227, 228, 232, 245, 250, 253, 255, 263, 264, 282, 323, 333, 340
cell, 59, 61, 64, 79, 80, 81, 85, 86, 94, 102, 155, 162, 165, 179, 192, 194, 195, 196, 197, 198, 203, 208, 209, 213, 214, 215, 223, 238, 240, 241, 242, 245, 246, 247, 253, 254, 256, 271, 303, 323, 328, 334
chloroplasts, 215
cold-blooded, 203, 204, 220
Ecosystem, 81, 155, 192, 193, 212, 214, 216, 218, 223, 224, 225, 229, 238, 244, 252, 253, 272, 335, 336
egg, 210, 211, 212, 221, 222, 286
endosymbiosis, 215
enzymes, 65, 161, 196, 210
eukaryotes, 215, 217, 218, 219, 243, 244, 334
exoskeleton, 158, 199, 200, 208, 209, 268, 310
genes, 212, 223, 244
hormone, 104, 105, 196, 197, 198, 199, 209, 239, 240, 241, 250, 254, 255, 256
microbe, 160, 194, 223, 224, 238, 239, 256, 323
mitochondria, 215, 246
multicellular life, 60, 153, 208, 215, 334
mycorrhizal fungi, 198, 208
organs, 56, 61, 79, 80, 81,

82, 92, 93, 95, 96, 101, 104, 107, 141, 153, 159, 160, 161, 162, 164, 165, 166, 174, 179, 180, 181, 182, 185, 187, 191, 192, 195, 198, 200, 206, 207, 208, 209, 210, 212, 213, 214, 215, 216, 226, 231, 234, 238, 239, 240, 242, 245, 247, 250, 251, 252, 253, 255, 256, 260, 261, 262, 263, 268, 269, 298, 303, 304, 306, 308, 323, 328, 333
Osmosis, 197, 201, 202
Photosynthesis, 155, 159, 183, 194, 195, 196, 198, 205, 214, 215, 217, 247, 265, 272, 334
sex, 209, 212, 251, 254, 258, 260
sexual, 60, 196, 199, 200, 208, 211, 212, 215, 241, 258, 260, 261, 323
skeleton, 174, 199
Soil, 81, 155, 156, 164, 187, 189, 190, 191, 192, 193, 195, 196, 200, 204, 205, 216, 253, 267, 301, 332, 333
sperm, 99, 222, 286
theory of evolution, 19, 217
vitamin, 104, 105, 210, 223, 238, 240, 241, 250, 254
warm-blooded, 203, 204, 216, 220, 333
zoology, 19

Chemistry
air, 18, 24, 59, 67, 76, 82, 104, 105, 107, 110, 118, 120, 129, 132, 141, 144, 153, 154, 155, 156, 160, 172, 179, 186, 187, 189, 190, 191, 193, 194, 195,

198, 203, 204, 205, 207, 221, 227, 228, 230, 232, 233, 234, 235, 250, 251, 253, 256, 265, 269, 273, 303, 305, 306, 311, 323, 324, 332, 340
ammonia, 92, 192, 204, 205, 206, 214, 253, 301
carbon dioxide, 155, 173, 175, 189, 191, 196, 197, 200, 203, 204, 205, 206, 230, 231, 245, 246, 250, 253, 254, 256, 301, 308, 340
Elements
  aluminum, 225, 289, 299
  argon, 203, 204, 230, 301
  arsenic, 142, 182
  Astatine, 104, 105, 109, 110, 112
  beryllium, 36
  bismuth, 105
  boron, 142, 182, 193
  Bromine, 105, 108, 109, 110, 111, 182
  cadmium, 182
  calcium, 172, 182, 189, 193, 241, 246
  carbon, 91, 92, 127, 155, 172, 173, 175, 182, 189, 191, 193, 194, 195, 196, 197, 200, 203, 204, 205, 206, 230, 231, 245, 246, 247, 250, 253, 254, 256, 265, 275, 276, 301, 308, 340
  cerium, 182, 289
  Chlorine, 105, 108, 109, 110, 111, 116, 182, 193, 289
  chromium, 182
  cobalt, 142, 182
  copper, 144, 148, 178, 182, 189, 193, 290
  Fluorine, 105, 106, 108, 109, 110, 111, 182
  gallium, 182
  germanium, 142, 182

Index

gold, 83, 178, 290
helium, 20, 22, 27, 30, 36, 105, 203, 206, 230, 275, 276, 289, 301
Hydrogen, 20, 21, 22, 27, 30, 36, 39, 40, 92, 104, 108, 109, 110, 111, 135, 182, 193, 196, 203, 206, 250, 275, 276, 289, 301, 308
Iodine, 104, 105, 108, 109, 110, 111, 182
iron, 16, 37, 88, 104, 105, 125, 142, 147, 170, 172, 178, 182, 183, 184, 189, 193, 217, 220, 265, 273, 275, 276, 294
krypton, 106, 203, 230
lanthanum, 182, 289
Lead, 17, 48, 112, 177, 178
lithium, 36, 108, 182
magnesium, 172, 182, 183, 193, 195, 217, 246, 265, 275, 276, 289
manganese, 182, 193
mercury, 142
molybdenum, 183, 193
neon, 123, 203, 301, 324
nickel, 142, 183, 193
nitrogen, 60, 91, 92, 108, 182, 189, 192, 193, 195, 196, 203, 204, 205, 230, 243, 250, 253, 275, 276, 301, 308, 331, 340
oxygen, 91, 92, 108, 135, 154, 155, 172, 174, 175, 182, 183, 192, 193, 194, 196, 197, 203, 204, 205, 206, 209, 214, 215, 217, 230, 243, 245, 246, 250, 253, 254, 256, 268, 275, 276, 307, 308, 331, 340
phosphorus, 182, 183, 193, 276, 289

polonium, 105, 142
potassium, 91, 108, 172, 183, 193, 289, 301
radon, 105, 106
rubidium, 17
silicon, 49, 142, 172, 173, 179, 182, 183, 191, 194, 200, 247, 265, 268, 275, 276, 289
silver, 11, 142, 178
sodium, 11, 108, 116, 172, 183, 189, 241, 289, 301
strontium, 17, 183
sulfur, 182, 183, 189, 193, 200, 204, 205, 219, 275, 276, 308
Tennesseeum, 104, 105, 109, 110, 112
thallium, 183
Thorium, 17
tin, 178, 183
titanium, 183, 289, 299
tungsten, 122, 183
Uranium, 14, 17, 182
vanadium, 183
xenon, 106
yttrium, 183
zinc, 183
zirconium, 183

halogens, 104, 105, 106, 108, 109, 110, 111, 113
heavy elements, 20, 177, 183
isotopes, 17, 105, 137, 289, 299
light elements, 20, 22, 37, 326, 333
metals, 40, 106, 107, 108, 121, 142, 147, 157, 159, 178, 179, 181, 182, 184, 188, 189, 193, 229, 265, 288, 289
methane, 92, 194, 203, 204, 206, 214, 230, 231, 301
periodic table, 105, 106, 107, 108, 109, 112, 113, 137, 142, 173, 177, 178, 179, 182, 247
water vapor, 185, 203, 230, 231, 253, 254, 308
wave length, 27, 28, 30

Cosmology

Cosmos, 23, 56, 59, 60, 61, 68, 69, 76, 101, 102, 103, 106, 113, 121, 153, 154, 160, 162, 231, 238, 252, 262, 263, 264, 265, 323
being, 58, 59, 60, 61, 63, 71, 77, 79, 80, 84, 99, 100, 101, 106, 111, 146, 153, 154, 156, 157, 159, 162, 164, 165, 168, 176, 177, 178, 179, 180, 181, 183, 187, 195, 201, 203, 207, 213, 221, 222, 227, 229, 233, 234, 242, 245, 248, 249, 251, 252, 258, 259, 260, 262, 265, 266, 267, 268, 269, 303, 304, 316, 323, 339, 343, 345
cosmic unit, 56, 59, 60, 61, 99, 100, 101, 153, 323
creature, 60, 84, 123, 147, 154, 156, 157, 158, 194, 195, 196, 200, 203, 204, 207, 209, 221, 227, 229, 238, 244, 245, 250, 253, 260, 263, 264, 267, 269, 309, 310, 316, 323, 332, 333
life-forms, 56, 59, 60, 61, 64, 81, 84, 96, 102, 109, 153, 154, 156, 157, 159, 164, 165, 167, 173, 175, 176, 177, 179, 180, 182, 183, 184, 187, 190, 192, 194, 195, 196, 200, 203, 207, 208, 213, 215, 217, 250, 258, 260, 265, 267, 268, 271, 272, 303,

309, 311, 312, 331, 333, 334, 340

Enneagram, 100, 160, 161, 322, 347, 368

evolution, 10, 19, 22, 38, 57, 71, 84, 93, 187, 188, 194, 208, 213, 214, 215, 217, 218, 219, 221, 223, 224, 225, 238, 243, 244, 245, 261, 262, 264, 265, 268, 279, 304, 307, 309, 316, 333, 334, 336

evolutionary, 63, 69, 70, 71, 208, 209, 213, 215, 217, 218, 222, 224, 226, 249, 264, 265

food, 59, 60, 64, 80, 82, 95, 96, 100, 101, 104, 105, 106, 110, 112, 153, 154, 155, 156, 157, 160, 161, 162, 164, 176, 179, 191, 193, 195, 196, 198, 203, 204, 209, 210, 211, 212, 216, 220, 223, 224, 230, 233, 238, 248, 250, 253, 254, 257, 263, 264, 267, 269, 294, 303, 323, 347

involution, 69, 98, 262, 268

involutionary, 66, 69, 70

material, 2, 1, 8, 9, 16, 24, 25, 61, 62, 67, 68, 69, 74, 75, 76, 83, 85, 87, 88, 89, 90, 110, 115, 127, 128, 134, 135, 136, 137, 141, 142, 143, 169, 174, 178, 179, 189, 191, 193, 229, 230, 253, 254, 268, 293, 299, 301, 306, 313, 314, 322, 324

point of stability, 95, 177, 310, 323, 326

Substance, 3, 4, 16, 45, 59, 60, 64, 65, 66, 67, 68, 69, 74, 77, 81, 82, 85, 86, 88, 91, 92, 94, 95, 96, 97, 98, 99, 100, 101, 102, 103, 104, 105,

108, 110, 111, 112, 113, 114, 115, 123, 126, 127, 130, 131, 134, 135, 136, 137, 141, 142, 143, 146, 147, 149, 151, 153, 154, 155, 156, 157, 160, 161, 162, 172, 175, 176, 177, 178, 179, 182, 183, 184, 185, 190, 198, 200, 201, 210, 214, 217, 230, 233, 240, 243, 245, 251, 253, 254, 256, 258, 263, 264, 268, 289, 303, 304, 306, 316, 317, 322, 323, 324, 331

universe, 4, 9, 12, 14, 20, 21, 22, 23, 24, 26, 28, 31, 32, 33, 34, 35, 36, 37, 38, 39, 40, 41, 42, 43, 48, 51, 52, 56, 58, 59, 61, 62, 63, 68, 69, 76, 77, 79, 80, 81, 85, 86, 87, 89, 90, 91, 94, 95, 97, 98, 99, 100, 101, 102, 106, 107, 113, 115, 121, 123, 127, 134, 137, 138, 144, 145, 149, 150, 156, 176, 177, 199, 200, 212, 233, 237, 249, 262, 271, 272, 273, 274, 277, 278, 279, 282, 287, 288, 293, 297, 304, 319, 320, 321, 322, 323, 324, 325, 327, 328, 329, 331, 333, 335, 337, 339, 341, 343, 345, 349, 351

Electricity

anode, 116, 117, 130, 280, 288

capacitor, 133, 141, 287

cathode, 116, 117, 130, 280, 288

conductors, 133, 142, 144, 184, 230

dielectric, 141, 143, 144

dynamo, 148

electric current, 116, 120, 121, 124, 126, 130, 131, 132, 133, 136, 139, 140, 144, 145, 147, 148, 166, 177, 202, 235, 236, 237, 255, 274, 275, 276, 277, 278, 279, 280, 281, 282, 283, 284, 285, 286, 288, 296, 305, 325, 326, 327

Birkeland currents, 177, 274, 275, 276, 277, 278, 279, 280, 281, 325, 327

electromagnetic force, 41, 43, 47, 277, 278, 295

frequency, 30, 36, 61, 67, 68, 72, 80, 85, 94, 98, 100, 121, 138, 139, 177, 300

Induction, 16, 147, 148

magnet, 19, 24, 30, 47, 48, 52, 53, 61, 81, 89, 104, 105, 110, 120, 122, 124, 125, 130, 138, 141, 142, 143, 144, 145, 147, 148, 197, 201, 231, 236, 250, 274, 276, 277, 281, 282, 292, 295, 302, 327

semiconductors, 142

Static Electricity, 133, 139, 140

voltage, 48, 123, 124, 130, 132, 133, 199, 235, 236, 246, 287, 288, 305, 307

potential difference, 124, 132, 133, 235, 236, 246, 305

Four Elements

Four Elements, 86, 91, 92, 166, 184, 245, 331

Air, 4, 86, 87, 88, 89, 91, 92, 149, 166, 167, 172, 190, 203, 204, 205, 207, 215, 229, 230, 264, 265, 267, 268, 269, 272, 279, 324, 331, 332, 334, 340

Index

Earth, 86, 87, 88, 89, 91, 92, 123, 135, 166, 167, 169, 172, 179, 180, 181, 182, 183, 184, 187, 190, 194, 200, 264, 267, 268, 269, 279, 324, 331, 332

Fire, 4, 86, 87, 88, 89, 91, 92, 114, 166, 167, 172, 190, 233, 234, 237, 264, 265, 267, 269, 272, 324, 331, 332, 334

Water, 4, 86, 87, 88, 89, 91, 92, 123, 135, 166, 167, 172, 180, 182, 185, 187, 190, 200, 202, 216, 230, 264, 265, 266, 268, 279, 324, 331, 332, 334, 340

Geology

Anthroposphere, 154, 190, 333

Asthenosphere, 333

atmosphere, 42, 63, 80, 81, 88, 89, 90, 99, 107, 186, 187, 189, 190, 191, 192, 194, 195, 196, 203, 204, 205, 206, 207, 214, 215, 219, 228, 230, 231, 232, 235, 236, 243, 255, 256, 265, 279, 290, 293, 294, 298, 301, 302, 305, 307, 308, 309, 323, 331, 333, 339

Biosphere, 314, 333

crystals, 82, 89, 94, 95, 98, 99, 108, 111, 143, 149, 150, 172, 173, 175, 179, 180, 181, 182, 183, 184, 186, 189, 191, 201, 213, 227, 228, 235, 237, 248, 249, 262, 270

Cycle of Rock, 170, 171, 174

earthquake, 46, 120, 169, 170, 180, 183, 184, 228, 229, 300, 311, 315, 316

Erosion, 171, 173, 174, 180, 182, 188, 189, 193, 295, 296

fossil, 14, 174, 189, 207, 208, 213, 214, 215, 216, 218, 238, 243, 244, 245, 291, 309, 312, 314, 316, 326, 334

geological, 17, 18, 58, 86, 169, 170, 172, 173, 184, 207, 208, 219, 244, 296, 308, 310, 312, 314, 326

geological dating, 17, 18

Hydrosphere, 333

ice age, 293

Igneous Rock, 171, 172, 173, 174, 180, 181, 182, 183, 184, 188, 189, 217, 291, 299, 300

lava, 17, 171, 172, 173, 181, 184, 297

Lithosphere, 333

Magma, 170, 171, 172, 175, 181, 182, 265, 299, 300, 333, 340

Metamorphic Rock, 171, 172, 174, 175, 180, 182

metamorphism, 171, 172, 173, 174, 175, 180, 181, 182, 188

Pedosphere, 333

rock, 17, 70, 85, 87, 88, 156, 164, 167, 170, 171, 172, 173, 174, 175, 178, 179, 180, 181, 182, 183, 184, 187, 188, 189, 190, 191, 192, 193, 195, 200, 207, 214, 217, 260, 267, 268, 291, 293, 295, 296, 298, 299, 300

sediment, 164, 171, 173, 174, 183, 184, 188, 194, 195, 208, 220, 267, 268, 310, 333

Sedimentary Rock, 171, 172, 173, 174, 175, 180, 182, 291, 298

Tectonic, 170, 171, 173, 175, 180, 184, 185, 188, 268, 300, 311

Volcano, 174, 181, 184, 185, 187, 192, 219, 260, 265, 268, 297, 309, 333

Gurdjieff

Gurdjieff, 2, 3, 5, 10, 24, 55, 57, 63, 66, 67, 69, 71, 72, 76, 77, 80, 83, 86, 87, 89, 90, 91, 94, 95, 98, 100, 101, 103, 104, 105, 106, 109, 110, 111, 113, 125, 134, 145, 153, 154, 155, 156, 157, 158, 159, 160, 162, 164, 172, 175, 176, 177, 183, 191, 194, 195, 197, 200, 207, 212, 227, 231, 232, 233, 245, 248, 251, 255, 256, 260, 261, 262, 263, 264, 265, 266, 290

Gurdjieffian Neologisms

Alillnofarab, 110

Djartklom, 99, 100, 289

Fulasnitamnian, 233, 234, 249

Heptaparaparshinokh, 98, 101

Hydro-oomiak, 110, 111

Itoklanoz, 233, 234

Khritofalmonofarab, 110

Klananoizufarab, 111

Krilnomolnifarab, 110

Kundabuffer, 234, 261, 262, 263

Obligolnian, 5

Okidanokh,, 94, 99, 100, 290

Omnipresent-Okidanokh, 94, 97, 98, 100, 117

Petrkarmak, 110, 111

Planekurab, 110

Polormedekhtic, 160, 195

Rascooarno, 248

Sirioonorifarab, 110, 111

Stopinders, 67, 68, 71, 98

357

Talkoprafarab, 110
Theomertmalogos, 77, 95, 98
Triamazikamno, 66, 97, 99, 100, 101
Trogoautoegocrat, 59, 60, 69, 100, 101, 102, 113, 155, 156, 201, 204, 212, 221, 233, 244, 252, 253, 260
Ouspensky, 2, 86, 94, 104, 164, 176, 239, 317
The Work, 1, 7, 10, 59, 71, 74, 82, 83, 209, 213, 214, 227, 237

Hydrogens

Hydrogens, 3, 65, 70, 82, 85, 91, 92, 94, 95, 96, 97, 98, 99, 100, 101, 102, 103, 104, 105, 109, 110, 112, 130, 155, 157, 159, 160, 161, 162, 163, 176, 210, 250, 253, 258, 265, 266, 267, 268, 271, 272
H1, 65, 105, 272
H3, 65, 103, 105, 268, 271
H6, 65, 96, 97, 103, 105, 110, 112, 157, 162, 250, 251, 253, 254, 258, 265, 266, 267, 268, 269, 272
H12, 65, 104, 105, 109, 110, 112, 157, 159, 162, 163, 228, 250, 253, 254, 257, 264, 266, 267, 268, 269, 271
H24, 65, 104, 105, 109, 110, 157, 159, 162, 210, 212, 228, 250, 254, 257, 265, 266, 267, 268, 269, 308
H48, 65, 104, 105, 109, 110, 157, 159, 160, 162, 210, 240, 250, 253, 254, 256, 257,
258, 263, 264, 266, 267, 268, 269, 308
H96, 65, 104, 105, 110, 155, 157, 240, 250, 254, 255, 257, 265, 266, 267, 268, 272, 306
H192, 65, 104, 105, 110, 154, 155, 156, 157, 160, 162, 240, 250, 253, 254, 263, 264, 265, 266, 267, 268, 308
H384, 105, 110, 112, 155, 161, 187, 250, 253, 254, 264, 267, 268
H768, 105, 110, 112, 154, 155, 157, 160, 250, 253, 254, 264, 267, 268, 308
H1536, 105, 155, 184, 250, 253, 254, 264, 267, 268
H3072, 104, 105, 155, 176, 268
H6144, 176, 177

Carbon, 91, 92, 96
C6, 162
C24, 162
C512, 176
C1024, 176
Nitrogen, 91, 92, 96
N1024, 176
N2048, 176
Oxygen, 91, 92, 96
O24, 162
O96, 162
O1536, 176
O3072, 176

Laws

laws, 3, 5, 8, 15, 22, 26, 32, 34, 36, 40, 41, 47, 50, 56, 58, 64, 65, 66, 67, 68, 70, 72, 74, 75, 76, 77, 78, 85, 86, 87, 88, 89, 90, 91, 92, 93, 96, 97, 98, 99, 100, 101, 103, 109, 110, 113,
134, 137, 140, 149, 155, 157, 160, 161, 162, 166, 167, 176, 177, 187, 188, 194, 213, 217, 232, 237, 240, 241, 242, 243, 264, 265, 266, 267, 268, 269, 271, 272, 304, 320, 321, 322, 332, 347
Boyle's Law, 8
Coulombs Law, 140
first law of thermodynamics, 40
he third law of thermodynamics, 41
Hubble's Law, 26
inverse square law, 36
Kepler's laws of planetary motion, 32
law of accident, 241, 242
Law of Catching up, 90, 101
law of cause and effect, 241, 242
Law of Falling, 89, 90, 101, 176, 177
law of fate, 232, 237, 241, 242, 243
law of immobility, 74
Law of Seven, 3, 22, 64, 65, 66, 67, 74, 77, 85, 86, 96, 160, 237, 322
Law of Three, 3, 64, 65, 66, 68, 72, 77, 85, 86, 91, 92, 93, 96, 100, 103, 113, 137, 155, 157, 160, 161, 162, 166, 217, 240, 265, 271, 322, 347
laws governing the universe, 22
laws of mathematics, 15
laws of World-creation and World-maintenance, 5, 58
Newton's first law of motion, 50
Newton's Laws of Motion, 15
orders of law, 74, 75, 86,

Index

88, 113, 137, 188, 264, 265, 266, 267, 268, 269, 272, 304
primordial sacred laws, 77
scientific laws, 321
the electro-magnetic force inverse square law, 47
the force of gravity inverse square law, 47

Mathematics

Mathematics, 2, 3, 14, 15, 16, 17, 18, 20, 28, 29, 32, 33, 34, 37, 41, 45, 46, 50, 52, 53, 58, 72, 120, 125, 126, 128, 138, 177, 271, 276, 313, 351
continuous, 15, 16, 20, 31, 37, 40, 43, 45, 79, 83, 99, 100, 101, 111, 138, 171, 180, 196, 218, 222, 236, 237, 262, 288, 295, 303, 319, 325
dimension, 23, 108, 125, 126, 219, 283
discrete, 15, 16, 44, 138
Extrapolation, 16, 17, 22, 27, 28, 29, 33, 39, 41, 48, 49, 316
map, 15, 16, 17, 36, 41, 218, 277, 293, 368
statistics, 13, 14, 45, 46, 58

Nature

Nature, 60, 61, 80, 102, 103, 106, 109, 165, 212, 214, 219, 220, 221, 222, 223, 230, 233, 234, 239, 244, 245, 260, 261, 277, 278, 316, 317, 323, 330, 338, 339, 340

Objective Science

Objective Science, 1, 2, 3, 4, 5, 7, 16, 20, 22, 23, 24, 29, 55, 56, 57, 58, 59, 60, 61, 62, 64, 68, 74, 75, 76, 85, 86, 87, 88, 89, 98, 105, 106, 107, 113, 115, 118, 129, 136, 137, 149, 153, 154, 199, 217, 236, 238, 243, 258, 272, 273, 284
Objective Science Theory information, 1, 3, 9, 11, 55, 57, 58, 81, 82, 83, 84, 120, 123, 145, 149, 157, 179, 180, 190, 197, 214, 218, 231, 247, 257, 270, 271, 297, 299, 322, 352
knowledge, 1, 3, 4, 7, 9, 10, 11, 19, 57, 58, 66, 82, 83, 84, 88, 104, 115, 126, 149, 156, 157, 160, 165, 176, 190, 225, 226, 231, 256, 258, 264, 265, 266, 267, 268, 269, 273, 274, 317, 322, 328, 333
music, 2, 70, 72, 73, 74, 242, 260, 264, 319
equal temperament scale, 72, 74
just scale, 72, 73, 74
reality, 11, 14, 15, 16, 25, 33, 41, 46, 61, 70, 83, 84, 86, 96, 106, 125, 126, 140, 142, 155, 181, 187, 261, 262, 274, 315
truth, 9, 10, 11, 12, 32, 41, 55, 57, 70, 134, 154, 222, 259, 261, 273, 304

Octave

Octave, 3, 4, 22, 66, 67, 68, 69, 70, 71, 72, 73, 74, 77, 80, 81, 82, 85, 92, 94, 95, 96, 97, 98, 99, 100, 101, 102, 105, 109, 110, 111, 112, 113, 137, 160, 161, 164, 165, 167, 168, 177, 194, 207, 214, 227, 228, 232, 239, 243, 244, 252, 253, 254, 257, 258, 259, 260, 263, 270, 322, 329, 330, 346
do, 4, 22, 66, 67, 68, 69, 70, 71, 72, 73, 77, 78, 80, 81, 82, 94, 95, 96, 99, 112, 160, 177, 214, 232, 244, 259, 260, 270, 322
do-si, 67, 70, 81, 95
fa, 4, 22, 66, 68, 69, 70, 71, 72, 73, 80, 81, 82, 92, 95, 97, 112, 164, 166, 167, 168, 194, 207, 219, 239, 243, 244, 322, 329
fa-mi, 70, 81, 95
interval, 3, 4, 22, 69, 70, 71, 77, 80, 81, 82, 92, 95, 96, 97, 139, 164, 166, 167, 168, 194, 207, 219, 239, 243, 255, 285, 329
la, 4, 22, 66, 69, 70, 73, 77, 80, 81, 92, 95, 96, 97, 164, 207, 219, 244, 253, 270, 322, 329, 330
mi, 4, 22, 66, 68, 69, 70, 71, 72, 73, 80, 81, 82, 90, 92, 94, 95, 97, 112, 160, 164, 166, 167, 168, 194, 207, 214, 219, 239, 243, 244, 252, 322, 329
mi-fa, 4, 68, 71, 72, 80, 81, 82, 164, 166, 167, 168, 207, 219, 243
notes, 4, 66, 67, 68, 69, 70, 71, 72, 73, 77, 80, 81, 82, 90, 92, 95, 96, 97, 99, 112, 125, 156, 160, 164, 167, 177, 183, 207, 214, 219, 239, 252, 253, 259, 260, 264, 270,

298, 329
re, 4, 22, 66, 68, 70, 71, 73, 81, 82, 90, 94, 112, 214, 239, 244, 246, 322
si, 4, 22, 66, 67, 69, 70, 72, 73, 77, 81, 95, 96, 112, 244, 270, 322
si-do, 4, 72
sol, 4, 22, 66, 69, 70, 73, 80, 81, 82, 90, 92, 94, 95, 97, 164, 207, 219, 239, 244, 259, 260, 322, 329
Ansapalnian-octave, 110, 111
Inner Ansapalnian octave, 109, 110, 112, 113
Lateral Octave, 80, 81, 92, 164, 165, 167, 177, 194, 207, 214, 239, 243, 253, 258, 259, 260, 270, 329, 330, 346

Physics
adjustable parameters, 18, 39, 40, 41, 49
aether, 24, 25, 26, 31, 50, 52, 53, 62, 118, 120, 125, 317, 347
closed system, 8, 12, 13, 41
Convection, 121, 173, 201, 230, 235, 275
Diffusion, 89, 198, 201, 204, 205, 230, 245, 280, 323
Doppler, 22, 26, 27, 28, 30
energy, 16, 20, 21, 30, 31, 33, 34, 37, 39, 40, 42, 43, 45, 61, 68, 87, 90, 104, 118, 119, 120, 121, 123, 124, 125, 128, 129, 130, 131, 132, 133, 134, 135, 136, 137, 138, 139, 140, 144, 145, 147, 148, 155, 162, 184, 192, 202, 210, 212, 215, 227, 228, 240, 241, 243, 245, 246, 251, 256, 261, 268, 272, 282, 285, 289, 298, 301, 302, 303, 304, 319, 322, 324
Chemical energy, 87, 130
Electrical energy, 87, 130, 131, 140
Gravitational energy:, 130
Kinetic energy, 87, 121, 129, 130, 131, 132, 140, 144, 145, 147, 148
Magnetic energy, 130, 131
Mechanical energy, 130, 131, 162
Nuclear energy, 130
Potential energy, 124, 129, 130, 131, 140, 145
Radiant energy, 45, 131
Sound energy, 131
Thermal energy, 121, 131
field, 19, 20, 30, 40, 43, 50, 52, 53, 89, 115, 116, 120, 122, 124, 125, 130, 141, 142, 143, 144, 145, 147, 148, 151, 190, 197, 231, 236, 274, 276, 277, 281, 282, 292, 294, 295, 302, 327
electric field, 52, 53, 122, 141, 142, 143, 144, 145, 147, 151, 294
magnetic field, 19, 30, 52, 89, 115, 116, 120, 122, 125, 130, 142, 143, 144, 145, 147, 148, 197, 231, 236, 274, 276, 277, 281, 282, 292, 295, 302, 327
force, 18, 23, 24, 30, 32, 37, 41, 43, 46, 47, 48, 50, 52, 61, 64, 65, 66, 67, 70, 74, 75, 76, 77, 78, 80, 85, 89, 90, 91, 92, 93, 96, 97, 98, 99, 104, 113, 119, 124, 125, 127, 128, 129, 130, 131, 132, 137, 140, 142, 143, 144, 145, 147, 162, 166, 176, 177, 181, 184, 188, 190, 201, 202, 212, 216, 223, 230, 238, 241, 247, 254, 261, 265, 273, 274, 275, 277, 278, 279, 281, 292, 294, 295, 311, 312, 321, 329, 334
active force, 66, 91, 92, 93, 96, 97, 129, 162, 166
Lorentz force, 145
neutralizing force, 65, 91, 92, 99, 104, 129, 166, 241
passive force, 91, 92, 93, 129, 162, 166
Friction, 127, 128, 129, 132, 133
gravity, 20, 23, 24, 32, 35, 37, 40, 46, 47, 48, 53, 67, 98, 124, 127, 131, 132, 140, 149, 150, 152, 159, 177, 181, 197, 209, 216, 227, 249, 263, 264, 265, 267, 278, 279, 281, 287, 292, 293, 300, 305, 306, 310, 311, 313, 326
gravitation, 15, 20, 23, 24, 28, 30, 31, 32, 41, 47, 50, 52, 90, 128, 130, 131, 140, 169, 202, 273, 278, 279, 293, 300, 307, 312
interference, 9, 34, 43, 44, 48, 49, 212
interferometer, 25, 27
light, 16, 21, 22, 24, 25, 26, 27, 28, 29, 30, 31, 34, 43, 44, 45, 48, 50, 61, 63, 68, 79, 88, 90, 94, 117, 118, 122, 131, 134, 136, 138, 146, 148, 152, 165, 196, 197, 198, 214, 231, 232, 236, 251, 253, 269, 270, 271, 282, 285, 286, 306, 324, 333
mass, 8, 24, 29, 32, 33,

Index

35, 37, 43, 50, 51, 52, 84, 87, 92, 93, 94, 95, 102, 126, 127, 129, 130, 132, 134, 139, 140, 153, 170, 172, 177, 188, 229, 242, 281, 284, 293, 310, 313, 314, 323

Matter, 4, 20, 21, 29, 30, 31, 32, 33, 34, 39, 40, 45, 68, 87, 88, 89, 90, 101, 104, 105, 107, 109, 113, 115, 121, 127, 134, 137, 150, 153, 164, 167, 169, 191, 195, 228, 236, 252, 253, 254, 255, 256, 257, 273, 275, 277, 279, 319, 322, 323, 324

Mechanics, 126, 127, 128, 129, 130, 131, 132, 133

model, 4, 15, 16, 17, 18, 19, 20, 21, 22, 23, 26, 28, 29, 31, 32, 33, 34, 35, 37, 38, 39, 41, 42, 43, 45, 47, 48, 49, 52, 57, 107, 108, 116, 120, 122, 130, 137, 138, 170, 181, 211, 225, 273, 287, 288, 289, 292, 297, 299, 300, 307, 311, 315, 319, 321, 323, 325, 327, 329, 331, 333, 335, 337, 339, 341, 343, 345

Newton's Cradle, 118

nuclear fusion, 20, 37, 130

radiation, 20, 21, 24, 30, 33, 34, 35, 36, 37, 42, 43, 45, 48, 68, 90, 95, 117, 118, 119, 120, 121, 122, 138, 145, 146, 147, 231, 255, 256, 301, 305, 308

black body radiation, 34

electromagnetic radiation, 24, 30, 34, 35, 37, 43, 48, 68, 90, 118, 120, 121

EMR, 118, 120, 121, 122, 123, 124, 125, 131, 134, 138, 139, 144, 146, 147,
231, 232, 236, 286, 317

Recession of Galaxies, 21

Redshift, 22, 26, 27, 28, 29, 30, 31, 36, 40, 349

resistance, 18, 24, 64, 66, 127, 128, 129, 130, 132, 133

SI unit, 126, 127, 128, 132, 133

speed of light, 28, 50, 120, 134, 138, 271, 272

Standard Models, 18, 19

Standard Model, 18, 19, 20, 21, 22, 26, 28, 31, 32, 33, 34, 41, 42, 43, 45, 47, 49, 287, 319

star, 20, 21, 29, 32, 35, 36, 37, 38, 40, 48, 59, 62, 63, 78, 79, 91, 94, 96, 100, 101, 125, 242, 270, 273, 279, 282, 283, 284, 285, 286, 287, 292, 317, 323

Surface tension, 89, 105, 201, 230

Temperature, 8, 11, 21, 34, 35, 39, 107, 109, 121, 126, 135, 141, 142, 151, 170, 171, 172, 173, 174, 175, 178, 181, 182, 184, 186, 188, 189, 197, 203, 204, 209, 216, 219, 231, 235, 236, 254, 267, 282, 283, 284, 285, 292, 293, 303, 305, 308, 324

theory of relativity, 8, 20, 23, 28

Time, 23, 28, 38, 39, 49, 66, 101, 126, 128, 129, 133, 134, 138, 139, 170, 171, 182, 184, 187, 198, 210, 231, 255, 270, 307, 312, 323, 347

vibration, 24, 49, 61, 66, 67, 68, 72, 73, 77, 80, 82, 85, 86, 87, 91, 94, 96, 98, 99, 100, 102,
112, 120, 131, 177, 184, 229, 234, 256, 267, 269, 289, 290

wavelength, 16, 27, 34, 35, 48, 68, 120, 121, 122, 123, 124, 125, 138, 139, 202, 231, 322

Waves, 24, 27, 28, 30, 37, 43, 44, 45, 49, 50, 68, 90, 117, 118, 119, 120, 121, 122, 123, 124, 125, 131, 139, 147, 174, 176, 202, 205, 230, 231, 257, 267, 268, 269, 286, 294, 317

Longitudinal wave, 118, 119, 131

Transverse waves, 118, 119, 131

Weight, 47, 105, 112, 113, 127, 134, 310, 316

atomic weight, 105, 112, 113

Plasma

Arc, 123, 124, 135, 136, 137, 282, 284, 286, 297, 324

glow mode, 91, 123, 124, 135, 137, 265, 269, 282, 284, 297, 324

Ionosphere, 231, 234, 235, 236, 305, 306, 333

Ions, 89, 115, 116, 117, 122, 127, 130, 131, 135, 139, 140, 144, 151, 175, 177, 190, 204, 205, 235, 236, 246, 256, 276, 281, 305, 306, 307, 324

Magnetosphere, 152, 206, 207, 236, 280, 295, 297, 298, 301, 302, 305, 306, 309, 325, 327, 332, 333

Marklund convection, 275

plasma, 3, 4, 30, 31, 61, 87, 88, 89, 90, 114,

115, 116, 117, 119, 120, 121, 123, 124, 125, 127, 128, 129, 130, 131, 133, 134, 135, 136, 137, 139, 140, 141, 143, 144, 145, 147, 149, 150, 151, 152, 190, 191, 230, 232, 234, 235, 236, 237, 245, 246, 247, 248, 250, 254, 256, 257, 258, 259, 260, 264, 265, 269, 272, 273, 274, 275, 276, 277, 278, 279, 281, 282, 283, 284, 285, 287, 288, 289, 293, 294, 297, 302, 303, 306, 307, 309, 316, 319, 324, 325, 332, 333, 334, 341, 345

Safire Project, 9, 287, 288, 289, 319

plasmasphere, 341

plasmoid, 277, 279

Z-pinch, 276, 277, 279, 325, 326

Psyche

emanations, 5, 77, 95, 98, 145, 146, 147, 231, 232, 234, 236, 255, 256, 258, 267

emotional, 110, 157, 159, 162, 191, 199, 210, 211, 212, 225, 226, 227, 228, 250, 251, 254, 255, 257, 263, 266, 267, 268, 303, 322, 333

ESP, 11

formatory apparatus, 227, 250, 257

impressions, 2, 3, 33, 59, 70, 82, 84, 95, 100, 110, 124, 153, 160, 161, 164, 179, 207, 209, 211, 212, 226, 228, 233, 247, 249, 250, 251, 253, 254, 256, 257, 258, 262, 263, 267, 303, 323

instinctive, 157, 159, 199, 209, 210, 211, 212, 225, 226, 227, 228, 248, 254, 257, 262, 263

intellectual, 29, 38, 41, 162, 211, 212, 225, 227, 229, 230, 240, 241, 250, 258, 267, 269, 273, 319

intentional suffering, 83, 304

moving center, 209, 210, 211, 212, 225, 226, 250, 254, 263, 266, 268, 269

personality, 1, 2, 237, 242, 248, 261, 269, 273, 303, 304, 306

self-observation, 71, 83, 254, 258

self-remembering, 71, 83, 254, 258

Suggestibility, 5, 10, 19, 55

Quantum Mechanics

atom, 7, 16, 20, 21, 30, 31, 37, 39, 40, 41, 44, 45, 47, 48, 52, 59, 60, 61, 64, 74, 75, 76, 79, 85, 87, 88, 89, 90, 92, 98, 102, 105, 106, 107, 108, 109, 112, 113, 115, 116, 120, 121, 122, 127, 130, 131, 134, 135, 136, 137, 144, 166, 172, 175, 183, 184, 196, 201, 204, 213, 247, 278, 294, 301, 322

Large Hadron Collider, 12, 48, 122

LHC, 12, 49, 122

particle, 20, 23, 24, 30, 36, 41, 42, 43, 44, 45, 48, 49, 50, 53, 55, 61, 68, 89, 90, 113, 116, 117, 122, 123, 127, 131, 134, 137, 139, 141, 150, 174, 175, 177, 178, 182, 189, 191, 236, 274, 278, 280, 292, 301, 302, 325

Antiparticle, 42

Boson, 43, 45, 47, 49

Higgs Boson, 43

leptons, 42, 43, 45

meson, 42, 43

neutrino, 30, 49, 285

neutron, 36, 41, 42, 43, 137, 273, 287

photon, 42, 43, 44, 47, 50, 122, 123, 125, 138, 139, 301, 306, 322

positron, 40, 42

quark, 36, 41, 42, 43, 45

tachyon, 42

quantum mechanics, 16, 19, 41, 42, 43, 45, 46, 48, 49, 50, 52, 85, 138

Copenhagen interpretation, 45

strong nuclear force, 41, 43

strong force, 47

weak nuclear force, 41, 43

weak force, 47

Ray of Creation

Ray of Creation, 3, 6, 22, 62, 63, 64, 66, 67, 68, 69, 74, 76, 77, 78, 80, 81, 82, 86, 87, 90, 92, 93, 94, 95, 96, 107, 109, 134, 156, 159, 164, 219, 243, 252, 259, 264, 265, 270

Absolute, 4, 22, 55, 61, 62, 63, 67, 74, 76, 77, 78, 80, 81, 86, 87, 90, 92, 94, 95, 96, 97, 98, 113, 134, 136, 137, 139, 159, 175, 236, 251, 270, 271

All Suns, 4, 74, 75

All Worlds, 4, 62, 74, 75, 76, 270

Earth, 4, 12, 14, 19, 25, 26, 27, 30, 35, 36, 38,

Index

58, 62, 63, 70, 71, 75, 76, 78, 80, 81, 82, 83, 84, 85, 86, 87, 88, 90, 92, 93, 94, 95, 96, 101, 102, 103, 120, 121, 122, 127, 142, 150, 153, 154, 156, 164, 165, 166, 169, 170, 171, 173, 175, 177, 179, 181, 183, 185, 191, 194, 200, 201, 202, 206, 207, 212, 213, 214, 217, 219, 220, 228, 229, 230, 231, 234, 236, 237, 238, 239, 241, 242, 243, 246, 248, 251, 252, 260, 262, 264, 265, 267, 268, 269, 272, 274, 280, 289, 291, 292, 293, 294, 295, 296, 297, 298, 299, 300, 301, 302, 303, 304, 305, 306, 307, 308, 309, 310, 311, 312, 313, 314, 315, 316, 317, 323, 325, 326, 327, 328, 329, 331, 333, 339, 340

Milky Way, 4, 26, 29, 30, 35, 36, 48, 62, 76, 90, 91, 102, 160, 270, 271, 278, 282, 284, 286

Moon, 4, 62, 63, 74, 75, 78, 80, 81, 82, 83, 86, 87, 88, 90, 93, 94, 95, 96, 97, 127, 131, 160, 164, 166, 167, 195, 202, 206, 207, 214, 219, 239, 241, 243, 248, 260, 261, 262, 264, 269, 272, 286, 291, 293, 294, 295, 296, 297, 298, 299, 300, 301, 302, 303, 304, 305, 306, 307, 308, 309, 311, 313, 314, 315, 316, 317, 323, 327, 341

Planet, 4, 62, 63, 75, 76, 78, 79, 81, 86, 88, 89, 92, 93, 94, 164, 165, 166, 167, 187, 232, 234, 237, 238, 239, 240, 241, 242, 243, 264, 265, 269, 272, 292, 294, 295, 298, 339

Sun, 4, 25, 35, 48, 62, 63, 75, 79, 80, 81, 86, 88, 90, 91, 92, 94, 95, 96, 102, 147, 164, 165, 167, 177, 194, 196, 202, 204, 206, 207, 209, 214, 220, 231, 234, 236, 237, 239, 241, 243, 250, 253, 259, 264, 265, 267, 269, 270, 271, 272, 274, 277, 278, 279, 280, 281, 282, 283, 284, 285, 287, 288, 289, 291, 292, 294, 295, 298, 299, 301, 302, 305, 306, 308, 316, 325, 327, 329, 330, 333

Sun Absolute, 4, 62, 76, 77, 78, 90, 91, 94, 95, 98, 100, 159, 162, 164, 265, 267, 270, 271, 272, 323

Science

cognitive bias, 17, 18

Contemporary science, 1, 5, 7, 10, 11, 14, 15, 18, 19, 20, 22, 24, 26, 31, 55, 56, 57, 58, 59, 60, 85, 88, 90, 107, 115, 199, 224, 287, 298, 317

engineering, 14, 15, 83, 120, 149, 215, 244, 246, 269, 310

Experiment, 2, 7, 8, 9, 11, 12, 13, 17, 23, 24, 25, 26, 31, 42, 43, 44, 46, 47, 49, 50, 51, 56, 57, 58, 68, 95, 107, 108, 119, 131, 149, 150, 154, 188, 201, 211, 214, 223, 224, 227, 232, 238, 248, 253, 257, 258, 267, 268, 271, 285, 287, 289, 290, 294, 296, 313, 347

Double Slit Experiment, 43, 44, 49

Michelson-Morley, 25

thought experiment, 24, 50, 51

hypothesis, 8, 9, 11, 12, 13, 14, 19, 25, 243, 272, 287, 310, 311, 315, 317

Modern science, 3, 4, 26, 87, 104, 113, 136, 137, 153, 195, 197, 203, 217, 256, 274, 319

Prediction, 8, 9, 15, 17, 23, 26, 31, 34, 35, 36, 37, 39, 41, 43, 45, 46, 179, 242, 297

repeatability, 11, 12, 13, 56

Roman Catholic Church, 9, 19

scientific establishment, 5, 10, 287, 293, 294, 315

scientific method, 7, 12, 18, 49, 258

scientist, 8, 9, 12, 13, 14, 17, 18, 19, 25, 35, 55, 56, 57, 58, 111, 126, 146, 179, 180, 194, 211, 213, 214, 236, 243, 274, 275, 277, 278, 281, 286, 289, 301, 308, 315, 323

Scientists

Anthony Peratt, 275, 287

Arno Penzia, 34

Bob Wilson, 34

Brian J. Ford, 296

C J Ransom, 297

Christiaan Huygens, 43

Clark Maxwell, 24

Copernicus, 9, 32

Darwin, 10, 214, 224

Dayton Miller, 25, 31

Dr. Alexander Volkov, 198

Dr. Yuri Shtessel, 198
E Silvertooth, 26
Einstein, 15, 23, 26, 41, 45, 138
Ernst Mach, 50, 52, 349
Fred Hoyle, 37, 213
G T Marklund, 275
Galileo, 9, 32, 273
Heisenberg, 50
Heraclitus, 12
Hermes Trismegistus, 5
Hubble, 26, 27, 28, 40, 286
J B Rhine, 11, 12
James Clerk Maxwell, 120
M Allais, 26
M G Sagnac, 26
Mesmer, 10, 31, 104
Michelson, 25
Newton, 15, 23, 32, 43, 51, 127, 129, 132, 324
Niels Bohr, 45
Planck, 39, 45, 138, 139
R DeWitte, 26
R Hazen, 214
Thomas Young, 43
Velikovsky, 10, 293, 294, 295, 349, 350
Werner Heisenberg, 45
Wilhelm Weber, 50
Y Galaev, 26

Step Diagram
Step Diagram, 3, 153, 157, 159, 162, 163, 164, 165, 166, 168, 169, 175, 179, 180, 182, 187, 194, 195, 239, 243, 250, 263, 264, 265, 268, 270, 272, 332, 333
Angels, 159, 162, 164, 166, 167, 237, 239, 244, 249, 250, 251, 263, 264, 267, 271
Archangels, 157, 159, 162, 163, 164, 165, 166, 167, 237, 239, 244, 249, 250, 251, 252, 260, 261, 264, 265, 269, 270, 272
Eternal Unchanging, 159, 164, 265, 269, 270, 271, 272
Invertebrates, 157, 158, 159, 162, 164, 167, 180, 187, 188, 199, 202, 207, 208, 209, 210, 211, 212, 213, 214, 226, 239, 244, 245, 250, 264, 265, 266, 268, 269, 333
Kernel, 159, 162, 164, 166, 167, 175, 176, 177, 183, 184, 214, 244, 250, 264, 267, 268, 272
Man, 159, 162, 163, 164, 167, 180, 190, 197, 209, 210, 213, 219, 223, 225, 226, 227, 228, 229, 239, 243, 244, 245, 249, 250, 251, 257, 259, 260, 261, 262, 264, 265, 266, 268, 271, 308, 333, 343
Metals, 4, 159, 162, 164, 166, 167, 176, 177, 178, 179, 182, 184, 188, 214, 239, 244, 250, 260, 263, 264, 265, 267, 268, 332, 333, 343
Minerals, 159, 162, 164, 166, 167, 180, 182, 183, 184, 187, 188, 190, 194, 214, 239, 244, 250, 264, 265, 267, 268, 333
Plants, 157, 159, 162, 164, 167, 187, 188, 190, 191, 193, 194, 195, 196, 197, 198, 199, 200, 203, 204, 207, 208, 209, 210, 213, 215, 239, 244, 245, 250, 253, 263, 264, 265, 266, 267, 268, 333
Sun Absolute, 159, 162, 164, 265, 267, 270, 271, 272
Vertebrates, 157, 158, 159, 162, 164, 167, 199, 209, 210, 213, 214, 218, 225, 226, 227, 228, 239, 244, 245, 250, 259, 260, 263, 264, 266, 267, 269, 333

Gurdjieff's Hydrogens: The Ray of Creation

To Fathom The Gist

Volume 1: Approaches to the writings of G.I. Gurdjieff
Volume II: The Arch-Absurd
Volume III: The Arousing of Thought

The first volume of To Fathom The Gist, provoked a renaissance of interest in Gurdjieff's literary masterpiece. Rather than presenting a compendium of thoughts and theories on the meaning of *The Tales*, the book provided a practical guide on how to approach it productively. It provides a clear and concise description, with abundant examples, of the techniques and tricks the reader needs to employ to better understand it.

The second volume in the series goes deeper, examining the text, and with the assistance of an enneagram, maps out how Gurdjieff wrote, reviewed and edited his masterpiece. If you appreciate how the book was created it is easier to understand. The third volume in the series provides a sentence-by-sentence analysis of the text of the first half of the first chapter of The Tales (The Arousing of Thought). It is a demonstration of the results of reading the book in the manner advocated by the first two volumes in the series.

These books have been described as: "Insightful and original," "Essential reading for anyone studying Gurdjieff's writings," "True companions to All and Everything," "Valuable to any student of The Tales," "An ultra-effort" and "Exceptional." They have had a profound impact on many readers and have inspired the organization of study groups focussing on Gurdjieff's literature in North America, South America and Europe.

THE 1931 MANUSCRIPT

This is an edited version of the original 1931 manuscript of *Beelzebub's Tales to His Grandson* by G. Gurdjieff. The text is, for the most part, unchanged from the original manuscript that was published in a limited edition in 1931 by A. R. Orage. The Karnak Press published two versions of the book, both of which are available. In this version Gurdjieff's neologisms are changed to match the words in the later 1950 published version of this classic literary work. In the alternative edition the spellings remain as they were in the original. The former version is usually preferred but latter is truer to the original.

The text of *The 1931 Manuscript* edition is significantly different to *The Tales* in some areas, including the appearance of a whole part-chapter on Termosses. Reading *The 1931 Manuscript* at times feels like reading a different book, albeit one written by Gurdjieff.

THE SEARCHABLE INDEX

This ebook provides a productive means for searching through, reading and pondering *Beelzebub's Tales To His Grandson* by G. I. Gurdjieff. It has been written and compiled by Robin Bloor, the author of *To Fathom The Gist, Volumes I, II, and III*. Its goal is to help readers apply some of the approaches to fathoming the gist of Beelzebub's Tales described in those books. It is likely to prove useful to anyone who studies *The Tales* in depth, since the indexes and lists it includes will, on their own, be thought provoking to the reader.

Words (including neologisms) that the compiler considered important or significant are organized into a comprehensive set of categories and arranged within those categories. There are about a hundred such categories covering everything from names of people and places to psychic states. It is a simple matter to search on any one of them and thus find all references in *The Tales* to one of these words. This ebook is a "digital concordance" as it includes the full text of *The Tales* and an index of every word used in *The Tales*, along with the frequency of its occurrence.

The book is currently available from two sources, either as an ebook which can be read on amazon's Kindle, or as a pdf. The majority of readers who have used this ebook prefer the PDF version. That version can be bought directly from the publisher by visiting the website LittleCrowPress.com.

The Herald of Coming Good

The Herald of Coming Good
First Appeal to Contemporary Humanity
[with notes]

G. Gurdjieff

The Herald of Coming Good was first published in 1933, apparently as a prelude to the publication of Gurdjieff's three series of books under the common title of All and Everything. It was written in the obtuse and difficult style of *The Tales*. As such, it is a mysterious work. Why did he write it like that?

In publishing it, Gurdjieff proclaimed, "Contrary to the established custom, I shall not only permit this first book of mine, as well as the books of the first series, to be reprinted in any country, but, if necessary, I am willing to subsidize it, on the condition of course that absolute accuracy is preserved." It pretends to be a marketing vehicle for attracting people to the Work, with registration blanks for readers to fill in, should they wish to subscribe to the books of the First Series. Some would be inclined to accuse it of being the least effective marketing document ever written.

The goal was clearly not marketing. Indeed it is difficult to know what Gurdjieff had in mind. The casual reader is unlikely to make much sense of it, but serious readers of Gurdjieff's writings will find its contents valuable, particularly in helping to understand *The Tales*.

Edited by Robin Bloor, This version of the book has been "translated" into American English. It also includes a rendering of the prospectus for Gurdjieff's Institute for the Harmonious Development of Man. It also includes some analytical notes on the text.

The Fragments Poster

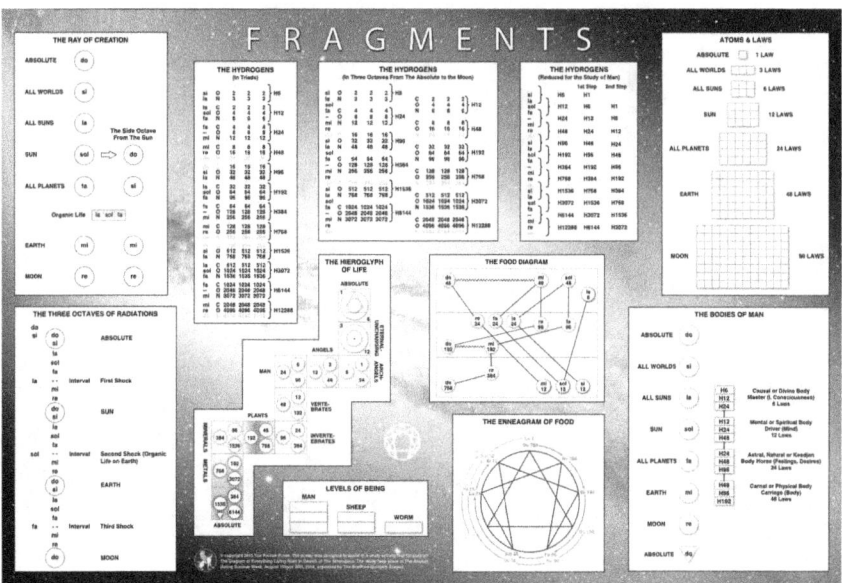

This colorful 30" x 20" poster combines many of the diagrams from *In Search of The Miraculous*. It can be thought of as a companion poster to that book, or indeed this one. It is available from the ToFathomTheGist.com website, as are all of the books advertised here.

Gurdjieff's Hydrogens: The Ray of Creation

www.ingramcontent.com/pod-product-compliance
Lightning Source LLC
Chambersburg PA
CBHW070526010526
44118CB00012B/1068